Principles
of
GROUNDWATER
ENGINEERING

WILLIAM C. WALTON

LEWIS PUBLISHERS

Library of Congress Cataloging in Publication Data

Walton, William Clarence

    Principles of Groundwater Engineering
      p.   cm.
    Includes bibliographical references and index.
    ISBN 0-87371-283-8
    1. Groundwater—Engineering. 2. Water—Analysis
I. Walton, William C.   II. Title.
GB1179.9.W43   1991
462.1'8643—dc20   90-90765
                                                  CIP

LEWIS PUBLISHERS, INC.
121 South Main Street, Chelsea, Michigan 48118

PRINTED IN THE UNITED STATES OF AMERICA

Preface

This book attempts to unify and bring together under one cover principles of groundwater engineering thereby providing a comprehensive manual for professionals working in the groundwater industry. Quantitative aspects of groundwater supply, drainage, and waste disposal are prime concerns. Although this book was specifically written for practicing engineers, geologists, and hydrogeologists it should be useful to water well contractors, educators and students, industry representatives, and others.

The application of theory and practical aspects of groundwater engineering are stressed rather than the derivation of equations. Ample references are cited to guide study of theoretical background, scientific principles, and technology topics. It is assumed that the reader has a working knowledge of the basic principles of geology, physics, chemistry, mathematics, and hydrogeology.

This book is aimed at bridging the gap between groundwater theory, covered in numerous textbooks, and groundwater problem solving. Towards this end, voluminous appendices are provided which contain quantitative data called for in groundwater flow and contaminant migration equations. Illustrative case histories describing the application of groundwater engineering principles to field conditions are omitted because they are readily available in other literature. Tables are restricted to the Appendix to avoid disruptions in the continuity of concept presentations. Because of the enormous recent growth of groundwater literature, particularly concerning contaminant migration, coverage but not reference support of certain topics is limited.

The computer has become a routine aid in the modeling of aquifer systems. Most professional workers have immediate access to a computer through a time-sharing terminal to a remote mainframe or minicomputer, or have a personal microcomputer sitting on their desk. Although this book does not contain computer program listings, the availability of pertinent groundwater programs and software is referenced throughout the text.

Unfortunately, easy access to computers often leads to unrealistic expectations concerning the precision of quantitative analyses. Because groundwater engineering modeling is only the art and science of approximation, precisions ranging from a few percent to orders of magnitude, depending largely on the degree of hydrogeologic and hydrogeochemical complexities, are considered to be realistic.

This book is a distillation of my training and experience over the past 44 years. Many colleagues over these years contributed in various ways to the writing of this book especially Bill Drescher, Frank Foley, Stanley Norris, Ed Schaefer, Ray Nace, Smitty, Bill Ackermann, Burt Maxey, Bill Guyton, John Harshbarger, Mahdi Hantush, Harvey Banks, Tom Prickett, Phil Davis, and Olin Braids. This book would have been impossible without the continuing support of my wife Ellen.

This book is dedicated to my three granddaughters: Alyse Solberg, Heidi Solberg, and Katy Kramer.

William C. Walton

Mahomet, Illinois

William C. Walton received his B.S. in Civil Engineering from Lawrence Technological University, Southfield, Michigan in 1948 and attended Indiana University, University of Wisconsin, Ohio State University, and Boise State University. Bill served for 10 years as Director of the Water Resources Research Center and Professor of Geology and Geophysics at the University of Minnesota. He taught groundwater courses at the University of Illinois on a part time basis for 2 years.

Bill's 44 years of experience in the water resources field include 2 years as a Water Well Contractor at Detroit, Michigan; 1 year with the U.S. Bureau of Reclamation at Cody, Wyoming; 8 years with the U.S. Geological Survey at Madison, Wisconsin, Columbus, Ohio, and Boise, Idaho; 6 years with the Illinois State Water Survey at Urbana, Illinois; 6 years with consulting firms including Shaefer and Walton at Columbus, Ohio; Camp Dresser and McKee, Inc. at Champaign, Illinois; and Geraghty and Miller at Champaign, Illinois; and 3 years as a self employed Consultant In Water Resources at Mahomet, Illinois. He was Executive Director of the Upper Mississippi River Basin Commission at Minneapolis, Minnesota for 5 years and has participated in water resources projects throughout the United States and Canada, and in Haiti, El Salvador, Libya, and Saudi Arabia. Bill is presently a self employed writer.

The positions he has held include: Vice President of the National Water Well Association: founding editor of the Journal Ground Water, Chairman of the Ground Water Committee of the Hydraulics Division, American Society of Civil Engineers; member of the U.S. Geological Survey Advisory Committee on Water Data for Public Use; Consultant to the Office of Science and Technology, Washington, D.C.; member of the Steering Committee of the International Ground Water Modeling Center; and advisor to the United States Delegation to the Coordinating Council of the International Hydrological Decade of UNESCO.

Bill served as a Visiting Scientist for the American Geophysical Union and the American Geological Institute and lectured at many universities throughout the United States. He also lectured at short courses sponsored for several years by the International Ground Water Modeling Center and presented papers at several professional society meetings in Europe. Bill is author of over 75 technical papers and 6 books: Groundwater Resource Evaluation, McGraw Hill; The World Of Water, Weidenfeld and Nicolson, London; Practical Aspects of Groundwater Modeling, National Water Well Association; and Groundwater Pumping Tests, Analytical Groundwater Modeling, and Numerical Groundwater Modeling, Lewis Publishers.

Contents

List of Figures

Chapter 1

Chapter 2

Chapter 8

Chapter 11

1

Introduction

Groundwater engineering is concerned with groundwater supply facilities (wells, well fields, collectors, etc.), dewatering facilities (trenches, mines, drains, etc.), waste disposal facilities (landfills, pits, basins, etc.), heat storage and extraction facilities (heat pump and geothermal energy), and facilities for land subsidence control. Protection of groundwater supplies, avoiding future groundwater contamination, and restoration of water quality in aquifers are prime concerns. Problems of municipal and industrial waste disposal, agricultural practices, petroleum spills, mining activities, and contamination of coastal groundwater supplies by saltwater intrusion fall within the purview of groundwater engineering. The short history of quantitative groundwater hydrology methods presented by Bredehoeft (1976, pp. 8-14) is relevant to the history of groundwater engineering.

In this book, radial flow towards production wells under nonleaky artesian, leaky artesian, water table, and fractured rock aquifer conditions is described and the general principles of groundwater flow are presented. Analytical methods for aquifer boundary and discontinuity and flow net analysis are summarized. Aquifer, tracer, and well production test design and analysis techniques are outlined. Contaminant migration from slug and continuous sources with advection, dispersion, sorption, and radioactive decay is described. The general principles of groundwater mass transport are presented. Analytical methods for simulating heat conduction and convection from heated water injection wells are summarized. Techniques for estimating ultimate and time dependent subsidence are presented. Production well design is described and methods for analytically simulating construction dewatering, mine drainage, and agricultural drainage are summarized. The principles of analytical and numerical computer modeling are reviewed. Popular mainframe and micro-computer numerical program and software availability is discussed. Features of a finite-difference numerical constant density 3-D flow model and a random walk constant density mass 3-D transport model are described in detail. Finally, graphics methods for display-ing computer model results are summarized.

The selection of subject matter included in this book is based on frequency of use criteria. Although an attempt has been made to provide a broad coverage of groundwater engineering subjects, de-tailed information on several topics are not included in this book but may be pursued through cited references. Among the topics not discussed in detail are: geologic principles, concepts, and processes that control the occurrence, movement, storage, and chemical char-acter of groundwater (see Back, et al. 1989 and Freeze and Cherry, 1979, pp. 144-166); methods of drilling; the detailed design, con-struction, and maintenance of wells; geophysical logging and sam-pling; groundwater monitoring; and groundwater law. In addition, most groundwater textbooks contain detailed discussions of the hydrological cycle including precipitation, evaporation, stream run-off, and soil moisture (see Fetter, 1988, pp. 1-114). Those detailed discussions are not repeated herein.

Although not discussed in detail in this book, groundwater engi-neering plays an important role in geotechnical problems and geo-logic processes (Freeze and Cherry, 1979, pp. 464-524). Some geotechnical problems such as leakage at dams (Harr, 1962, pp. 101-248 and Strack, 1989, pp. 379-380) and inflows into tunnels (Goodman, et al., 1965) and excavations arise from groundwater flow. Others such as landslides, slope stability, and subsidence are a consequence of excessive fluid pressures in the groundwater. There is a close relationship between groundwater flow and fluid pressures and geologic processes such as faulting, thrusting, migra-tion and accumulation of petroleum (Toth, 1970), geothermal sys-tems, karst landscapes, and hydrothermal deposits.

In general, several techniques are applied to groundwater exploration and environmental impact analysis which are not discussed in detail in this book including the use of aerial photographs and other remote sensing data and surface, subsurface, and borehole geophysical methods (Fetter, 1988, pp. 479-524 and Todd, 1980, pp. 409-457). Fracture-trace analysis has been successfully used to locate high yielding production wells and groundwater monitoring well sites. Surface geophysical methods including direct-current, seismic refraction, gravity, and magnetics are useful for determining the extent and nature of geological deposits beneath the surface and the extent of a contaminant plume. Magnetometer surveys and ground-penetrating radar are used to locate areas of waste disposal. Borehole geophysical data can help in locating zones of high porosity, hydraulic conductivity, and salinity encountered by a well.

Groundwater development and management occurs within the framework of water law (Goldfarb, 1988) consisting of myriad legal obligations, rights, and constraints at Federal, State, and local levels which are not discussed in detail in this book. The riparian, prior appropriation, Winters, and mutual prescription doctrines and the English and American Rules regulate water quantity in the United States. Each individual State has its own body of water law. In general, State laws can be more but not less strict than related federal laws. There are a large number of laws regulating water quality including the: National Environmental Policy Act of 1969 (P.L. 91-190); Federal Water Pollution Control Act of 1972 (P.L. 92-500); Clean Water Act Amendments of 1977 (P.L. 95-217); Safe Drinking Water Act of 1974 (P.L. 93-523) and amendments; Resource Conservation and Recovery Act of 1976 (RCRA) (P.L. 94-580); Comprehensive Environmental Response- Compensation and Liability Act of 1980 (CERCLA) (P.L. 96-510); Superfund Amendments and Reauthorization Act of 1986 (SARA); Surface Mining Control and Reclamation Act (SMCRA) (P.L. 95-87); Uranium Mill Tailings Radiation and Control Act of 1978 (UMTRCA) (P.L. 95-604 as amended by P.L. 95-106 and P.L. 97-415); Toxic Substances Control Act (TOSCA) (P.L. 94-469 amended by P.L. Law 97-129); and Federal Insecticide, Fungicide and Rodenticide Act (FIFRA) (P.L. 92-516, amended by P.L. 94-140, P.L. 95-396, P.L. 96-539, and P.L. 98-201).

Factors to be considered in the design, construction, and sampling of wells to monitor groundwater quality are are not covered in this book but are described by Everett (1989), Fetter (1988, pp.375-389), and Driscoll (1986, pp. 702-728). Chemical analysis of water samples is typically done in a specialized analytical laboratory, however, the hydrogeologist is usually involved in the collection of water samples. Sampling protocols have been developed by the U.S. EPA (Ford, et al., 1983) and techniques used in the sampling process are described by Ford, et al. (1983), Dunlap, et al. (1977), Fenn, et al. (1977), Gibb, et al. (1981), Scalf, et al. (1981), Barcelona, et al. (1983), and Barcelona, et al. (1985).

There is a large number of sources (see Driscoll, 1986, pp. 903-904) of information on groundwater engineering including books, journals, and governmental publications. The most frequently cited textbooks and journals are listed in Appendix A. Bi-monthly bibliographic listings of new books, journal papers, and reports appear in the New Publications, New from the USGS, New from the NTIS, and New Books sections of Ground Water, Journal of the National Water Well Association. This comprehensive bibliography, if properly used, can greatly assist readers in keeping up-to-date with current groundwater engineering literature.

A comprehensive groundwater library is maintained by the Ground Water Information Center of the National Water Well Association, 6375 Riverside Drive, Dublin, Ohio 43017. This library contains key reference works, periodicals, and data bases. It offers technical assistance, expert referrals, and document copying and loans. Many key reference works may be purchased from the National Water Well Association Ground Water Bookstore, P.O. Box 182039, Dept. 017, Columbus, Ohio 43218.

Several Universities offer curricula with broad coverage in the science of groundwater. The National Water Well Association maintains and distributes statistical information concerning educational facilities and course work. To supplement formal educational opportunities, many short courses, workshops, meetings, and conferences are sponsored by the National Water Well Association, Universities, the International Ground Water Modeling Center, professional organizations, and consulting firms. An extensive listing of these continuing education opportunities is presented bimonthly in the Meeting Calendar section of Ground Water, Journal of the National Water Well Association.

DEFINITIONS

Definitions of several important terms used throughout this book follow. An aquifer is a saturated bed, formation, or group of formations which yield water in sufficient quantity to be of consequence as a source of supply. An aquitard is a saturated bed, formation, or group of formations which yields inappreciable quantities of water to drains, wells, springs, and seeps compared to an aquifer, but, through which there is appreciable leakage of water. An aquiclude is a saturated bed, formation, or group of formations which yields inappreciable quantities of water to drains, wells, springs, and seeps and through which there is inappreciable movement of water. A formation may be classified as an aquifer in one area but only as an aquitard or aquiclude in a different area depending on the availability of water. An aquifer acts as a transmission conduit and storage reservoir.

A water table (unconfined) aquifer is one in which groundwater possesses a free surface open to the atmosphere. The upper surface of the zone of saturation is called the water table. When the water table declines, water is released from aquifer storage by the compaction of the aquifer skeleton, expansion of the water itself, and the gravity drainage of aquifer pores.

An artesian aquifer is one in which groundwater is confined under pressure by overlying and underlying aquitards or aquicludes and water levels in wells rise above the aquifer top. Artesian aquifers are classified as leaky or nonleaky depending on whether aquitards or aquicludes overlie and underlie the aquifer, respectively. The imaginary surface to which water rises in wells tapping artesian aquifers is called the potentiometric surface. When water levels in artesian aquifers decline, water is released by the compaction of the aquifer skeleton and the expansion of the water in the aquifer pores until water levels decline below the aquifer top at which time gravity drainage of aquifer pores occurs.

The porosity of a rock is a measure of the pore space of the rock and is expressed as the percentage or fraction of the total volume of the rock occupied by the pores. Representative porosity ranges for selected rocks are listed in Appendix B. The specific yield or effective pore porosity of a rock is a measure of the water-yielding capacity of rock pores and it is the percentage of the total volume of rock occupied by the ultimate volume of water released from or added to aquifer storage in a water table aquifer per unit (horizontal) area of a rock per unit decline or rise in the water table. Representative specific yield ranges for selected rocks are listed in Appendix B.

The storativity of an aquifer is defined as the volume of water the aquifer releases from or takes into storage per unit surface area of the aquifer per unit decline or rise of head in the aquifer. In Figure 1.1 (from Ferris, et al., 1962, p. 77), the volume of water released from storage in the aquifer prism divided by the product of the prism's cross-sectional area and the change in head results in a dimensionless number called the storativity. Under water table conditions, storativity is equal to the aquifer specific yield, provided gravity drainage is complete. The specific storativity is the storativity divided by the aquifer thickness. Representative ranges of storativity are listed in Appendix B.

Hydraulic conductivity is a measure of the ease of movement of groundwater through aquifers and aquitards. The horizontal hydraulic conductivity is defined as the rate of flow of groundwater through a unit cross-sectional area of the aquifer (see opening A in Figure 1.2 from Ferris, et al., 1962, p. 72) under a unit hydraulic gradient at the prevailing temperature and density of the groundwater. Representative horizontal hydraulic conductivities for selected rocks are listed in Appendix B. A related term, transmissivity, indicates the capacity of an aquifer to transmit water through its entire thickness and is equal to the hydraulic conductivity multiplied by the aquifer thickness. Transmissivity is defined as the rate of flow of groundwater through a vertical strip of the aquifer (see opening B in Figure 1.2) one unit wide and extending the full aquifer thickness under a unit

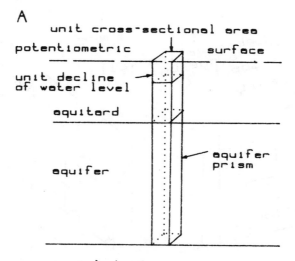

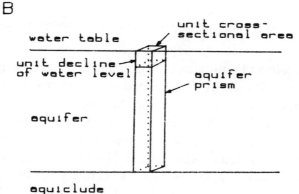

FIGURE 1.1   Diagrammatic representation of storativity under artesian (A) and water table (B) conditions.

hydraulic gradient at the prevailing temperature and density of the groundwater.

The vertical hydraulic conductivity is defined as the rate of vertical flow of groundwater through a unit horizontal cross sectional area of the aquifer or aquitard under a unit vertical hydraulic gradient at the prevailing temperature and density of the groundwater (see opening C in Figure 1.2). Representative vertical hydraulic conduc-

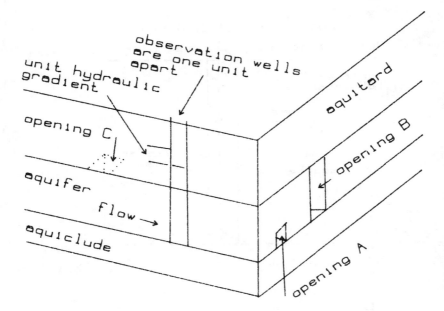

FIGURE 1.2   Diagrammatic representation of transmissivity and hydraulic conductivity.

tivities for selected aquitards and horizontal to vertical aquifer hydraulic conductivity ratios are listed in Appendix B. Commonly used unit abbreviations are listed in Appendix C.

Ranges in aquifer hydraulic conductivities and transmissivities for major groundwater regions in the United States presented by Heath (1982) are listed in Appendix B. Hydraulic conductivity in the United States ranges from 0.001 to 10000 ft/day (0.0003 to 3000 m/day) and transmissivity ranges from 5 to 5000000 ft²/day (0.5 to 500000 m²/day).

Hydraulic conductivity depends on the characteristics of the aquifer skeleton and the groundwater, whereas, intrinsic permeability (see Marsily, 1986, pp. 59-60) depends only on the characteristics of the aquifer skeleton and is related to hydraulic conductivity as follows (Bear, 1979, p. 67):

$$P = P_i \, \rho_w g / \mu_w = P_i g / \nu_w \qquad (1.1)$$

where P is the aquifer hydraulic conductivity, P_i is the aquifer intrinsic permeability, ρ_w is the groundwater density, g is the gravitational acceleration, μ_w is the groundwater dynamic viscosity, and v_w is the groundwater kinematic viscosity.

The Darcy velocity and average pore velocity of groundwater flow are derived from Darcy's Law which may be stated as follows (Darcy, 1856):

$$Q = -PIA \qquad (1.2)$$

where Q is the rate of flow of groundwater through an aquifer cross-sectional area A with a hydraulic conductivity P and a hydraulic gradient I.

The negative sign indicates that the flow of groundwater is in the direction of decreasing head.

The Darcy velocity from equation 1.2 is as follows:

$$V = Q/A = -PI \qquad (1.3)$$

Equation 1.3 assumes that flow occurs through the entire aquifer cross-sectional area without regard for the aquifer skeleton and pores. Actually flow occurs only in a portion of the aquifer pore space so that the average pore velocity is defined as follows:

$$v = Q/(An_e) = V/n_e = -PI/n_e \qquad (1.4)$$

where n_e is the aquifer effective porosity.

An aquifer is said to be homogeneous when hydraulic conductivity is independent of space location; if hydraulic conductivity is dependent on space location, the aquifer is said to be heterogeneous. An aquifer is said to be isotropic when hydraulic conductivity is independent of space direction. If hydraulic conductivity is dependent on space direction, the aquifer is said to be anisotropic.

The specific weight of a substance at a stated reference temperature is defined as its density multiplied by the local acceleration due to gravity. It is the weight per unit volume that takes into account the magnitude of the local gravitational force. The mass of any substance is the weight of the substance divided by the local acceleration due to gravity. Inasmuch as weight is dependent on the local gravitational force, it varies with location. However, mass is an absolute property that does not change with location. The mass of a contaminant source is the product of the solubility (or density in the case of infinite solubility) and the volume of the contaminant. Density is defined as mass per unit volume and does not change with location.

Dispersivity is a measure of the mechanical contaminant dispersion property of porous materials. It is defined as a characteristic mixing length (Bear, 1979, p. 234).

Thermal conductivity is defined as the rate of flow of heat through a unit cross-sectional area of the aquifer under a unit temperature gradient. Thermal diffusivity is analogous to aquifer transmissivity divided by aquifer storativity and is equal to the aquifer thermal conductivity divided by the product of the aquifer specific heat and density. The specific heat of an aquifer is the quantity of heat necessary to raise the temperature of a unit mass of the aquifer by 1°. The calorie is the quantity of heat required to raise the temperature of 1 gm of water by 1°C.

Equations in this book will accept any consistent set of units. Commonly used unit conversion factors are presented in Appendix C. Difficulty with the calculation of the constant for a particular groundwater flow or contaminant migration equation with a desired system of units can be minimized by writing an equation of units (dimensional analysis) compatable with the particular equation. For example, suppose the particular equation is $s = QW(u)/(4 T)$ where s is drawdown at time t, Q is production well discharge rate, and T is aquifer transmissivity. With s in foot, Q in gallons per minute, and T in gallons per day per foot the equation of units is: foot = [(gallons/minute) × 1440(minutes/day)]/[(gallons/day)/foot × (4 × 3.1416)] or 114.6QW(u)/T.

NATURAL AND ARTIFICIAL RECHARGE

Natural recharge from precipitation is irregularly distributed in time and space and depends on many factors. Among these are the:

geological character and thickness of the soil and underlying deposits; topography; vegetal cover; land use; soil moisture content; depth to the water table; intensity, duration, and seasonal distribution of precipitation; occurrence of precipitation as rain or snow; and air temperature and other meteorological factors (humidity, wind, etc.).

There are several methods for estimating natural recharge including streamflow hydrograph separation studies, regional hydrologic and groundwater budget studies, and regional flow net analysis. In streamflow hydrograph separation methods, total streamflow is separated into its two major components, surface and groundwater runoff. Surface runoff reaches streams rapidly and is discharged from a drainage basin usually within a few days. Groundwater runoff percolates slowly towards streams and is discharged from a drainage basin gradually. Commonly, a few (3 to 5) days after precipitation ceases there is no surface runoff and streamflow is derived entirely from groundwater runoff. Groundwater runoff is the portion of natural recharge that is not removed by evapotranspiration and constitutes the lower limit of natural recharge.

Groundwater runoff several days after precipitation ceases is readily determined from streamflow hydrographs, however, groundwater runoff under flood hydrographs is subjective. During floods, hydrographs may be approximated in the following manner (Linsley, et al., 1958). A straight line (line 1) is drawn from the hydrograph point of rise to the hydrograph N days after the peak. N is defined as the time after the peak of the streamflow hydrograph at which surface water runoff terminates and is equal to $A^{0.2}$ where A is the drainage basin area in square miles. Another line (line 2) is drawn by projecting the recession of the streamflow after the storm back under the hydrograph to a point under the inflection point of the falling limb. An arbitrary rising limb (line 3) is drawn from the point of rise of the hydrograph to connect with the projected streamflow recession (line 2). Groundwater runoff is estimated as the average of lines 1 and 3 taking into consideration frozen ground and climatic conditions. These procedures give reasonably accurate estimates of groundwater runoff even though the lines do not describe the exact sequence of events occurring during storms.

The streamflow hydrograph separation method and streamflow

data for years of near-normal, below-normal, and above-normal precipitation were used to determine annual groundwater runoff from 109 drainage basins in Illinois and 38 drainage basins in Minnesota by Walton (1965) and Ackroyd, et al. (1967). Groundwater runoff in Illinois and Minnesota ranged from 0.14 inch (0.04 m) per year with below-normal precipitation to 5.7 inches (1.74 m) per year with near-normal precipitation (about 18% of precipitation) to 11 inches (3.35 m) per year with above-normal precipitation. 1 inch (0.3 m) of water over 1 square mile of drainage basin per year is equivalent to 47700 gpd (185.68 m/day). In general, groundwater runoff in Illinois and Minnesota is greatest from glaciated and unglaciated drainage basins having considerable surface deposits of sand and gravel and underlain by permeable bedrock. Groundwater runoff in Illinois and Minnesota is least from glaciated drainage basins with surface lakebed sediments and underlain by impermeable bedrock. Groundwater runoff in Illinois and Minnesota is increased by the presence of buried bedrock valleys and lakes and wetlands.

A hydrologic budget is a conservation of mass statement expressing the balance between the total water gains and losses and changes in water storage in a drainage basin. Hydrologic budget factors include: precipitation, surface and groundwater evapotranspiration, surface water inflow, imported water, groundwater inflow, surface water outflow, groundwater outflow, reservoir evaporation, exported water, changes in soil moisture, and changes in groundwater storage. Precipitation, streamflow, transported water, changes in soil moisture, and reservoir evaporation are measured directly. Evapotranspiration is usually estimated with equations derived by Thornthwaite and Hare (1965, pp. 163-180) or Penman (1948, pp. 120-145). Groundwater inflow, outflow, and changes in storage are usually estimated from aquifer hydraulic characteristics and measured water levels in wells.

A groundwater budget (sub-hydrologic budget) is a statement expressing the balance between the total groundwater gains and losses and changes in storage in a subsurface drainage basin. Groundwater budget factors include: natural and artificial groundwater recharge; groundwater evapotranspiration; groundwater runoff; groundwater discharge to wells, springs and other man-made devices; subsurface inflow; subsurface outflow; and changes in groundwater storage. Usually, groundwater discharge, artificial recharge, and changes in storage are measured directly. Groundwater runoff is frequently estimated with streamflow hydrograph separation methods and subsurface inflow and outflow are estimated from aquifer hydraulic characteristics and measured water levels. Double groundwater rating curves showing the relation between the water table and streamflow, one for times when evapotranspiration is high, and the other for times when evapotranspiration is low, are frequently used to estimate groundwater evapotranspiration. In some basins and for some inventory periods, several groundwater budget factors are negligible and the complexity of the groundwater budget is reduced considerably.

Schicht and Walton (1961) prepared hydrologic and groundwater budgets for 3 small watersheds in central Illinois during years of near-normal, below-normal, and above-normal precipitation. Budget factors for a year of near-normal precipitation in central Illinois are listed in Appendix D. Groundwater recharge in central Illinois ranges from 10 to 28% of precipitation in the year of near-normal precipitation and is 4.5% of precipitation during a year of below-normal precipitation. Groundwater recharge in central Illinois generally is at a maximum during the spring and most recharge occurs prior to July. In many years, very little recharge occurs during the five-month period July through November.

Natural recharge is an important factor in groundwater modeling studies. The precision of most estimates of natural recharge is not high. Natural recharge may increase or decrease with lowered water levels as a result of development. Ranges in recharge rates for years of near-normal precipitation and major groundwater regions in the United States presented by Heath (1982) are listed in Appendix D. These rates range from 0.001 to 40 inches (0.0003 to 12.19 m) per year or from about 2 to 40% of annual precipitation.

Commonly, natural recharge, except in desert and especially humid areas, is estimated to be about 10 to 20% of annual precipitation. Some natural recharge rates for years of near-normal precipitation and selected locations are: (Walton, 1970, pp. 360-438)

Langlade County Wisconsin — 5.4 inch (1.65 m) per year or 18% of precipitation; Goshen County, Wyoming — 0.7 inch (0.21 m) per year or 5% of precipitation; Clay County, Nebraska — 1.56 inch (0.48 m) per year or 6.5% of precipitation; Olds, Alberta, Canada — 0.48 inch (0.15 m) per year or 3% of precipitation; Dayton, Ohio — 12 inches (3.66 m) per year or 33% of precipitation; and Mason and Tazewell Counties, Illinois — 10.5 inches (3.20 m) per year or 28% of precipitation.

Artificial recharge methods (Huisman and Olsthoorn, 1983) include water spreading, recharging through pits and wells, and pumping to induce recharge from surface water bodies. Techniques of artificial recharge are described in detail by Todd (1980, pp. 458-493). Basin and stream channel spreading are the most widely used methods. Spreading recharge rates usually range from 0.1 to 2.9 m/day (0.328 to 9.51 ft/day). Because recharge wells easily clog and lose their efficiency, well recharging has been limited to a few areas where extensive experience in water treatment and redevelopment of wells is available. Average recharge well rates range from 200 to 5600 m³/day (706.2 to 197736 ft³/day). Wastewater recharge for nonpotable reuse has been practiced with success at some places in recent years. Kazmann (1948, pp. 404-424) and Klaer (1953, pp. 620-624) describe streambed induced infiltration facilities.

QUALITY REQUIREMENTS

Groundwater quality is determined by the solutes and gases dissolved in the water and the matter suspended in and floating on the water. Common units of measurement of groundwater quality are milligrams per liter (mg/L) and micrograms per liter (µg/L). It is usually assumed that 1 part per million (ppm) is equal to 1 mg/L. The principles of chemical thermodynamics provide a framework for interpreting chemical analyses of groundwaters and for defining the processes by which groundwater attains its observed chemical character (Hem, 1970; Drever, 1982). The law of mass action and the concepts of activity coefficients, equilibrium constants, free energy, and hydrochemical facies (Freeze and Cherry, 1979, pp. 89-

133) lead to the description of the groundwater chemistry in relation to the geohydrologic environment and to the prediction of changes in chemical character of groundwater over space and time.

Pictorial diagrams of chemical analyses of natural groundwaters are useful for display purposes, comparing analyses to emphasize differences and similarities, detecting and identifying mixtures of groundwaters of difference composition, and identifying chemical processes resulting from the circulation of groundwaters. Widely used pictorial graphing techniques are (Matthess, 1982, pp. 299-321): bar graphs, circular diagrams, radial diagrams, trilinear diagrams, four-coordinate diagrams, two-coordinate diagrams, parallel scale diagrams, horizontal scale diagrams, and vertical scale diagrams.

The microcomputer program PLOTCHEM developed by TECSOFT and distributed through the Scientific Software Group produces Piper trilinear, Stiff, and radial diagrams; pie charts; and bar charts from epm, ppm, or meq/l input units. A data base file created with AQUABASE may be imported by PLOTCHEM. AQUABASE is a microcomputer three-dimensional water quality data storage and retrieval system developed by TECSOFT. FORTRAN programs for plotting pictorial diagrams are developed by Morgan and McNellis (1969) and McNellis and Morgan (1969).

The trilinear diagram is commonly used because it forms the basis for the classification of natural groundwaters. Sources of dissolved constituents in natural groundwaters, modifications in the character of groundwater as it passes through an area, mixing of groundwaters, and related geochemical problems are studied with the Piper (1944, pp. 914-923) trilinear diagram. In this diagram, groundwater is treated as though it contained three cation constituents (Mg, Na, and Ca) and three anion constituents (Cl, SO_4, and HCO_3). Less abundant constituents than these are summed with the major constituents to which they are related in chemical characteristics.

The Piper trilinear diagram (Figure 1.3) combines three distinct fields for plotting, two triangular fields at the lower left and right and an upper diamond-shaped field (Piper, 1953). All three fields have scales ranging from 0 to 100%. In the triangular field at the lower left, the percent of equivalents per million values of the cation group

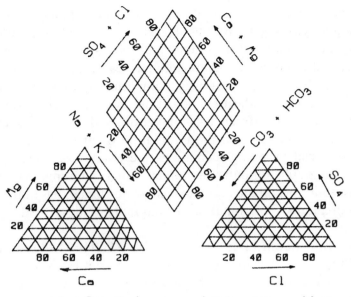

percent of total equivalents per million

FIGURE 1.3   Piper trilinear diagram.

(Ca, Mg, and Na) for a groundwater sample are plotted as a single point according to conventional trilinear coordinates. The anion group (Cl, SO_4, and HCO_3) is plotted likewise in the triangular field at the lower right. Thus, two points on the diagram, one in each of the two triangular fields, indicate the relative concentrations of the several dissolved constituents of a groundwater. The subtotal of all cation or anion equivalents per million is taken as the 100% base for calculating the percent of equivalent per million value of individual cations or anions.

The upper diamond-shaped field is used to show the overall chemical character of the groundwater by a third single-point plotting, which is at the intersection of rays projected from the lower left and right plots. The position of this plotting indicates the relative composition of a groundwater in terms of the cation-anion pairs that correspond to the four vertices of the field. The three trilinear plots show the essential chemical character of a groundwater according to the relative concentration of its constituents, but not according to the

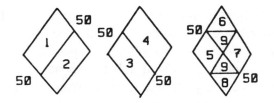

FIGURE 1.4   Subdivisions of diamond-shaped field of Piper trilinear diagram.

absolute concentrations. Because the absolute concentrations commonly are decisive in many problems of interpretation, it is convenient to indicate the plotting in the upper field by a circle whose area is proportional to the absolute concentration of the groundwater. The chemical character of groundwaters containing free acid in substantial quantities (hydrogen is present as a cation) cannot be fully represented in the trilinear diagram.

Distinct groundwater types can be quickly determined by their plots in certain subareas of the central diamond-shaped field (Piper, 1944, pp. 914-928) as indicated in Figure 1.4 and the following explanation: area 1, alkaline earths exceed alkalies; area 2, alkalies exceed alkaline earths; area 3, weak acids exceed strong acids; area 4, strong acids exceed weak acids; area 5, carbonate hardness (secondary alkalinity) exceeds 50% and chemical properties of the groundwater are dominated by alkaline earths and weak acids; area 6, noncarbonate hardness (secondary salinity) exceeds 50%; area 7, noncarbonate alkali (primary salinity) exceeds 50% and the chemical properties of the groundwater are dominated by alkalies and strong acids; area 8, carbonate alkali (primary alkalinity) exceeds 50% and the groundwater is inordinately soft in proportion to its content of dissolved solids; area 9, no one cation-anion pair exceeds 50%.

Hydrochemical facies (see Freeze and Cherry, 1979, pp. 250-254) describe zones of groundwater that differ in their chemical composition as a result of the interaction of water with the aquifer lithologic framework. Hydrochemical facies are commonly classified by means of the trilinear diagram.

Interpretations of groundwater quality data are often made to determine if the water is satisfactory for a proposed use. Whether a groundwater is suitable for a proposed use depends on the criteria, classifications, or standards of acceptable quality for that use. Water quality standards are regulations that set specific limitations on the quality of water that may be applied to a specific use. Water quality criteria are values of dissolved substances and their toxicological and ecological meaning. Public, industrial, and agricultural supplies have different quality requirements. A classification of saline groundwaters developed by Carroll (1962) and a groundwater hardness classification developed by Sawyer and McCarty (1967) are presented in Appendix E.

The U.S. Environmental Protection Agency (EPA) has been directed to establish drinking water standards under the provisions of Public Law 93-523, the Safe Drinking Water Act of 1974, and its amendments. In a continuing process, maximum contaminant-level goals (MCLGs), nonenforceable health goals, are set at a level to prevent known or anticipated adverse effects with an adequate margin of safety. Maximum contaminant levels (MCLs), enforceable standards, are set as close to the MCᵀ Gs as is feasible based on water-treatment technologies and cost, Primary MCLs are promulgated for substances with a health risk, and secondary maximum contaminant levels (SMCLs) are promulgated for substances which can affect the aesthetic quality of water by imparting taste and odor and staining fixtures. National Interim Drinking Water Standards (Federal Register, Feb. 1978, No. 266) are listed in Appendix E.

Periodically, Congress assigns additional compounds for which MCLGs and MCLs will be established by EPA and the short list of Interim Drinking Water Standards is expanded into a longer list of Final Drinking Water Standards. Contaminants regulated under the Safe Drinking Water Act, 1986 amendments (Federal Register, Nov. 13, 1985) are listed in Appendix E. The status of MCLGs and MCLs for selected contaminants as of 1987 is presented in Appendix E. MCLs are enforced for all public water-supply systems by the various states which may set MCLs that are more strict than the federal standards, but not less stringent. Public water-supply systems that rely on groundwater are required to perform a complete analysis of the water for the drinking water standards prior to the time a drinking water supply well is put into service and periodically thereafter.

Chemical requirements of groundwaters used in different industrial processes vary widely. For example, makeup water for high-pressure boilers must meet extremely exacting criteria whereas water of as low a quality as sea water can be satisfactorily employed for cooling of condensers. Even within each industry, criteria cannot be established, instead, only recommended limiting values or ranges can be stated. Of equal importance for industrial purposes as quality is the relative constancy of the various constituents.

In some areas, groundwaters have been used extensively for industrial purposes and heat-pumps (Gass and Lehr, 1977, pp. 42-47) because of their low and relatively constant temperature. Seasonal groundwater temperatures are usually damped out below depths of 32.81 ft (10 m) in the tropics and 65.62 ft (20 m) in polar regions. Groundwater temperatures at these depths are about 1° to 2° C higher than local mean annual air temperature (see Todd, 1980, pp. 307-310). In the United States, these temperatures range from about 4°C in the northern part to 10°C in the northwest to 20°C in the south. Below depths of 32.81 to 65.62 ft (10 to 20 m), groundwater temperatures normally increase from 1 to 5°C (2.9°C on the average) for each 328 ft (100 m) of depth increase (Bouwer, 1978, pp. 378-380).

It is technically possible to treat any water to give it a desired quality, however, if the water requires extensive treatment, it may not be economically feasible to utilize the supply. Suggested water quality criteria for selected industries developed by the American Water Works Association (1971) are presented in Appendix E.

Range cattle seem to be able to use water containing a high amount of dissolved solids, but a high proportion of sodium or magnesium and sulfate is very undesirable for stock use. A high concentration of selenium in water can be very poisonous to animals. Maximum concentrations of total and specific ions in drinking water for farm animals, as recommended by the National Academy of Sciences and the National Academy of Engineering (1972), are presented in Appendix E.

Factors to be considered in evaluating the usefulness of groundwaters for irrigation are the: total concentration of dissolved solids, concentration of some individual constituents, relative proportions of some constituents, nature and composition of the soil and subsoil, topography, depth of the water table, amounts of groundwater used and method of application, kinds of crops grown, and climate of the area. Most of the soluble matter in irrigation groundwater remains behind in the soil. In order to maintain the productivity of irrigated soil on a permanent basis, excess soluble matter left in the soil from irrigation must be removed by leaching the topsoil. If the water table rises excessively because of irrigation, leaching may not be effective. Lack of proper drainage of the water table to remove excess soluble matter and groundwater will cause the abandonment of irrigation and land. In many irrigated areas, open ditches, buried drain tile, or drainage wells are used to provide drainage and the necessary leaching of soluble matter from the soil.

Boron is essential for proper plant nutrition, however, a small excess over the needed amount is toxic to some plants (see Hem, 1970). Permissible boron concentrations suggested by Wilcox (1955) are presented in Appendix E. The relative tolerances of selected plants to boron are listed in Appendix E (Salinity Laboratory, 1954).

The sodium adsorption ratio (SAR), defined as follows (Salinity Laboratory, 1954), is frequently used to determine the suitability of groundwaters for irrigation purposes:

$$SAR = Na/[(Ca + Mg)/2]^{0.5} \qquad (1.5)$$

where Na, Ca, and Mg refer to the concentrations of these ions expressed in meq/L units. Factors for converting concentrations expressed in mg/L to concentrations expressed in milliequivalents per liter meq/L are listed in Appendix C.

A soil high in exchangeable sodium is very undesirable for agriculture because it can become deflocculated and tends to have a relatively impermeable crust. This condition is promoted by waters of high SAR and is reversed by waters containing a high proportion of calcium and magnesium (Hem, 1970). Soil amendments such as gypsum or lime may correct this situation. A diagram for evaluation of the suitability of irrigation waters based on SAR and developed by the Salinity Laboratory (1954) is shown in Figure 1.5. The

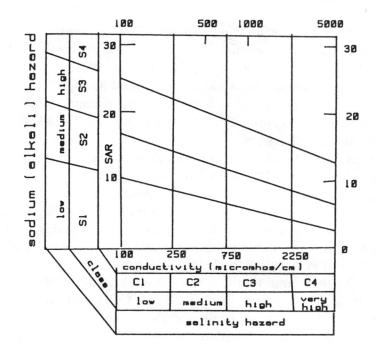

FIGURE 1.5   Diagram for use in determining the suitability of groundwater for irrigation purposes.

relative tolerances of selected crops to the salinity of water are presented in Appendix E (Salinity Laboratory, 1954).

CONTAMINANT SOURCES AND REMEDIAL MEASURES

Contaminants are solutes reaching aquifer systems as a result of man's activities. Pollution occurs when contaminant concentrations reach objectionable levels. Man's interference with natural flow patterns and his introduction of chemical and biological material into the ground usually results in undesirable groundwater quality changes. Contaminant sources include (Todd, 1980, pp. 316-337 and Fetter, 1988, pp. 406-421): municipal sewer leakage, liquid waste disposal, and solid waste disposal (Page, et al., 1987); urban runoff; lawn fertilizer application; industrial liquid waste disposal, tank, and pipeline leakage; mining activities; chemical spills; cooling water, process water, and oilfield brine disposal; agriculture irrigation return flows; animal feed lot and waste disposal; fertilizer, herbicide, and pesticide application (D'Itri and Wolfson, 1987 and Fairchild, 1987); domestic wastewater disposal through septic tanks and cesspools (Canter and Knox, 1985 and Kaplan, 1987); salt-water intrusion; spreading of highway deicing salts; stockpile run-off; and surface discharges.

The general characteristics of contaminants from these sources are listed by Miller (1980). Common point and nonpoint sources are seepage pits and trenches, percolation ponds, lagoons, waste disposal facilities, streambeds, landfills, deep disposal wells, injection wells, surface spreading and irrigation areas, and farming areas.

The largest component of municipal land disposal of solid wastes is paper, but, substantial food wastes, yard wastes, glass, metals, plastics, rubber, and liquid wastes are also included. Landfill leachates can contain high levels of BOD, COD, iron, manganese, chloride, nitrate, hardness, heavy metals, stable organics, and trace elements. Gases such as methane, carbon dioxide, ammonia, and hydrogen sulfide are by-products of municipal landfills. Many municipal waste disposal sites receive industrial process residuals and pollution control sludges. Radioactive, toxic, and hazardous wastes have been disposed of in some municipal landfills.

Municipal waste water may reach aquifers by leakage from collecting sewers, leakage from the treatment plant during processing, land disposal of the treatment plant effluent, effluent disposal to surface waters which recharge aquifers, and land disposal of sludge. Sewer leakage can introduce high concentrations of BOD, COD, nitrate, organic chemicals, bacteria, and heavy metals into groundwater. Potential contaminants from sludge include nutrients, heavy metals, and pathogenic organisms. Potential contaminants from industrial waste disposal sites cover the full range of inorganic and organic chemicals including phenols, acids, heavy metals, and cyanide. Chemicals known to have contaminated groundwater and their industrial uses are listed in Appendix F (OTA, 1984). Common industrial solvents such as trichloroethylene, 1,1,1-trichloroethane, tetrachlorethane, benzene, and carbon tetrachloride have been detected in some groundwaters. Major industrial categories with significant contamination potential include the: pulp and paper industry; petroleum refining industry; steel industry; organic chemical industry; inorganic chemicals, alkalies, and chlorine industry; plastic material and synthetics industry; nitrogen fertilizer industry; and phosphate fertilizer industry.

The regulated disposal of hazardous and toxic industrial wastes is sometimes accomplished by the installation of deep injection wells (Warner and Lehr, 1981). Kimbler, et al. (1975) presents equations and lists FORTRAN computer programs pertaining to the injection of wastes with dispersion and segregation of two fluids due to density differences (interface is inclined with respect to the vertical).

Water associated with oil-field brines and containing high levels of sodium, calcium, ammonia, boron, chloride, sulfate, trace metals, and dissolved solids has contaminated some aquifers. The principle groundwater contaminants from mine waste are acidity, dissolved solids, metals, radioactive materials, color, and turbidity. The most prevalent contamination problem associated with coal mining is the formation and discharge of large volumes of acid.

Irrigation return flows in arid and semiarid regions with high levels of calcium, magnesium, sodium, bicarbonate, sulfate, chloride, and nitrate can be a major cause of groundwater contamination. Animal wastes may transport salts especially nitrate-nitrogen, organic loads, and bacteria into and below the soil. Nitrogen in solution is the primary agricultural fertilizer contaminant.

Septic tank effluent contains bacteria and viruses. Significant amounts of nitrogen can be added to groundwater through the use of septic tanks and drain tile fields. Suburban areas frequently have groundwater with high levels of nitrate due to the use of lawn fertilizers.

Seller and Canter (1980) describe six empirically based site-specific prevention methodologies for assessing the potential for groundwater contamination from sources. The six methodologies involve a landfill or hazardous waste disposal facility and include the: U.S. EPA methodology; criteria listing method; water balance methods; Hagerty, Pavoni, and Heer system; Phillips, Nathwane, and Mooij matrix; and LeGrand methods. Other methodologies are described by the Michigan Department of Natural Resources (1983) and Hutchinson and Hoffman (1983)

The popular LeGrand method (LeGrand, 1983) assigns a score to each of four disposal site conditions in addition to assigning a score to characterize the waste itself. Summing the five scores then gives an assessment of the groundwater contamination potential of the site. One of the primary advantages of the method is its simplicity; its major disadvantage is the subjectiveness of the scoring process and the need for an experienced evaluator to compile the scores. Factors considered in the method include soil hydraulic conductivity, soil sorption, depth to the water table, groundwater flow gradient, distance from contaminant source to point of use, and groundwater quality.

A standardized system for evaluating groundwater contamination potential using hydrogeologic settings is described by Aller, et al. (1987). The system has two major portions: the designation of mappable units, termed hydrogeologic settings, and the superposition of a relative rating system called DRASTIC. Hydrogeologic settings incorporate factors which affect and control groundwater movement including: depth to water, net recharge, aquifer media, soil media, topography, impact of the unsaturated zone, and aquifer hydraulic conductivity. These factors are incorporated into a relative ranking scheme that uses a combination of weights and ratings to produce a numerical value called the DRASTIC index.

The decision tree approach, commonly used by regulatory agencies, is a logical step by step process for assessment of the site contamination potential. This approach begins with the most important question, followed by a hierarchy of decreasing critical questions. In this manner, a "no" answer to an early important question can eliminate the site from further consideration and, from a practical standpoint, prevent the expenditure of unnecessary money for additional site investigations. A "no" answer may also indicate that an alternative type of waste disposal site or disposal method should be utilized (Cal. Dept. of Health, 1975).

The configuration of contamination is unique for each source because of variable hydrogeological conditions and mechanisms of contaminant migration. Most contaminant plumes tend to be elliptical in shape and plumes expand or contract generally in response to changes in the rate of waste discharge. Plumes may become stable with constant waste discharge because of attenuation mechanisms or the contaminant reaches a groundwater discharge boundary such as a stream. Contaminant plume dimensions commonly encountered are: width — 100 to 2000 ft (30.48 to 609.6 m), length — 200 to 10000 ft (60.96 to 3048 m), and depth — 25 to 200 ft (7.62 to 60.96 m). Associated water table hydraulic gradients usually range from .5 to 500 ft/mile (0.094 to 94.72 m/km).

A large number of test borings and monitoring wells are required to delineate and monitor the horizontal and vertical movement of a contaminant plume. The number of test borings required for plume delineation commonly ranges from 10 to 50 and the number of installed monitor wells usually ranges from 5 to 20. Frequently groundwater quality at different depths must be sampled and multi-level sampling devices are often installed.

Whenever possible, the input of contaminants into an aquifer should be controlled or eliminated. Source control measures reduce the volume of wastes to be handled or reduce the threat a certain waste poses by altering its physical or chemical makeup. Potential source control strategies include: recycling, resource recovery, centrifugation, filtration, sand drying beds, chemical fixation, de-toxification, degradation, encapsulation, waste segregation, co-disposal, and leachate recirculation (Canter and Knox, 1985).

Nonpoint fertilizer, herbicide, and pesticide losses to groundwater may be reduced by decreasing application rates and by applying fertilizers, herbicides, and pesticides at the time they are needed and in forms which are available to the crop but not subject to excessive leaching. Surface contamination from a feedlot area can be controlled through the use of appropriate land disposal procedures.

If source control measures are not feasible then other appropriate measures must be taken to confine contaminant damage, protect drinking water supply wells, and decontaminate the aquifer. These remedial measures are time-consuming (months to years) and very expensive (tens of thousands to millions of dollars). It often takes longer to decontaminate an aquifer then it took to contaminate the aquifer.

Remedial measures remove or isolate point sources and/or pump and treat contaminated groundwater (JRB Associates, 1982). Remedial measures include: changing the surface drainage so that runoff does not cross the source, using source subsurface drains and ditches, constructing low-permeability caps above the source, installing a low-permeability vertical barrier (slurry wall, grout curtain, or sheet piling) around the source, lowering the water table where it is in contact with the source, chemical or biological in situ treatment of the source plume, modifying nearby production well discharge patterns, changing water table hydraulic gradients through the installation of injection wells, and extracting contaminated groundwater via production wells.

Caps and top liners reduce infiltration into a waste site and thereby reduce the quantity of leachate generated at a source. Subsurface drains, ditches, and bottom liners installed in the unsaturated zone capture leachate before it reaches the water table. Vertical barriers divert uncontaminated groundwater around a waste site or limit the migration of contaminated groundwater. Injection wells improve the efficiency of contaminant plume capture by modifying groundwater flow patterns and flush contaminants toward production wells.

Extraction wells are designed to capture the contaminant plume by removing as little uncontaminated water as possible. Extraction wells are usually located within the plume and sized to control water table hydraulic gradients towards extraction wells. Plume-stabilization wells may be designed to reverse the water table hydraulic gradient beyond the edge of the plume and prevent further movement of the plume. Locating an extraction well outside the plume tends to expand the plume boundaries. The positions of extraction wells and screened zones are designed so that a line of equal travel time is created which encloses the plume to be removed as narrowly as possible (Kinzelbach, 1986, pp. 239-252).

Removal of contaminated water through extraction wells with aquifer advective and dispersion mechanisms but without aquifer sorption mechanisms requires that a volume of groundwater about twice the volume of the contaminant plume be removed from the aquifer. Ten to 100 times more groundwater needs to be removed with aquifer advection, dispersion, and desorption mechanisms (Gilham, 1982). Contaminated water that has been removed from the plume must be treated before being discharged. Types of treatment include: air stripping, activated carbon adsorption, and biological processes for organics and chemical precipitation for inorganics (Canter and Knox, 1985).

Design features of remedial measures are usually determined with analytical, semi-analytical, and numerical models as described by Boutwell, et al. (1986). Many of the analytical model equations and numerical model techniques presented in this book are applicable to the evaluation of the effectiveness of remedial measures. Data on the costs of remedial measures is presented by SCS Engineers (1982).

2

Nonleaky Artesian Radial Flow

When a water supply production well in a uniformly porous nonleaky artesian aquifer (Figure 2.1) is pumped, water is continuously withdrawn from storage within the aquifer to balance pumpage as the cone of depression progresses radially outward from the well. Water levels decline (drawdown) and there is no stabilization of the head in the aquifer. However, the rate of the decline in head continuously decreases as the cone of depression spreads. Water is released instantaneously from storage by the compaction of the aquifer skeleton and by the expansion of the water itself. The cone of depression is described by the following equation (Theis, 1935, pp. 519-524):

$$s = QW(u)/(4\pi T) \tag{2.1}$$

with

$$W(u) = -0.577216 - \ln(u) + u - u^2/2.2! \\ + u^3/3.3! - u^4/4.4! + ... \tag{2.2}$$

$$u = r^2S/(4Tt) \tag{2.3}$$

where s is the drawdown at time t, Q is the constant production well discharge rate, W(u) is a well function [sometimes designed as $-E_i(-u)$ or $E_1(u)$], T is the aquifer transmissivity, r is the distance between the production and observation wells, S is the aquifer storativity, t is the time after pumping started, and $\pi = 3.141592654$.

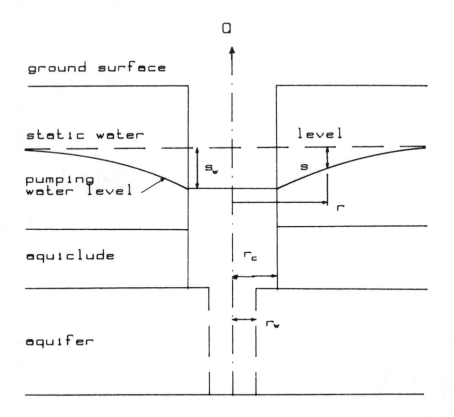

FIGURE 2.1   Cross section through production well in nonleaky artesian aquifer.

The drawdown inside the production well s_w is obtained at $r = r_w$ with r_w being the effective production well radius.

Values of W(u) for the practical range of u (Ferris, et al., 1962, pp. 96-97) are listed in Appendix G. Logarithmic and semilogarithmic graphs of W(u) vs. 1/u and W(u) vs. u are shown in Figures 2.2 to 2.5. The shapes of time-drawdown curves are defined by the graphs in Figures 2.2 and 2.4 and the shapes of cones of depression are defined by the graphs in Figures 2.3 and 2.5.

Often, a particular value of u is not tabled and W(u) must be interpolated based on available u and W(u) values. The precision of interpolation depends upon the spacing of tabled values and the

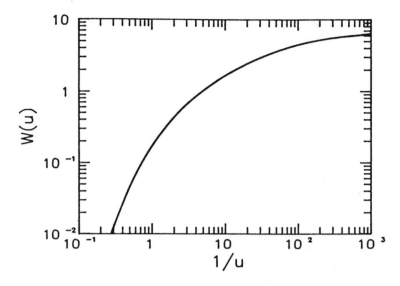

FIGURE 2.2   Logarithmic graph of W(u) vs 1/u.

method of interpolation. Linear interpolation is less precise than curvilinear interpolation. Linear interpolation (see Figure 2.6), which has a precision acceptable for many field applications, is based on the following equation (Davis, 1986, p. 147):

$$y_0 = [(y_2 - y_1)(x_0 - x_1)]/(x_2 - x_1) + y_1 \qquad (2.4)$$

where y_0 is the calculated W(u) value for the interpolation point, x_0 is the known u value for the interpolation point, x_1 and x_2 are the known u values for tabled points immediately surrounding the interpolation point, and y_1 and y_2 are the known W(u) values for tabled points immediately surrounding the interpolation point.

Tabled u values are searched to identify the u values immediately surrounding the interpolation point. This is accomplished by comparing the u value of the interpolation point with tabled u values and flagging the two u values (x_1 and x_2) which satisfy the relation $x_1 < x_0 < x_2$. The tabled W(u) values (y_1 and y_2) corresponding to the flagged u values are then identified. Finally, values of x_0, x_1, x_2, y_1, and y_2 are substituted into Equation 2.4 to calculated y_0.

Lagrangian curvilinear interpolation is based on the following equation (Clark, 1987, p. 5.3):

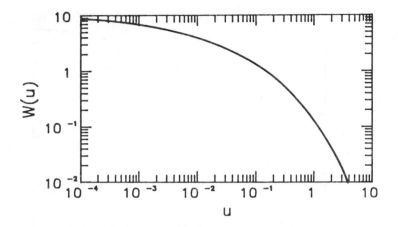

FIGURE 2.3   Logarithmic graph of W(u) vs u.

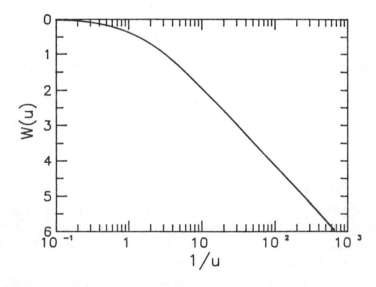

FIGURE 2.4   Semilogarithmic graph of W(u) vs 1/u.

$$y_0 = \{[(x_0 - x_2)(x_0 - x_3)...(x_0 - x_n)]/[(x_1 - x_2)$$
$$(x_1 - x_3)...(x_1 - x_n)]\}y_1 + \{[(x_0 - x_1)(x_0 - x_3)...(x_0 - x_n)]/$$
$$[(x_2 - x_1)(x_2 - x_3)...(x_2 - x_n)]\}y_2 + ... + \{[(x_0 - x_1)$$
$$(x_0 - x_2)...(x_0 - x_{m-1})]/[(x_n - x_1)$$
$$(x_n - x_2)...(x_n - x_{m-1})]\}y_n \tag{2.5}$$

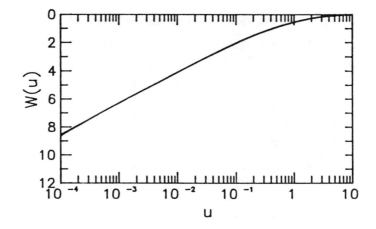

FIGURE 2.5   Semilogarithmic graph of W(u) vs u.

where x_1, x_2,...x_n are the known u values for tabled points; y_1, y_2,...y_n are the known W(u) values for tabled points; x_0 is the u value for the interpolation point; y_0 is the calculated W(u) value for the interpolation point; and n is the number of known points.

A BASIC subroutine for Lagrangian interpolation between known points on a user-defined curve is listed by Poole, et al (1981, pp. 84-85). A BASIC subroutine for Lagrangian interpolation of user-defined well functions in one- and two-dimensions is listed by Clark (1987, pp. 5.1-5.14).

Interpolation between pairs of tabled points may also be accomplished with a spline function (see Press, et al, 1986, pp. 86-89). A spline is a low-level polynomial, normally a cubic, fitted to adjacent points whose coefficients are determined slightly nonlocal. A line may be extended through points to produce a continuous smooth curve with a spline function. A BASIC spline subroutine is listed by Fowler (1984, pp. 301-305) and a FORTRAN spline subroutine is listed by Press, et al (1986, pp. 88-89).

When $u \leq 0.01$ then $W(u) = -0.5772 - \ln(u) = \ln(0.562/u)$ (Hantush, 1964, p. 321). According to Huisman and Olsthoorn (1983, p. 68), if $u < .25$ then $W(u) = \ln(0.78/u)$ has an error of less than 1 percent. If $u > 1$ then $W(u) = \exp(-1.2u - 0.60)$.

Values of W(u) are commonly calculated with the following polynomial approximation (Abramowitz and Stegun, 1964):

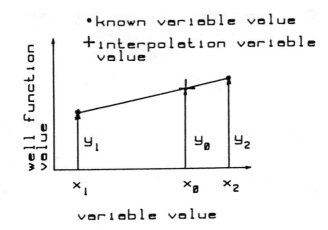

FIGURE 2.6   Graph showing well function linear interpolation symbols .

when $0 < u \leq 1$

$$W(u) = -\ln u + a_0 + a_1u + a_2u^2 + a_3u^3 + a_4u^4 + a_5u^5 \quad (2.6)$$

where

$$
\begin{array}{ll}
a_0 = -.57721566 & a_3 = .05519968 \\
a_1 = .99999193 & a_4 = -.00976004 \\
a_2 = -.24991055 & a_5 = .00107857
\end{array}
$$

when $1 < u < \infty$

$$W(u) = [(u^4 + a_1u^3 + a_2u^2 + a_3u + a_4)/$$
$$(u^4 + b_1u^3 + b_2u^2 + b_3u + b_4)]/[u\exp(u)] \quad (2.7)$$

where

$$
\begin{array}{ll}
a_1 = 8.5733287401 & b_1 = 9.5733223454 \\
a_2 = 18.0590169730 & b_2 = 25.6329561486 \\
a_3 = 8.6347608925 & b_3 = 21.0996530827 \\
a_4 = .2677737343 & b_4 = 3.9584969228
\end{array}
$$

BASIC microcomputer programs for generating values of W(u) with polynomial approximations are listed by Walton (1988, p. 341) and Clark (1987, pp. 1.9-1.10). A FORTRAN program for generating values of W(u) is listed by Reed (1980, p. 65).

The volume of the cone of depression depends on the time after pumping started, aquifer hydraulic characteristics, and the discharge rate. The radius of the cone of depression is often called the radius of investigation (Streltsova, 1988, p. 78). It is commonly defined as the distance from the production well beyond which drawdown is <0.01 ft or 0.003 m. The radius of investigation may be calculated by first substituting assumed values of Q and T and a value of 0.01 ft or 0.003 m in Equation 2.1 and solving for W(u). Next, the value of u corresponding to the calculated value of W(u) is obtained from Appendix G. This value of u is substituted in Equation 2.3 with assumed values of S, T, and t to obtain r which is the radius of investigation. A BASIC program listed by Clark (1987, pp. 3.1-3.12) calculates u when W(u) is known.

Equation 2.1 assumes that the aquifer is homogeneous, isotropic, infinite in areal extent (there are no aquifer boundaries within the cone of depression), overlain and underlain by aquicludes, and constant in thickness throughout. It is further assumed that the production well fully penetrates the aquifer and has an infinitesimal diameter, no wellbore storage, and no well loss; and the observation well fully penetrates the aquifer.

Equation 2.1 describes drawdown distribution with a production well having a finite diameter and negligible wellbore storage when (Streltsova, 1988, p. 49)

$$t \geq 50 \ r_w^2 S/T \ \text{(for production well)} \tag{2.8}$$

$$r \geq 20r_w \ \text{(for observation well)} \tag{2.9}$$

where t is the time after pumping started, r_w is the effective production well radius, S is the aquifer storativity, T is the aquifer transmissivity, and r is the distance between the production and observation wells.

Equation 2.1 describes drawdown distribution with a production well having a finite diameter and wellbore storage when (Papadopulos and Cooper, 1967, p. 242)

$$t \geq 2.5 \times 10^3 (r_c^2 - r_d^2)/T \quad \text{(observation well)} \qquad (2.10)$$

$$t \geq 250(r_c^2 - r_d^2)/T \quad \text{(production well)} \qquad (2.11)$$

where t is the time after pumping started, r_c is the radius of the production well casing in the interval over which the water level declines, r_d is the pump-column pipe radius, and T is the aquifer transmissivity.

Equation 2.1 describes drawdown distribution with a partially penetrating production well when (Hantush, 1964, p. 351)

$$r \geq 1.5m(P_h/P_v)^{0.5} \qquad (2.12)$$

where r is the distance from the production well, m is the aquifer thickness, P_v is the aquifer vertical hydraulic conductivity, and P_h is the aquifer horizontal hydraulic conductivity.

Equation 2.1 describes drawdown distribution under water table conditions with delayed gravity drainage when (Boulton, 1954)

$$t \geq 5S_y m/P_v \qquad (2.13)$$

where t is the time after pumping started, S_y is the aquifer gravity yield, m is the aquifer thickness, and P_v is the aquifer vertical hydraulic conductivity.

Hantush (1962) presents equations for calculating drawdown distribution in a nonleaky artesian aquifer having a variable thickness which increases exponentially in the x-direction and is uniform in the y-direction. Kanwar, et al. (1979) presents equations for calculating drawdown distribution in a nonleaky artesian aquifer when a production well with a hemispherical bottom just penetrates the aquifer top.

FINITE WELL DIAMETER

Drawdown with a finite production well diameter and no wellbore storage during early times when finite diameter impacts are appreciable is defined by the following equation (Streltsova,1988, pp. 45-

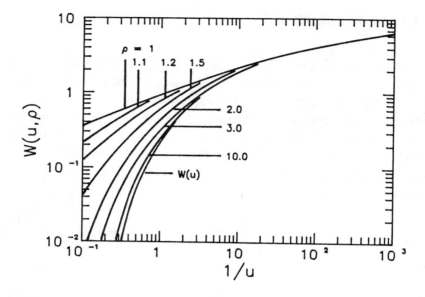

FIGURE 2.7   Logarithmic graphs of W(u,ρ) vs 1/u for selected values of ρ.

49):

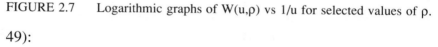

$$s = QW(u,\rho)/(4\pi T) \qquad (2.14)$$

with

$$u = r^2S/(4Tt) \qquad (2.15)$$

$$\rho = r/r_w \qquad (2.16)$$

where s is the drawdown at time t, Q is the constant production well discharge rate, W(u,ρ) is a well function, T is the aquifer transmissivity, r is the distance between the production and observation wells, S is the aquifer storativity, t is the time after pumping started, and r_w is the production well effective radius.

Values of W(u,ρ) for a production well (ρ = 1) and the practical range of 1/u as listed by Streltsova (1988, p. 48) are presented in Appendix G. When 1/u ≥ 50 then W(u,ρ) = W(u). A family of logarithmic graphs of values of W(u,ρ) vs. 1/u for selected values of ρ is shown in Figure 2.7.

WELLBORE STORAGE

If the production well has a finite diameter and wellbore storage is appreciable, the discharge rate is the sum of the aquifer flow rate and the rate of wellbore storage depletion. The aquifer flow rate increases exponentially with time to the discharge rate and the wellbore storage depletion rate decreases in a like manner to zero (Streltsova, 1988, pp. 49-55). Drawdown is defined by the following equation (Papadopulos and Cooper, 1967, pp. 241-244 and Papadopulos, 1967, pp. 157-168):

$$s = QW(u,\rho,\alpha)/(4\pi T) \tag{2.17}$$

with

$$u = r^2 S/(4Tt) \tag{2.18}$$

$$\alpha = (r_w^2 - r_d^2)S/r_c^2 \tag{2.19}$$

$$\rho = r/r_w \tag{2.20}$$

where s is the drawdown at time t, Q is the production well constant discharge rate, $W(u,\rho,\alpha)$ is a well function, T is the aquifer transmissivity, r_w is the effective production well radius, S is the aquifer storativity, r_d is the pump-column pipe radius, r_c is the radius of the production well casing in the interval over which the water level declines, t is the time after pumping started, and r is the distance between the production and observation wells ($r = r_w$ and $\rho = 1$ for the production well).

Values of $W(u,\rho,\alpha)$ for the practical range of u,ρ, and α presented by Reed (1980, pp. 41-43) are listed in Appendix G. An observation well family of logarithmic graphs of $W(u,\rho,\alpha)$ vs. $1/u$ for selected ρ and α values is shown in Figure 2.8. A production well family of logarithmic graphs of $W(u,\rho,\alpha)$ vs. $1/u$ for selected α values and a ρ value of 1 is shown in Figure 2.9.

$W(u,\rho,\alpha)$ approaches $W(u)$ as time becomes large. When $t > 2.5 \times 10^3 r_c/T$ or $\alpha\rho^2/u > 10^4$ then $W(u,\rho,\alpha) = W(u)$ and when $t > 2.5 \times$

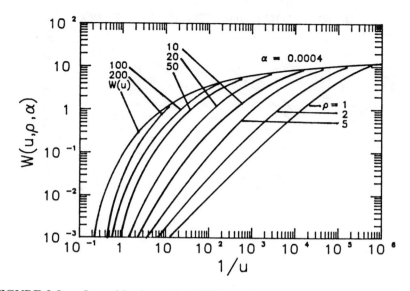

FIGURE 2.8   Logarithmic graphs of $W(u,\rho,\alpha)$ vs $1/u$ for observation wells and selected values of ρ and α.

$10^2 r_c^2/T$ or $\alpha/u_w > 10^3$ then $W(u,1,\alpha) = W(u_w)$ where $u_w = r_w^2 S/(4Tt)$ (Papadopulos, 1967, p. 161). A FORTRAN program for generating values of $W(u,\rho,\alpha)$ is listed by Reed (1980, pp. 92-96).

WELL PARTIAL PENETRATION

Production wells and piezometers or observation wells often do not completely penetrate aquifers (Figure 2.10). Production well partial penetration affects the drawdown in nearby piezometers or observation wells (Hantush, 1961, pp. 83-98). The cone of depression is appreciably distorted within a distance $r = 1.5m(P_h/P_v)^{0.5}$ of the production well where m is the aquifer thickness, P_h is the aquifer horizontal hydraulic conductivity, and P_v is the aquifer vertical hydraulic conductivity. Within this distance, observed drawdowns in piezometers or observation wells differ from those defined in Equation 2.1 for full well penetration conditions depending on the vertical position of the screen in the production well and piezometer or observation well.

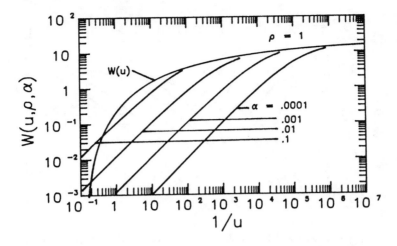

FIGURE 2.9 Logarithmic graphs of $W(u,\rho,\alpha)$ vs $1/u$ for production wells and selected values of α and $\rho = 1$.

For example, if the production and observation wells are both open in either the top or bottom portion of the aquifer, the observed drawdown in the observation well is greater than for Figure 2.10. If the production well is open to the top of the aquifer and the observation well is open to the bottom of the aquifer, or visa versa, the observed drawdown in the observation well is smaller than for full penetrating conditions. Thus, partial penetration impacts may be negative or positive depending on well geometry. Several methods are available for calculating partial penetration impacts (Kruseman and Ridder, 1970, pp. 146-155).

Commonly, the drawdown due to well partial penetration which must be added to s to obtain the total drawdown with well partial penetration is calculated with one of the following equations (Hantush, 1961a, pp. 85 and 90; Reed, 1980, pp. 8-10):

for piezometer

$$s_p = QW(u, ar/m, L/m, d/m, z/m)/(4\pi T) \qquad (2.21)$$

for observation well

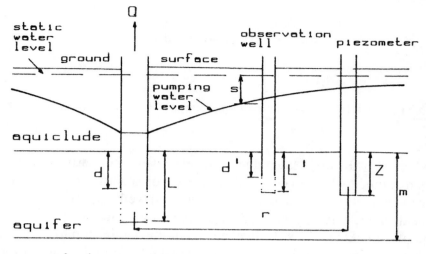

FIGURE 2.10   Cross section through partially penetrating wells in nonleaky artesian aquifer.

$$s_p = QW(u,ar/m,L/m,d/m,L'/m,d'/m)/(4\pi T) \qquad (2.22)$$

with

$$a = (P_v/P_h)^{0.5} \qquad (2.23)$$

$$u = r^2S/(4Tt) \qquad (2.24)$$

$$W(u,ar/m,L/m,z/m) = 2m/[\pi(L-d)]\sum_{n=1}^{\infty}(1/n)$$

$$[\sin(n\pi L/m) - \sin(n\pi d/m)\cos(n\pi z/m)W(u,b) \qquad (2.25)$$

$$b = n\pi\, ar/m \qquad (2.26)$$

$$W(u,ar/m,L/m,d/m,L'/m,d'/m)=$$

$$2m^2/[\pi^2(L-d)(L'-d')]$$

$$\sum_{n=1}^{\infty}(1/n^2)[\sin(n\pi L/m)-\sin(n\pi d/m)][\sin(n\pi L'/m-\sin(n\pi d'/m)]W(u,b)$$

(2.27)

where s_p is the drawdown due to the effects of well partial penetration at time t which may be either positive or negative depending on well geometry, Q is the constant production well discharge rate, r is the distance between the production and observation wells, m is the aquifer thickness, L is the distance from the aquifer top to the base of the production well screen, d is the distance from the aquifer top to the top of the production well screen, z is the distance from the top of the aquifer to the base of the piezometer, T is the aquifer transmissivity, L' is the distance from the top of the aquifer to the base of the observation well screen, d' is the distance from the top of the aquifer to the top of the observation well screen, P_v is the aquifer vertical hydraulic conductivity, P_h is the aquifer horizontal hydraulic conductivity, t is the time after pumping started, and S is the aquifer storativity.

The partial penetration impact in the production well may be calculated by substituting r_w for r and 0.5(L + d) for z in Equation 2.21 or by substituting r_w for r, 0.5(L + d) for L' and 0.5(L + d) − 1 for d' in Equation 2.22 (Hantush, 1964, p. 352).

Assuming production well partial penetration, Equation 2.1 may be rewritten as:

$$s = QW_t(u)/(4\pi T)$$

(2.28)

with

$$W_t(u) = W(u) + W(u,ar/m,L/m,d/m,L'/m,d'/m)$$

(2.29)

W(u,b) in Equations 2.25 and 2.27 is the so called leaky artesian well function which can be calculated with the following recursive relationships (see Streltsova, 1988, p. 86):

$$W(u,b) + 2K_o(b) - \sum_{m=0}^{\infty} [(-u)^m / m!] E_{m+1}[b^2 / (4u)] \qquad (2.30)$$

and

$$W(u,b) + E_1(u) - \sum_{m=0}^{\infty} (1/m!)[-b^2 / (4u)]^m E_{m+1}(u) \qquad (2.31)$$

with

$$E_1(u) = W(u) \qquad (2.32)$$

$$E_{m+1}(u) = (1/m)[\exp(-u) - uE_m(u)] \quad m = 1,2, \dots \qquad (2.33)$$

Values of $W(u,b)$ over the practical range of u and b are listed in Appendix G. For large values of time $[t > m^2S/(2a^2T)$ or $t > mS/(2P_v)]$, partial penetration impacts are constant in time and (Reed, 1980, p. 9)

$$W(u,b) = 2K_0(b) \qquad (2.34)$$

$$W(u,ar/m,L/m,d/m,z/m) = W(ar/m,L/m,d/m,z/m) \qquad (2.35)$$

Values of $W(ar/m,L/m, d/m,z/m)$ for selected values of ar/m, L/m, d/m, and z/m presented by Weeks (1969, pp. 196-214) are listed in Appendix G.

A FORTRAN program for calculating values of $W(u,b)$ is listed by Reed (1980, pp. 66-72) and BASIC programs for calculating values of $W(u,b)$ are listed by Kinzelbach (1986, pp. 225-226), Clark (1987), and Walton (1987, pp. 123-127). Values of $K_0(b)$ listed in Appendix G were calculated with the following polynomial approximations presented by Abramowitz and Stegun (1964) which have been incorporated into a FORTRAN program listed by Reed (1980, pp. 62-64) and a BASIC program listed by Walton (1987, p. 115):

when $0 < X \leq 2$

$$K_0(X) = -\ln(X/2)I_0(X) - .57721566 + .42278420(X/2)^2$$
$$+ 23069756(X/2)^4 + .03488590(X/2)^6 + .0026298(X/2)^8$$
$$+ .00010750(X/2)^{10} + .00000740(X/2)^{12} \qquad (2.36)$$

when $2 < X < \infty$

$$K_0(X) = [1.25331414 - .07832358(2/X) + .02189568(2/X)^2$$
$$- .01062446(2/X)^3 + .00587872(2/X)^4 - .00251540(2/X)^5$$
$$+ .00053208(2/X)^6]/X^{1/2}exp(X) \qquad (2.37)$$

when $-3.75 \leq X \leq 3.75$

$$I_0(X) = 1 + 3.5156229(X/3.75)^2 + 3.0899424(X/3.75)^4$$
$$+ 1.2067492(X/3.75)^6 + .2659732(X/3.75)^8$$
$$+ .0360768(X/3.75)^{10} + .0045813(X/3.75)^{12} \qquad (2.38)$$

There is a $W(u,ar/m,L/m,d/m,L'/m,d'/m)$ vs. $1/u$ logarithmic graph for each value of ar/m, $L/m,d/m$, L'/m, d'/m for each observation well. A family of $W_t(u)$ vs. $1/u$ logarithmic graphs for selected values of a, r, m, L, d, L', d' is shown in Figure 2.11.

A FORTRAN program for calculating values of $W(u,ar/m,L/m,d/m,L'/m,d'/m)$ is listed by Reed (1980, pp. 57-65) and a BASIC program for calculating values of $W(u,ar/m, L/m,d/m,L'/m,d'/m)$ is listed by Walton (1987, pp. 113-116).

AQUIFER ANISOTROPHY

Aquifers can be homogeneous but anisotropic in the sense that the transmissivity varies in two different directions. The transmissivity in the major direction of anisotrophy is greater than that in the minor direction. With the x and y axes oriented parallel to the principal directions of anisotrophy as shown in Figure 2.12, the equation

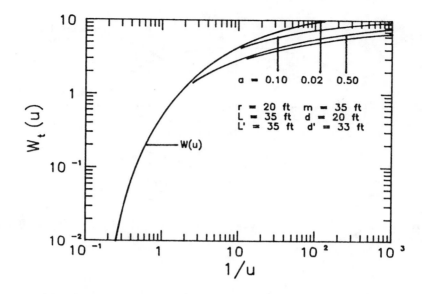

FIGURE 2.11   Logarithmic graphs of Wt(u) vs 1/u for selected values of a, r, m, L, d, L', and d'.

defining drawdown distribution is as follows (Hantush, 1966, pp. 421-426):

$$s = QW(u)/[4\pi(T_xT_y)^{0.5}] \tag{2.39}$$

with

$$u = r^2S/(4T_rt) \tag{2.40}$$

$$T_r = T_x/[\cos^2\phi + (T_x/T_y)\sin^2\phi] \tag{2.41}$$

where s is the drawdown at time t, Q is the constant production well discharge rate, W(u) is a well function defined earlier, T_x is the aquifer transmissivity in the x-direction, T_y is the aquifer transmissivity in the y-direction, r is the distance between the production and observation wells, S is the aquifer storativity, t is the time after pumping started, and ϕ is the angle between the x-axis and a line connecting the production and observation wells.

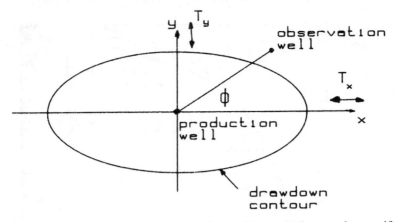

FIGURE 2.12 Plan view of wells in anisotrophic nonleaky artesian aquifer.

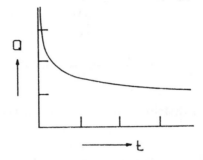

FIGURE 2.13 Graph of exponentially declining well discharge.

VARIABLE DISCHARGE

Often the discharge of a production well declines exponentially during the pumping period (see Figure 2.13). The rate of discharge decline is generally greatest near the beginning of the pumping period and decreases thereafter. Frequently, the discharge attains a constant rate towards the end of the pumping period. The period of declining discharge may range from less than a minute to several days. Discharge variation is usually due to the self-adjustment of a constant-speed pump to water level declines. The time-discharge curve may be fitted to a simple curve described by a variable discharge function and drawdowns may be calculated with exact equations expressing the relation between the variable discharge

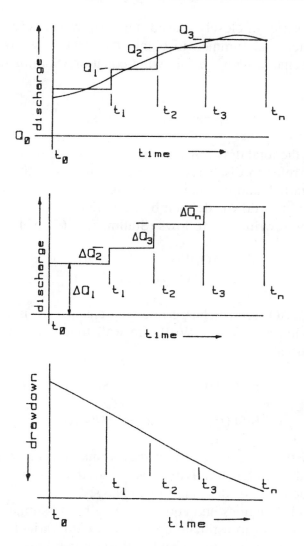

FIGURE 2.14   Selected variable discharge graphs.

function, aquifer conditions, and drawdown. However, approximation methods are most commonly used to estimate drawdowns due to variable discharge.

Drawdowns due to the complicated continuous discharge variations shown in Figure 2.14 may be estimated with equations developed by Stallman (1962, p. 120). The discharge curve is approximated by a series of closely spaced steps. Drawdowns for each step are calculated and summed to obtain the total drawdown due to variable discharge with the following equation (Stallman, 1962, p. 120):

$$s_T = s_1 + s_2 + s_3 \ldots + s_n \tag{2.42}$$

where s_T is the total drawdown at time t, $s_1 \ldots s_n$ are step drawdowns (subscripts refer to discharge rate increments), t is the time after pumping started, and n is the number of steps.

With step drawdowns defined by Equations 2.1 to 2.3, Equation 2.42 may be rewritten as follows (Stallman, 1962, p. 120):

$$s_T = [\Delta Q_1 W(u)_1 + \Delta Q_2 W(u)_2 \ldots \\ + \Delta Q_n W(u)_n]/4\pi T \tag{2.43}$$

where $\Delta Q_1 \ldots \Delta Q_n$ are discharge rate step changes, T is the aquifer transmissivity, and $W(u)_1 \ldots W(u)_n$ are well function values for the following u values:

$$u_1 = r^2 S/[4T(t_1 - t_0)], \quad u_2 = r^2 S/[4T(t_2 - t_1)] \tag{2.44}$$

$$u_3 = r^2 S/[4T(t_3 - t_2)], \quad \ldots u_n = r^2 S/[4T(t_n - t_{n-1})] \tag{2.45}$$

where r is the distance between the production and observation wells, S is the aquifer storativity, T is the aquifer transmissivity, and $t_0 \ldots t_n$ are the times of discharge rate changes.

Other well functions and equations may be substituted in Equations 2.43 to 2.45 to estimate drawdowns with variable discharge under leaky artesian or water table aquifer conditions.

Drawdown equations for certain variable discharge functions were presented by Abu-Zied and Scott (1963, pp. 119-132) and Werner (1946, pp. 687-708) for nonleaky artesian aquifer conditions, and by Hantush (1964, pp. 343-345) for both nonleaky and leaky artesian aquifer conditions and exponentially, hyperbolically, and inverse square root decreasing discharge functions. Moench (1971, pp. 4-8) developed an equation with a convolution integral for calculating drawdown with any arbitrary variable discharge function. Reed (1980, pp. 102-106) lists a FORTRAN program for solving that convolution integral. The program approximates the convolution integral by summing the trapezoidal rule applied to a sequence of segments. Drawdown is calculated with the following five variable discharge functions:

$$Q = Q_s[1 + a \exp(-t/t^*)] \text{ (exponential function)} \qquad (2.46)$$

$$Q = Q_s[1 + a/(1+t/t^*)] \text{ (hyperbolic function)} \qquad (2.47)$$

$$Q = Q_s[1 + a/(1+t/t^*)^{.5}] \qquad (2.48)$$
(inverse square root function)

$$Q = \sum_{i=0}^{5} c_i t^i$$

$$(2.49)$$

(5th degree polynomial function)

$$Q = c_j + b_j(t - t_{j-i}) \qquad (2.50)$$
(piecewise linear function of eight line segments)

$$\text{for } t_{j-1} < t \leq t_j, \, j = 1, 2, ..., 8 \qquad (2.51)$$

where Q is the production well discharge rate at time t, Q_s is the ultimate steady state production well discharge rate, a and t^* are parameters defining a particular function obtained empirically and depend on the aquifer and pump characteristics, c_i is a coefficient of the polynomial, c_j and b_j are parameters defining the jth line segment, and t is the time after pumping started.

Streltsova (1988, pp. 101-152) developed drawdown solutions for nonleaky artesian aquifer conditions and linear, parabolic, polynomial, exponential, and harmonic functional forms of discharge rate variance as well as for piecewise approximation of the discharge rate by step and linear functions.

Sternberg (1968, pp. 177-180) developed the following simplified equation for calculating the specific drawdown with variable discharge from a nonleaky artesian aquifer when u ≤ 0.01 (Kruseman and De Ridder, 1970, pp. 142-145):

$$s_k / Q_k = [1 / (4\pi T)]\{\sum_{j=0}^{k-1}[(Q_{j+1} - Q_j)\ln(t_k - t_j) / Q_k]$$

$$+ \ln[2.25T / (r^2S)]\} \qquad (2.52)$$

where T is the aquifer transmissivity, S is the aquifer storativity, r is the distance between the production and observation wells, t_k is the time after pumping started, t_n is the total pumping period, Q_k is the discharge rate at time t_k, s_k is the drawdown at time t_k.

Clark (1987, pp. 7.1-7.16) lists a BASIC program for analyzing data with Equation 2.52.

Another simplified equation for calculating approximate values of drawdown in a nonleaky artesian aquifer with variable discharge when u ≤ 0.01 is as follows (Aron and Scott, 1965, pp. 1-12; Kruseman and De Ridder, 1970, pp. 140-142):

$$s_t = \{[2.30Q_t(t)/(4\pi T)]\log[2.25Tt/(r^2S)]\} + s_e \qquad (2.53)$$

with

$$s_e = (Q_{ta} - Q_t)/(2.25\pi T) \qquad (2.54)$$

where s_t is the drawdown at time t, Q_t is the discharge rate at time t, s_e is the excess drawdown caused by the earlier higher discharge, Q_{ta} is the average discharge from time 0 to time t, $Q_{ta} - Q_t$ is the total excess volume pumped causing the excess drawdown s_e, T is the aquifer transmissivity, S is the aquifer storativity, t is the time after pumping started, and r is the distance from the production well.

FLOWING WELLS

A flowing well with a finite diameter is shown in Figure 2.15.

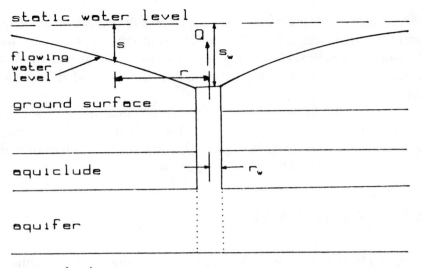

FIGURE 2.15   Cross section through flowing well in nonleaky artesian aquifer.

Initially, the well is capped. The cap is removed and the water level in the flowing well declines instantaneously to a constant stage lower than the initial water level. As a result, the well discharge decreases with time and drawdowns in observation wells increase with time as defined by the following equations (Jacob and Lohman, 1952, p. 560 and Hantush, 1964, p. 343):

$$Q = 2\pi T s_w W(\lambda) \tag{2.55}$$

$$s = s_w W(\lambda, \rho) \tag{2.56}$$

with

$$\lambda = Tt/(Sr_w^2) \tag{2.57}$$

$$\rho = r/r_w \tag{2.58}$$

where Q is the flowing well discharge rate at time t, T is the aquifer transmissivity, s_w is the constant drawdown in the flowing well,

$W(\lambda)$ and $W(\lambda,\rho)$ are well functions, s is the drawdown in an observation well at time t, S is the aquifer storativity, r_w is the effective flowing well radius, t is the time after the cap was removed from the well, and r is the distance between the flowing and observation wells.

Values of $W(\lambda)$ and $W(\lambda,\rho)$, over the practical range of λ and ρ (Reed, 1980, pp. 19 and 20), are listed in Appendix G. Logarithmic graphs of $W(\lambda)$ vs. λ and $W(\lambda,\rho)$ vs. λ/ρ^2 for selected values of ρ are shown in Figures 2.16 and 2.17, respectively.

MULTI-AQUIFER WELL

A production well with infinitesimal diameter and no wellbore storage can fully penetrate two uniformly porous nonleaky artesian aquifers each homogeneous, isotropic, infinite in areal extent, and separated by an aquiclude as shown in Figure 2.18. Equations defining the composite head in the multi-aquifer production well and the discharge (or recharge) rate from each aquifer are as follows (Papadopulos, 1966, pp. 4791-4797):

$$h = h_1 - \{[(h_1 - h_2)/(1 + a)] \; W(\lambda,\rho) + Q/[4\pi(T_1 + T_2)]$$
$$\{W(u) - [\ln(b^2)/(1 + a)] \; W(\lambda,\rho)\}\} \qquad (2.59)$$

and

$$Q_1 = \{[2\pi T_1(h_1 - h_2)/(1 + a)] \; W(\lambda) + Qa/[2(1 + a)]\}$$
$$\{2e^{(-1/4c)} - [\ln(b^2)/(1 + a)] \; W(\lambda)\} \qquad (2.60)$$

with

$$b = (d_1/d_2)^{0.5} \qquad (2.61)$$

$$a = T_1/T_2 \qquad (2.62)$$

$$f = b^{[a/(1+a)]} \qquad (2.63)$$

$$d_1 = T_1/S_1 \qquad (2.64)$$

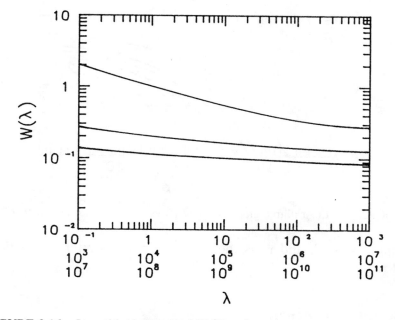

FIGURE 2.16   Logarithmic graph of $W(\lambda)$ vs λ.

$$d_2 = T_2/S_2 \tag{2.65}$$

$$c = d_1 t/r_w^2 \tag{2.66}$$

$$u = 1/(4c) \tag{2.67}$$

$$\lambda = c/f^2 \tag{2.68}$$

$$\rho = 1/f \tag{2.69}$$

where h is the composite head in the multi-aquifer production well at time t, h_1 is the initial head in aquifer 1, h_2 is the initial head in aquifer 2, Q is the constant discharge rate from the multiaquifer production well, T_1 is aquifer 1 transmissivity, T_2 is aquifer 2 transmissivity, $W(u)$, $W(\lambda)$, and $W(\lambda,\rho)$ are well functions defined earlier and listed in Appendix G, Q_1 is the constant discharge rate to the multi-aquifer production well from aquifer 1 at time t, S_1 is aquifer 1 storativity, S_2 is aquifer 2 storativity, t is the time after

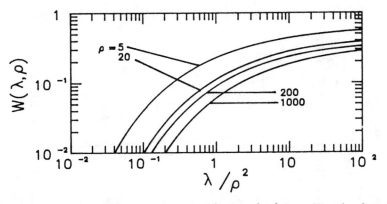

FIGURE 2.17   Logarithmic graphs of $W(\lambda,\rho)$ vs λ/ρ^2 for selected values of ρ.

pumping started, and r_w is the effective multi-aquifer production well radius.

FRACTURED ROCK

Flow behavior in fractured rock aquifers differs from that in uniformly porous aquifers such as sand and gravel deposits. Fractured rock aquifers possess, in addition to void spaces between mineral grains of rock and vesicular openings, fissures (cracks, crevices, joints, etc.) which make the pattern of porosity and hydraulic conductivity complex (Streltsova, 1988, pp. 357-364). There are two major fractured rock conceptual models: single fissure and double-porosity. In the single fissure model, fractured rock aquifers respond to pumping as a system of pipes or horizontal plates representing fissures with no significant contribution from the rock matrix. In the double-porosity model shown in Figure 2.19 (Streltsova, 1988, p. 377), fractured rock aquifers respond to pumping as two interconnected layers of porosity and hydraulic conductivity: the relatively high porosity and low hydraulic conductivity of unfractured but porous blocks of matrix rock and the relatively low porosity and high hydraulic conductivity of fissures.

Methods of analysis of the flow in fractured rock aquifers are reviewed by Streltsova-Adams (1978, pp. 357-423), Gringarten

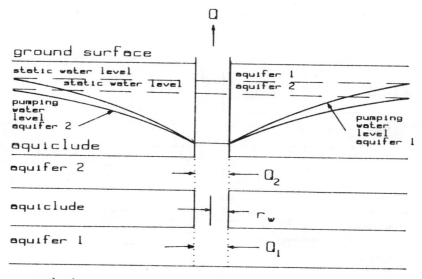

FIGURE 2.18 Cross section through multi-aquifer production well in nonleaky artesian aquifers.

(1984, pp. 549-564), Sauveplane (1984, pp. 171-206), and Streltsova (1988, pp. 357-400). Flow in fractured rock aquifers is very complex and is the subject of much debate. No method of analysis is universal in application and convincing field examples are scarce in the literature. Double-porosity models are often favored by the groundwater industry.

In the double-porosity model, flow in a fractured rock aquifer is due almost entirely to the presence of fissures, while porosity and therefore storativity is mainly associated with the porous blocks. Fissures have an immediate elastic response to a sudden change in water levels, while porous blocks have an induced subsequent elastic response. Commonly, the actual irregular network of interconnected blocks and fissures is simulated by a regular network of interconnected horizontal block and fissure units. Due to vertical symmetry, the fractured rock aquifer may be further simplified to the two layered model shown in Figure 2.19. The block unit has a thickness equal to the average thickness of individual blocks in the actual fractured rock aquifer and the fissure has a thickness equal to

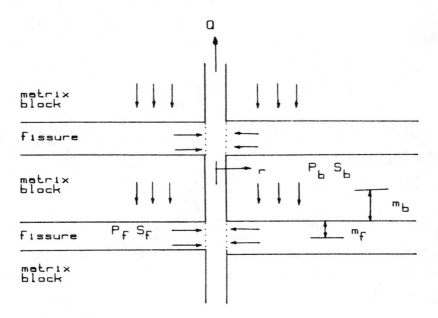

FIGURE 2.19 Cross section through fractured rock aquifer model.

the average thickness of the fissures in the actual fractured rock aquifer. Both the block and fissure average thicknesses and hydraulic characteristics are assumed to be constant in space.

Three time-drawdown segments in fractured rock aquifers have been identified. The first segment, representing the response of fractures to pumping, exists only at very early times and is often masked by wellbore storage impacts. The effective storativity during the first segment is the storativity of the fissure. The second segment represents the period during which the cone of depression slows in its rate of expansion (a quasi-steady state) as water stored in blocks reaches fractures. Block contribution is delayed because of the low block hydraulic conductivity. The third segment, approached asymptotically, represents the combined response of fractures and blocks to pumping as the cone of depression continues to expand. The effective storativity during the third segment is the fissure storativity plus the block storativity.

Moench (1984, pp. 831-846) developed an analytical model and equations for fractured rock aquifers with a production well of finite diameter having wellbore storage and skin. The effects of wellbore storage are shown to be dominant and to overshadow double-porosity impacts at early stages of production. Well function values for the model are generated numerically with the inversion procedure of Stehfest (1970).

Analytical model analysis can be significantly reduced in complexity by using a packer to isolate the fractured rock aquifer and remove the impacts of wellbore storage from analysis.

The Boulton and Streltsova (1977, pp. 257-269) double-porosity model and method of analysis presented by Streltsova (1988, pp. 313-319, 381-391) assume wellbore storage to be negligible and are commonly used in groundwater studies. The Boulton-Streltsova model equation is as follows:

$$s_f = QW(u_f,r_D,S_f/S_b)/(4\pi T_f) \tag{2.70}$$

with

$$u_f = r^2S_f/(4T_ft) \tag{2.71}$$

$$r_D = r(P_b/P_f)^{0.5}/m_b \tag{2.72}$$

$$T_f = P_fm \tag{2.73}$$

where s_f is the drawdown in the fissure portion of the fractured rock aquifer at time t, Q is the constant rate of discharge from the production well, $W(u_f,r_D,S_f/S_b)$ is a well function, S_f is the fissure storativity, S_b is the block storativity, T_f is the transmissivity of the fissured portion of the fractured rock aquifer, r is the distance between the production and observation wells, t is the time after pumping started, P_b is the vertical hydraulic conductivity of the block, P_f is the horizontal hydraulic conductivity of the fissures, m is the fractured rock thickness, and m_b is the half thickness of the average block unit.

Exact values of $W(u_f,r_D,S_f/S_b)$ in the Boulton-Streltsova model equation are calculated numerically by applying the inversion procedure of Stehfest (1970) to the flow equation in Laplace space. Logarithmic and semilogarithmic graphs of values of $W(u_f,r_D,S_f/S_b)$

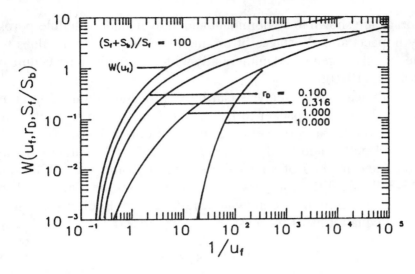

FIGURE 2.20   Logarithmic graphs of $W(u_f,r_D,S_f/S_b)$ vs $1/u_f$ for selected values of r_D and $(S_f+S_b)/S_f = 100$.

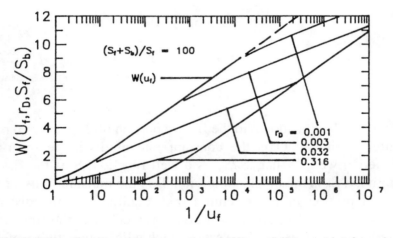

FIGURE 2.21   Semilogarithmic graphs of $W(u_f,r_D,S_f/S_b)$ vs $1/uf$ for selected values of r_D and $(S_f+S_b)/S_f = 100$.

vs. $1/u_f$ when the ratio $(S_b+S_f)/S_f = 100$ (Streltsova, 1988, pp. 315 and 316) are presented in Figures 2.20 and 2.21, respectively. Sauveplane (1984a, pp. 187-192) developed the following approximation for $W(u_f,r_D,S_f/S_b)$ and judged the accuracy of values calculated with the approximation to be sufficient for practical purposes during intermediate and long times:

$$W(u_f, r_D, S_f/S_b) = 2K_o(A^{0.5}) \tag{2.74}$$

with

$$A = 2u_f + [rc(2u_f)^{0.5}/(m_b b^{0.5})]\tanh(D) \tag{2.75}$$

$$D = m_b(2u_f)^{0.5}/(rb^{0.5}) \tag{2.76}$$

$$\tanh(D) = \sinh(D)/\cosh(D) \tag{2.77}$$

$$\sinh(D) = [\exp(D) - \exp(-D)]/2 \tag{2.78}$$

$$\cosh(D) = [\exp(D) + \exp(-D)]/2 \tag{2.79}$$

$$b = P_b S_f/(S_b P_f) \tag{2.80}$$

$$c = P_b/P_f \tag{2.81}$$

Drawdown distribution in fractured rock aquifers during all of the first time-drawdown segment and part of the second time-drawdown segment when $t \le 0.1 m_b^2 S_b/T_b$ is described by the following equation (Streltsova, 1988, pp.318):

$$s_f = QW(u_f, \tau)/(4\pi T_f) \tag{2.82}$$

with

$$\tau = 0.25 r_D(S_b/S_f) \tag{2.83}$$

$$r_D = r(P_b/P_f)^{0.5}/m_b \tag{2.84}$$

$$T_f = P_f m \tag{2.85}$$

where s_f is the drawdown in the fissure portion of the fractured rock aquifer at time t, Q is the constant rate of discharge from the production well, $W(u_f, \tau)$ is a well function, S_f is the fissure storativity, S_b is the block storativity, T_f is the transmissivity of the

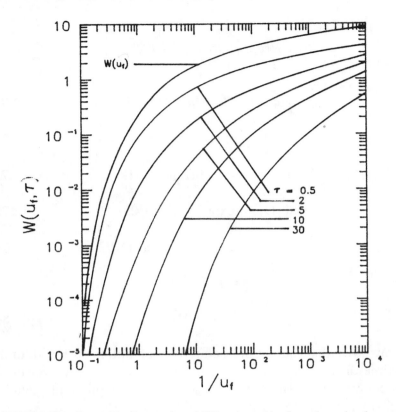

FIGURE 2.22   Logarithmic graphs of $W(u_f,\tau)$ vs $1/u_f$ for selected values of τ.

fissured portion of the fractured rock aquifer, m_b is the half thickness of the average block unit, t is the time after pumping started, P_b is the vertical hydraulic conductivity of the block, P_f is the horizontal hydraulic conductivity of the fissures, and m is the fractured rock thickness.

Values of $W(u_f,\tau)$, in terms of the practical range of u_f and τ, are presented in Appendix G and Figure 2.22. A FORTRAN program for calculating values of $W(u_f,\tau)$ is listed by (Reed, 1980). Sauveplane (1984b, p. 215) developed the following approximate equation for calculating values of $W(u_f,\tau)$:

$$W(u_f,\tau) = 2K_0(A)^{0.5} \tag{2.86}$$

with

$$A = u_f/2 + [4\tau/(2/u_f)^{0.5}]\coth(C) \qquad (2.87)$$

$$C = (4\tau/B^2)[1/(2/u_f)^{0.5}] \qquad (2.88)$$

$$B = (r/m_b)(P_b/P_f)^{0.5} \qquad (2.89)$$

$$\cotanh(C) = \cosh(C)/\sinh(C) \qquad (2.90)$$

S_f is sometimes determined from core data and the ratio S_b/S_f often is assumed equal to the thickness ratio m_b/m_f where m_f is the half thickness of a fissure in an average fissure-block unit and m_b is the half thickness of the average block unit. m_b is often determined from well logs.

3

Leaky Artesian Radial Flow

Horizontal radial flow in the uniformly porous leaky artesian aquifer in Figure 3.1 is augmented by downward vertical leakage of water through an aquitard. The discharge of water from the production well is balanced by aquifer storage depletion and leakage through the aquitard as the cone of depression spreads horizontally and vertically. The rate of water level decline continuously decreases until the entire discharge is derived from leakage and water levels stabilize. The rate of vertical leakage through an aquitard is proportional to the difference in heads in the source bed above the aquitard and the aquifer until the aquifer head declines below the aquitard base when maximum leakage rates are assumed to occur.

The most widely used equation to describe drawdown in a leaky artesian aquifer is as follows (Hantush and Jacob, 1955, p. 98):

$$s = QW(u,b)/(4\pi T) \qquad (3.1)$$

with

$$u = r^2 S/(4Tt) \qquad (3.2)$$

$$b = r/(Tm'/P')^{0.5} \qquad (3.3)$$

$$W(u,b) = 2K_o(b) - I_o(b)W(b^2/4u) + \exp(-b^2/4u)$$
$$\{0.5772 + \ln u + W(u) - u + u[I_o(b) - 1]/(b^2/4)$$

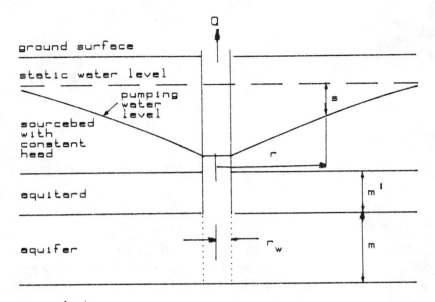

FIGURE 3.1   Cross section through production well in leaky artesian aquifer with lower aquiclude.

$$-u^2 \sum_{n=1}^{\infty} \sum_{m=1}^{n} [(-1)^{n+m}(n-m+1)!(b^2/4)^m u^{n-m}/(n+2)!^2]\}$$

(3.4)

when $b/2 \leq 0.1$ and $u \leq 1$ then

$$W(u,b) = 2K_o(b) - I_o(b)W(b^2/4u) + \{\exp(-b^2/4u)[0.5772 + \ln(u)$$

$$+ W(u) + (u/4)(b^2/4)(1 - u/9)]\}$$

(3.5)

where s is the aquifer drawdown at time t, Q is the constant production well discharge rate, W(u,b) is a well function, T is the aquifer transmissivity, t is the time after pumping started, r is the distance between production and observation wells, S is the aquifer storativity, m' is the aquitard thickness, and P' is the aquitard vertical hydraulic conductivity.

The series of Equation 3.4 is rapidly convergent for the values b/2 ≤ 1 and u ≥ 1. Only a few terms are required to obtain results accurate to four decimal places. Polynomial approximations of $K_o(b)$ and $I_o(b)$ are given in Equations 2.36 to 2.38. Values of $K_o(b)$ and $I_o(b)$, for the practical range of b, are listed in Appendix G. The factorials in Equation 3.4 are calculated using the following form of Stirling's approximation of the Gamma function (see Miller, 1981, p. 291):

$$n! = n(2\pi/n)^{0.5}n^n e^y \qquad (3.6)$$

with

$$y = 1/(12n) - 1/(360n^3) - n \qquad (3.7)$$

$$e = 2.71828183 \qquad (3.8)$$

A BASIC program for calculating factorials is listed by Clark (1987, p. 1.15). A FORTRAN program for calculating values of W(u,b) is listed by Reed (1980, pp. 66-72). Clark (1987, pp. 1.12-1.15) and Kinzelbach (1986, pp. 225-226) list BASIC programs for calculating values of W(u,b).

Values of W(u,b) may also be calculated with the following approximation developed by T.A. Prickett (oral communication, 1981) wherein the well function curve is divided into several convenient segments (Sandberg, et al. 1981):

when b = 0

$$W(u,b) = W(u) \qquad (3.9)$$

when b > 0 and $b^2/4u$ > 5

$$W(u,b) = 2K_o(b) \qquad (3.10)$$

when b > 0 and $b^2/4u$ ≤ 5 and u < 0.05 and u > 0.01 and b < 0.1

$$W(u,b) = W(u) - [(b/(4.7u^{0.6})]^2 \qquad (3.11)$$

when b > 0 and $b^2/4u \leq 5$ and $u \leq 0.9$ and $u \geq 0.05$ and u > b/2

$$W(u,b) = W(u) - [b/(4.7u^{0.6})]^2 \qquad (3.12)$$

when b > 0 and $b^2/4u \leq 5$ and $u \leq 0.01$ and $b \geq 0.1$

$$W(u,b) = 2K_o(b) - W(b^2/4u)I_o(b) \qquad (3.13)$$

when b > 0 and $b^2/4u \leq 5$ and $u \geq 0.05$ and $u \leq b/2$

$$W(u,b) = 2K_o(b) - 4.8 \times 10^E \qquad (3.14)$$

where

$$E = -(1.75u)^{-0.448b} \qquad (3.15)$$

when b > 0 and $b^2/4u \leq 5$ and u > 0.9

$$W(u,b) = 1.5637 \exp(-a - c/a)/a + 4.54\exp(-d - c/d)/d \qquad (3.16)$$

with

$$a = u + 0.5858 \quad c = b^2/4 \quad d = u + 3.414 \qquad (3.17)$$

when b>2

$$W(u,b) = [\pi/(2b)]^{0.5}\exp(-b)erfc[-(b - 2u)/(2u^{0.5})] \qquad (3.18)$$

with

$$\exp(b) = e^b \qquad (3.19)$$

$$\exp(-b) = e^{-b} \qquad (3.20)$$

Values of exp(b) and exp(-b), for the practical range of b, are listed in Appendix G. Polynomial approximations for the error functions erfc(b) and erf(b) are as follows (see Abramowitz and Stegun, 1964):

erfc(b)

$$erfc(b) = 1/(1 + a_1b + a_2b^2 + \dots a_6b^6)^{16} \tag{3.21}$$

$$erfc(-b) = 1 + erf(b) \tag{3.22}$$

with

$$erf(b) = 1 - 1/(1 + a_1b + a_2b^2 + \dots a_6b^6)^{16} \tag{3.23}$$

$$erf(-b) = -erf(b) \tag{3.24}$$

where

$$a_1 = .0705230784 \qquad a_4 = .0001520143$$
$$a_2 = .0422820123 \qquad a_5 = .0002765672$$
$$a_3 = .0092705272 \qquad a_6 = .0000430638$$

Walton (1988) lists BASIC programs for calculating values of $W(u,b)$, erfc(b), and erf(b) using Equations 3.9 to 3.17. Values of $W(u,b)$, for the practical range of u and b, are presented in Appendix G. Values of erfc(b) and erf(b), for the practical range of b, are presented in Appendix G. A family of logarithmic $W(u,b)$ vs. $1/u$ graphs for selected values of b is presented in Figure 3.2.

When $t > (30r_w^2S/T)[1 - (10r_w/(Tm'/P')^2]$ and $r_w/(Tm'/P') < 0.1$, the rate of yield from aquifer storage is as follows (Hantush, 1964, pp. 335, 338):

$$q_s = Qexp[-(Tt/S)/(Tm'/P')^2] \tag{3.25}$$

where q_s is the rate of yield from aquifer storage at time t, Q is the constant production well discharge rate, t is the time after pumping started, T is the aquifer transmissivity, r_w is the production well effective radius, S is the aquifer storativity, P' is the aquitard vertical hydraulic conductivity, and m' is the aquitard thickness.

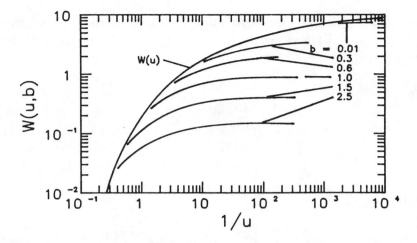

FIGURE 3.2   Logarithmic graphs of W(u,b) vs 1/u for selected values of b.

Equations 3.1 to 3.3 are based on the following assumptions: wells fully penetrate a uniformly porous artesian aquifer overlain by an aquitard and underlain by an aquiclude; overlying the aquitard is a sourcebed in which there is a water table; the aquifer is homogeneous, isotropic, infinite in areal extent, and constant in thickness throughout; the aquitard is more or less incompressible so that water released from storage therein is negligible; drawdown in the sourcebed is negligible; flow in the aquifer is two-dimensional and radial in the horizontal plane; flow in the aquitard is vertical; and the production well has an infinitesimal diameter and no wellbore storage.

These assumptions are valid provided $m/B < 0.01$ and $P/P' > 100$ m/m' (Hantush, 1967, p. 587), $\beta < 0.01$, $T_s > 100T$, and $Tt/(r^2S) < 1.6B^2/(r/B)^4$ (Neuman and Witherspoon, 1969, pp. 810,811) where m is the aquifer thickness, $B = (Tm'/P')^{0.5}$, T is the aquifer transmissivity, m' is the aquitard thickness, P' is the aquitard vertical hydraulic conductivity, P is the aquifer horizontal hydraulic conductivity, $\beta = (r/4m)[P'S'm/(PSm')]^{0.5}$, S' is the aquitard storativity, S is the aquifer storativity, and T_s is the sourcebed transmissivity.

Equations 2.10 or 2.11 are used to determine the time that must elapse before wellbore storage effects are negligible. Rushton and Redshaw (1979, pp. 242-244, 249) list a Fortran program which describes drawdown distribution under leaky artesian conditions with fully penetrating wells and wellbore storage.

When discharge is balanced by leakage and water levels stabilize (steady state conditions) drawdown in the leaky artesian aquifer system is described by the following equation (Marino and Luthin, 1982, p. 197):

$$s = QK_o(b)/(2\pi T) \tag{3.26}$$

or when $b < 0.05$

$$s = Q \ln(1.123/b)/(2\pi T) \tag{3.27}$$

with

$$b = r/(Tm'/P')^{0.5} \tag{3.28}$$

where s is the aquifer stabilized drawdown, Q is the constant production well discharge rate, $K_o(b)$ is the zero-order modified Bessel function of the second kind (see Equations 2.36 or 2.37), T is the aquifer transmissivity, r is the distance between production and observation wells, m' is the aquitard thickness, and P' is the aquitard vertical hydraulic conductivity.

Equations 3.26 to 3.28 may be combined with Equations 2.21 or 2.22 to describe drawdown with partially penetrating wells in a leaky artesian aquifer. The distance beyond which partial penetration effects are negligible may be ascertained with Equation 2.12. A FORTRAN program for calculating drawdown in a leaky artesian aquifer system with partially penetrating wells is listed by Reed (1980, pp. 75-83).

FLOWING WELL

A flowing well in a leaky artesian aquifer is illustrated in Figure 3.3. Initially the well is capped and the drawdown is zero throughout the aquifer. The cap is removed, the water level declines instantaneously to a lower position, and a constant drawdown and variable discharge by natural flow from the well is maintained. The equation governing the variable discharge from the flowing well is as follows (Hantush, 1959a, pp. 1043-1052):

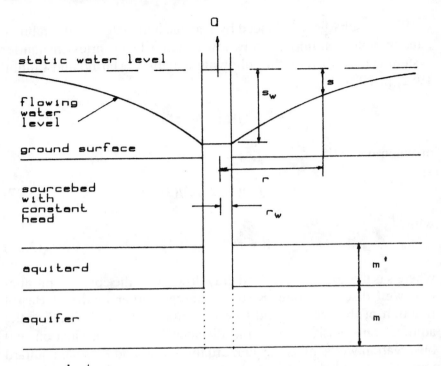

FIGURE 3.3   Cross section through flowing well in leaky artesian aquifer with lower aquiclude.

$$Q_f = 2\pi T s_w W(\lambda, b) \tag{3.29}$$

with

$$\lambda = Tt/(r^2_w S) \tag{3.30}$$

$$b = r_w/(Tm'/P')^{0.5} \tag{3.31}$$

where Q_f is the flowing well discharge at time t, T is the aquifer transmissivity, t is the time after the cap was removed from the well, s_w is the constant drawdown in the flowing well, $W(\lambda, b)$ is a well function, r_w is the flowing well effective radius, m' is the aquitard thickness, S is the aquifer storativity, and P' is the aquitard vertical hydraulic conductivity.

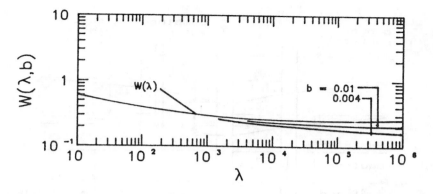

FIGURE 3.4   Logarithmic graphs of W(λ,b) vs λ for selected values of b.

The equation governing the drawdown in observation wells due to the flowing well is as follows (Hantush, 1964, p. 341):

$$s = s_w W(\lambda, \rho, b) \qquad (3.32)$$

$$\rho = r/r_w \qquad (3.33)$$

When discharge is balanced by leakage and steady state conditions exist Equation 3.32 becomes (Hantush, 1964, p. 341):

$$s = s_w K_o(r/B)/K_o(r_w/B) \qquad (3.34)$$

with

$$B = (Tm'/P')^{0.5} \qquad (3.35)$$

where s is the aquifer drawdown at time t, s_w is the constant drawdown in the flowing well, W(λ,ρ,b) is a well function, $K_o(r/B)$ is a Bessel function defined in Equations 2.36 or 2.37, r_w is the flowing well effective radius, and r is the distance between the flowing and observation wells.

The assumptions on which Equations 3.1 to 3.3 are based apply to Equations 3.29 to 3.35. Values of W(λ,b) listed by Reed (1980, p. 36), for the practical range of λ and b, are presented in Appendix G. A family of logarithmic graphs of W(λ,b) vs. λ for selected values of b is shown in Figure 3.4. A FORTRAN program for calculating values of the well function W(λ,ρ,b) is listed by Reed

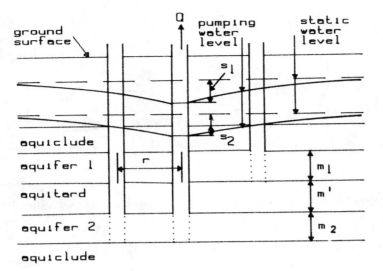

FIGURE 3.5   Cross section through production well in two mutually leaky artesian aquifers.

(1980, pp. 83-90). The function is not tabulated. Approximations for the function are presented by Hantush (1964, p. 325).

TWO MUTUALLY LEAKY AQUIFERS

Drawdown distribution in the two mutually leaky artesian aquifers in Figure 3.5 is governed by the following equations (Hantush, 1967b, pp. 1709-1720):

$$s_1 = Q_2[W(u) - W(u,b)]/[4\pi(T_1 + T_2)] \qquad (3.36)$$

$$s_2 = Q_2\{W(u) + a_1W(u,b) + \{[(c - 1)$$

$$(1 + a_1)]/[(c + 1)q]\}\ e^{-u}(1 - e^{-q})\}/[4\pi(T_1 + T_2)]\} \qquad (3.37)$$

with

$$u = r^2/(4dt) \qquad (3.38)$$

$$b^2 = b_1^2(1 + a_1) = b_2^2(1 + a_2) \tag{3.39}$$

$$a_1 = T_1/T_2 \tag{3.40}$$

$$a_2 = T_2/T_1 \tag{3.41}$$

$$q = dt(1/B_1^2 + 1/B_2^2) \tag{3.42}$$

$$d = 2d_1d_2/(d_1 + d_2) \tag{3.43}$$

$$c = d_2/d_1 \tag{3.44}$$

$$d_1 = T_1/S_1 \tag{3.45}$$

$$d_2 = T_2/S_2 \tag{3.46}$$

$$B_1 = (T_1m'/P')^{0.5} \tag{3.47}$$

$$B_2 = (T_2m'/P')^{0.5} \tag{3.48}$$

$$b_1 = r/B_1 \tag{3.49}$$

$$b_2 = r/B_2 \tag{3.50}$$

where s_1 is the drawdown in aquifer 1 at time t, s_2 is the drawdown in aquifer 2 at time t, Q_2 is the constant production well discharge in aquifer 2, t is the time after pumping started, T_1 is aquifer 1 transmissivity, T_2 is aquifer 2 transmissivity, W(u) and W(u,b) are well functions defined by equations 2.2 and 3.4 or 3.5, r is the distance between production and observation wells, S_1 is aquifer 1 storativity, S_2 is aquifer 2 storativity, P' is the aquitard vertical hydraulic conductivity, and m' is the aquitard thickness.

The assumptions on which Equations 3.36 to 3.50 are based are as follows (Hantush, 1967b, pp. 1709-1720): the diffusivities (T/S) of the two uniformly porous aquifers are unequal; the pumping period (t) is long [$t > 3.3(d_2 - d_1)B_1B_2/d_1d_2$]; the two aquifers are homogeneous, isotropic, infinite in areal extent, and of the same thickness throughout; the production well has an infinitesimal diameter, no wellbore storage, and fully penetrates the aquifer; aquitard storativity is negligible; and flow lines are refracted a full right angle as they cross the aquitard-aquifer interface.

A BASIC program for calculating the drawdown distribution in two mutually leaky artesian aquifers is listed by Walton (1984a).

When u is very small and pseudo steady-state conditions prevail, the drawdown distribution in the two aquifers is as follows (Hantush, 1967b, p. 1714):

$$s_1 = Q_2[\ln(r_e/r) - K_o(b)]/[2\pi(T_1 + T_2)] \tag{3.51}$$

$$s_2 = Q_2[\ln(r_e/r) + a_1K_o(b)]/[2\pi(T_1 + T_2)] \tag{3.52}$$

with

$$r_e = 1.5(dt)^{0.5} \tag{3.53}$$

AQUITARD STORATIVITY

An equation which describes the drawdown distribution in a leaky artesian aquifer with release of water from storage within the overlying aquitard (Figure 3.1) is presented by Streltsova (1988, pp. 313-323). The solution of that equation in the Laplace plane requires the use of the numerical procedure of Stehfest (1970). Hantush (1960, pp. 3713-3725) developed the following analytical solutions of that equation for short and long pumping periods. In case 1, with aquitards overlying and underlying the aquifer (Figure 3.6) and for small values of time [$t < m'S'/10P'$ and $t < m''S''/10P''$] drawdown is governed by the following case 1 equation (Hantush, 1960, p. 3716; Reed, 1980, p. 26):

$$s = QW(u,\tau)/(4\pi T) \tag{3.54}$$

with

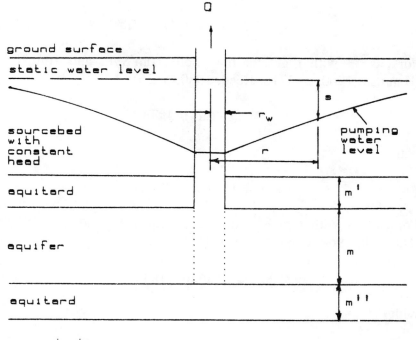

FIGURE 3.6   Cross section through production well in leaky artesian aquifer with lower aquitard.

$$u = r^2S/(4Tt) \tag{3.55}$$

$$\tau = (r/4)\{[P'S'/(m'TS)]^{0.5} + [P''S''/(m''TS)]^{0.5}\} \tag{3.56}$$

For case 2 with t greater than both $5m'S'/P'$ and $5m''S''/P''$ (Hantush, 1960, p. 3716; Reed, 1980, p. 26):

$$s = QW(u\delta_1,b)/(4\pi T) \tag{3.57}$$

with

$$\delta_1 = 1 + (S' + S'')/3S \tag{3.58}$$

$$b = r[(P'/m')/T + (P''/m'')/T]^{0.5} \qquad (3.59)$$

For case 3 with t greater than both $10m'S'/P'$ and $10m''S''/P''$ (Hantush, 1960, p. 3716; Reed, 1980, p. 26):

$$s = QW(u\delta_2)/(4\pi T) \qquad (3.60)$$

with

$$\delta_2 = 1 + (S' + S'')/S \qquad (3.61)$$

For case 4 with t greater than both $5m'S'/P'$ and $10m''S''/P''$ (Hantush, 1960, p. 3716; Reed, 1980, p. 26):

$$s = QW(u\delta_3,b)/(4\pi T) \qquad (3.62)$$

with

$$\delta_3 = 1 + (S'' + S''/3)/S \qquad (3.63)$$

$$b = r[(P'/m')/T]^{0.5} \qquad (3.64)$$

where s is the aquifer drawdown at time t, Q is the constant production well discharge rate, t is the time after pumping started, $W(u\delta_1,b)$ and $W(u\delta_3,b)$ are the well function defined in Equations 3.4 or 3.5, $W(u\delta_2)$ is the well function defined in Equation 2.2, T is the aquifer transmissivity, r is the distance between the production and observation wells, S is the aquifer storativity, P' is the vertical hydraulic conductivity of the aquitard overlying the aquifer, m' is the thickness of the aquitard overlying the aquifer, P'' is the vertical hydraulic conductivity of the aquitard underlying the aquifer, m'' is the thickness of the aquitard underlying the aquifer, S' is the storativity of the aquitard overlying the aquifer, and S'' is the storativity of the aquitard underlying the aquifer.

Values of $W(u,\tau)$, for the practical range of u and τ, are listed in Appendix G. A family of logarithmic graphs of $W(u,\tau)$ vs. 1/u for

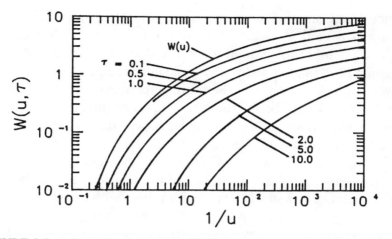

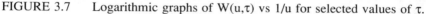

FIGURE 3.7   Logarithmic graphs of W(u,τ) vs 1/u for selected values of τ.

selected values of τ is shown in Figure 3.7. A FORTRAN program for calculating values of W(u,τ) is listed by Reed (1980, pp. 72-75). Values of W(u), for the practical range of u, are presented in Appendix G.

When $t < S'm'/10P'$ the rate of yield from storage in the aquifer is as follows (Hantush, 1964, p. 336):

$$q_s = Qexp(at)erfc[(at)^{0.5}] \tag{3.65}$$

with

$$a = P'S'/(S^2m') \tag{3.66}$$

where q_s is the rate of yield from aquifer storage at time t, Q is the constant production well discharge rate, erfc(at) is an error function defined in Equation 3.21, P' is the aquitard hydraulic conductivity, S is the aquifer storativity, S' is the aquitard storativity, and m' is the aquitard thickness.

The assumptions on which Equations 3.54 to 3.66 are based are as follows: the uniformly porous aquifer is homogeneous, isotropic, infinite in areal extent, and constant in thickness throughout; the hydraulic properties of the aquitards overlying and underlying the aquifer are constant in space and time; flow in the aquifer is two-dimensional and radial in the horizontal plane and flow in the aquitards is vertical; in case 1, there is no flow across the top of the upper aquitard or the base of the lower aquitard; in case 2, there are constant head plane sources above the upper aquitard and below the lower aquitard; in case 3, there are impermeable beds above the upper aquitard and below the lower aquitard; in case 4, there is a constant head plane source above the upper aquitard and an impermeable bed below the lower aquitard; the production well discharge rate is constant; and wells fully penetrate the aquifer.

Equations 2.21, 2.22, 3.54, 3.57, 3.60, and may be combined to describe drawdown in leaky artesian aquifers with partially penetrating wells. Sauveplane (1984, pp. 204-206) developed approximate equations describing drawdown distribution in two mutually leaky artesian aquifers with aquitard storativity using Schapery's (1961) technique of approximate inversion of derived functions in the Laplace plane. Moench and Ogata (1984, pp. 146-170) developed Laplace transform solutions for the following complex aquifer conditions: two interconnected leaky artesian aquifers with aquitard storativity, leaky artesian aquifer overlain by water table aquitard, partially penetrating well in leaky artesian aquifer with aquitard storativity, and large diameter production well with wellbore storage in leaky artesian aquifer with aquitard storativity.

Laplace transforms and the inversion theorem for Laplace transformation are briefly described by Sauveplane (1984b, pp. 197-199). In many cases, the inversion is difficult or impossible analytically and some form of numerical inversion is required. Many methods for numerical inversion are computer-time intensive. A simple numerical inverter which involves the Stehfest (1970) algorithm and requires relatively little computation is widely used in the groundwater field to evaluate well functions.

As described by Moench and Ogata (1981, p. 250), the inverse of a Laplace transform may be obtained with the following equation (Stehfest, 1970):

$$F_a = [(\ln 2)/T] \sum_{i=1}^{N} V_i P[i(\ln 2)/T] \qquad (3.67)$$

where F_a is the approximate value of the inverse $F(t)$ at T and $P(s)$ is the Laplace transformed function to be inverted. The coefficients V_i are given by the following equation (Stehfest, 1970):

$$V_i = (-1)^{(N/2)+1} \sum_{k=(i+1/2)}^{\min(i,N/2)} k^{N/2}(2k)! / [(N/2-k)!k!$$

$$(k-1)!(i-k)!(2k-i)!] \tag{3.68}$$

where N is an even number and k is calculated using integer arithmetic.

For a given N the V_i sum to zero and as N increases V_i tends to increase in absolute value. The value of N to use for maximum accuracy is approximately proportional to the number of computer dependent significant figures. Using 64 bit variables in computations, the value of N is usually between 18 and 20. The function $F(t)$ must have no discontinuities or rapid oscillations.

A FORTRAN program based on equations 3.67 and 3.68 and a companion diskette are described by Dougherty (1989, pp. 564-569). A FORTRAN program for a semianalytical solution to the radial dispersion in porous media with a Stehfest inverter subroutine is listed by Javandel, et al. (1984, pp. 167-170).

SUBSIDENCE

Subsidence is defined as the vertical and horizontal movement of the ground surface as a result of water, brine, steam, gas, or petroleum withdrawal. Subsidence ranging from 1 to 30 feet (0.30 to 9.14 m) has been measured due to artesian pressure decline (Helm, 1982, pp. 103-139) at such places as Aska, Japan; Tokyo, Japan; Mexico City, Mexico; Taipei Basin, Taiwan; Shanghai, China; Central Arizona; Santa Clara Valley, California; San Joaquin Valley, California; Las Vegas, Nevada; Houston-Galveston, Texas; and Baton Rouge, Louisiana. As a result of subsidence, damage has occurred at places to buildings, bridges, tunnels, streets, highways, railroads, water and sewer lines, power lines, well casings, dams, and irrigation and drainage canals. Subsidence has increased flood hazards in low areas and accompanied horizontal movement of the ground surface has produced cracks and fissures. Remedial or protective measures taken include: reduced groundwater withdrawal; construction of dikes, drainage pumping stations, water supply reservoirs, or recharge wells; adoption of a groundwater management code; and relocation of water supply well fields. Organic soil, coal mine, and sinkhole subsidence is described in Holzer (1984).

The ratio of subsidence to artesian pressure decline depends on the number, thickness, compressibility, and hydraulic conductivity of fine-grained sediments within the aquifer system; clay mineralogy and geochemistry; initial porosity; previous loading history; and degree of consolidation and cementation of sediments (Poland and Davis, 1969, pp. 187-269). The range in ratio where known is about 0.005-0.1 and stored water released by compaction of fine-grained sediments may be as much as 50 times greater than the water released by elastic expansion of groundwater (Poland, 1961, pp. B52-B54).

Fine-grained sediments (silty clay, clay, or shale) commonly are highly compressible compared to coarse-grained sediments. Compressibility in coarse-grained sediments is instantaneous but not appreciable in magnitude, whereas, compressibility in fine-grained sediments is slow but appreciable in magnitude. Coarse-grained sediments are largely elastic, whereas, fine-grained sediments are only partially elastic. Rebound (expansion) caused by decreases in groundwater withdrawals and resulting increases in artesian pressure is small or inappreciable in fine-grained sediments. The compressibility and expandability of coarse-grained sediments tend to be equal in magnitude. In contrast, the expandability of fine-grained sediments is generally 1 to 2 orders of magnitude smaller than compressibility. Thus, subsidence in coarse-grained sediments is largely recoverable and subsidence in fine-grained sediments is largely nonrecoverable. As a result of compaction, the hydraulic conductivity and storativity of fine-grained sediments decrease.

Subsidence mostly results from primary consolidation. However, secondary consolidation may be appreciable in fine-grained sediments (Brutsaert and Corapcioglu, 1976, pp. 1663-1675). Primary consolidation is commonly assumed to be the deformation resulting from the elastic properties of the aquifer system skeleton. Secondary consolidation or creep refers to the continuous readjustment of the aquifer system skeleton structure under a constant load.

Several factors complicate the prediction of subsidence (Helm, 1975, pp. 465-478). The compressibilities of sediments, particularly fine-grained sediments, are not the same and vary with the magnitude and duration of artesian pressure changes. Compressibility and expandability depend on whether the current artesian pressure is larger or smaller than past artesian pressure (prior history of artesian pressure). When artesian pressure increases before the vertical distribution of artesian pressure in fine-grained sediments reaches equilibrium, a transient internal discontinuity develops. Part of the fine-grained sediments continues to be stressed in response to declines in artesian pressure (and hence is compacting in part irreversible) and part of the fine-grained sediments is stressed in response to increases in artesian pressure (and hence is expanding elastically). Such conditions exist during cyclic artesian pressure changes.

Compressibility and expandability are generally assumed to be different but constant in time and with artesian pressure changes. The aquifer system is divided into volume segments which are assumed to be vertically homogeneous. Compactions and expansions in each segment are accumulated to determine total subsidence or rebound. Changes in artesian pressure are assumed to be instantaneous at time $t = 0$ and then constant (step changes). The magnitude and duration of artesian pressure changes and the vertical hydraulic conductivity, thickness, elasticity, and depth of sediments are required to estimate subsidence or rebound.

Artesian pressure decline causes an equivalent increase in grain-to-grain pressure on an aquifer system skeleton. The magnitude of the resulting compaction depends largely on the elasticity of sediments. The bulk modulus of elasticity of sediments is determined from void ratio e vs. effective pressure p' curves (Bouwer, 1978, pp. 317-319). A small sample of the sediments is placed in a consolidometer and stressed in the laboratory. Shortening of the freely draining saturated sample with each load increase is measured.

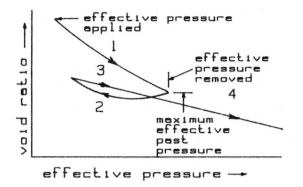

FIGURE 3.8   Conceptual graph of void ratio vs effective pressure .

An idealized e vs. p′ curve is presented in Figure 3.8. Void ratio decreases in response to the applied effective pressure as shown by arrow 1. If the effective pressure is removed, the void ratio will increase (rebound) at a rate less than the rate of decrease measured before the effective pressure was removed, but not to the ratio value at the start of the test (arrow 2). If the effective pressure is then re-applied, the void ratio will again decrease but a rate less than the rate of decrease measured before the effective pressure was removed until the previous maximum effective pressure is reached (arrow 3). Thereafter, the decrease in void ratio will accelerate (arrow 4) approaching the rate measured before the effective pressure was removed.

The bulk modulus of elasticity varies with the effective pressure but within the narrow changes in effective pressure usually encountered in groundwater situations it may be assumed to be constant. The bulk modulus of elasticity for coarse-grained sediments under normal conditions is commonly assumed to be the same for compaction (increasing effective pressure) as for expansion (rebound with decreasing effective pressure). In the case of fine-grained sediments, the bulk modulus of elasticity for effective pressure increases greater than past effective pressures (compaction) is generally 1 or 2 orders of magnitude larger than the bulk modulus of elasticity for effective pressure increases less than past effective pressure decreases (rebound). Representative ranges of bulk modulus of elasticity for selected rocks and effective pressure increases

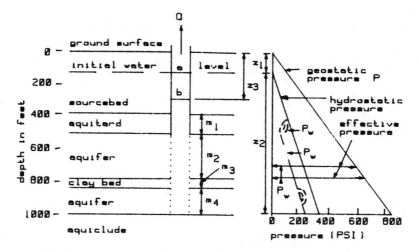

FIGURE 3.9   Cross section through aquifer system with depth-effective pressure diagram.

greater than past effective pressures are presented in Appendix B (Bouwer, 1978, p. 321).

The principle of effective pressure is expressed as (Terzaghi, 1925, pp. 874-878):

$$p = p' + p_w \tag{3.69}$$

where p is the total overburden load or geostatic pressure, p' is the effective overburden pressure of grain-to-grain load, and p_w is the water pressure.

A two-dimensional depth-pressure diagram for a selected aquifer system developed by Poland and Davis (1969, pp. 187-269) is presented in Figure 3.9. It is assumed that the artesian pressure head in the leaky artesian aquifer system is at the same level as the water table prior to withdrawal of water from the aquifer. As the result of pumping, the artesian pressure head in the aquifer system declines from position a to position b and the drawdown is equal to $z_3 - z_1$. The water table remains stationary. Artesian pressure head declines in the aquitard and clay interbed lag the artesian pressure head decline in the aquifer. Eventually, at equilibrium, the artesian pressure head decline is zero at the top of the aquitard (source bed is assumed unaffected by pumping) and $z_3 - 'z_1$ at the bottom, and the artesian pressure head decline above and below the clay interbed is $z_3 - z_1$.

The geostatic pressure prior to pumping equals the unit weight of moist sediments above the water table (γ_m) times their thickness, plus the unit weight of saturated sediments below the water table (γ) times their thickness; that is, at depth $z_1 + z_2$ (Poland and Davis, 1979, pp. 187-269):

$$p = z_1\gamma_m + z_2\gamma \tag{3.70}$$

or

$$p = z_1[\gamma_s(1 - n) + n_w\gamma_w] + z_2[\gamma_s - n(\gamma_s - \gamma_w)] \tag{3.71}$$

also prior to pumping (Poland and Davis, 1979, pp. 187-269):

$$p_w = z_2\gamma_w \tag{3.72}$$

and from Equation 3.70, the effective pressure is (Poland and Davis, 1979, pp. 187-269):

$$p' = z_1\gamma_m + z_2\gamma' \tag{3.73}$$

where γ_s is the unit weight of the sediment skeleton, n is the sediment porosity, n_w is the moisture above the water table in percent of total volume, γ_w is the unit weight of water, and $\gamma' = \gamma - \gamma_w$ is the submerged unit weight of the sediment skeleton.

Assuming negligible water expandability, lowering of the artesian pressure head in the aquifer does not change the geostatic pressure because the water table remains stationary. However, the water pressure is reduced by the amount $(z_3 - z_1) \gamma_w$, consequently from equation 3.69, the effective pressure increases by the amount $(z_3 - z_1) \gamma_w$ thereby causing consolidation and subsidence.

Two-dimensional depth-effective pressure head increase diagrams for an aquitard, coarse-grained sediments, and a clay interbed are presented in Figure 3.10 (Domenico and Mifflin, 1965, pp. 570,572). Areas of the two-dimensional depth-effective pressure

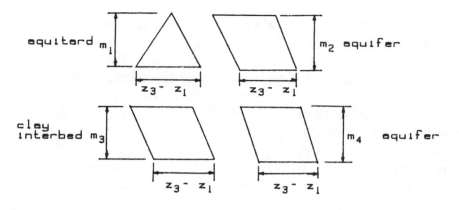

FIGURE 3.10   Two-dimensional depth-effective pressure head increase diagrams.

head increase diagrams in Figure 3.10 for t = ∞ are as follows (Domenico and Mifflin, 1965, p. 571):

Aquitard

$$[(z_3 - z_1)/2]m_1 \qquad\qquad (3.74)$$

Aquifer coarse-grained sediments immediately below aquitard

$$(z_3 - z_1)m_2 \qquad\qquad (3.75)$$

Clay interbed

$$(z_3 - z_1)m_3 \qquad\qquad (3.76)$$

Aquifer coarse-grained sediments immediately above aquiclude

$$(z_3 - z_1)m_4 \qquad\qquad (3.77)$$

   Areas of two-dimensional depth-effective pressure head increase diagrams for other subsidence or rebound conditions may be determined keeping in mind other water table and artesian pressure head changes and relationships. In some cases, geostatic pressure changes will occur that must be taken into account.

Consider the situation where the artesian pressure head in an aquifer coarse-grained sediments above and below a clay interbed is suddenly lowered and maintained at a constant level. Accompanying effective pressure increase in the aquifer coarse-grained sediments is instantaneous. However, the accompanying effective pressure increase in the clay interbed is time-dependent. Prior to equalization of artesian pressure heads in the aquifer coarse-grained sediments and the clay interbed, the equation governing effective pressure head decline in any particular horizontal element Δz of the clay interbed is as follows (Domenico, 1972, pp. 360-367):

$$s_c = s_a W_c(u, z/m_c) \qquad (3.78)$$

with

$$u = [(m_c/2)^2 \gamma_w]/(P't E_c) \qquad (3.79)$$

when there is no artesian pressure decline above the claylike materials

$$u = [m_c^2 \gamma_w]/(P't E_c) \qquad (3.80)$$

where s_c is the artesian pressure head decline at time t within any particular horizontal element z of the clay interbed, s_a is the artesian pressure head decline at time t in the aquifer coarse-grained sediments, $W_c(u, z/m_c)$ is a consolidation function, m_c is the clay interbed thickness, γ_w is the unit weight of water, t is the time after the abrupt artesian pressure head decline in the aquifer coarse-grained sediments occurred, P' is the vertical hydraulic conductivity of the clay interbed, E_c is the bulk modulus of elasticity of the clay interbed skeleton, and z is the depth below the top of the clay interbed to Δz.

Values of $W_c(u, z/m_c)$ based on a graph presented by Domenico and Mifflin (1965, p. 568), in terms of the practical range of $1/u$ and z/m_c, are presented in Appendix G. An arithmetic graph of $W_c(u, z/m_c)$ vs. z/m for selected values of $1/u$ is presented in Figure 3.11.

The average time-dependent effective pressure head increase in the clay interbed may be determined by noting that $s_c(t)\gamma_w$ is equal to the water pressure (p_w) at time t.

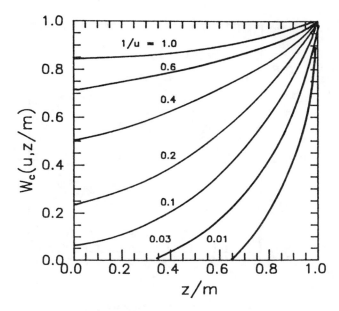

FIGURE 3.11   Arithmetic graphs of Wc(u,z/m) vs z/m for selected values of 1/u.

The equation governing the average effective pressure head decline in the clay interbed is as follows (Domenico, 1972, pp. 360-367):

$$s_{ca} = s_a W_c(u) \tag{3.81}$$

where s_{ca} is the average artesian pressure head decline at time t in the clay interbed, s_a is the artesian pressure head decline at time t in the aquifer coarse-grained sediments, t is the time after the artesian pressure head decline started, $W_c(u)$ is a consolidation function, and u is defined in Equations 3.79 or 3.80.

Values of $W_c(u)$ based on a graph presented by Domenico and Mifflin (1965, p. 574), in terms of the practical range of 1/u, are presented in Appendix G. An arithmetic graph of $W_c(u)$ vs. 1/u is presented in Figure 3.12.

With a single artesian pressure head decline, the equations governing the ultimate and time-dependent ground surface subsidence assuming water expandability and secondary consolidation are negligible are as follows (Domenico and Mifflin, 1965, pp. 563-576):

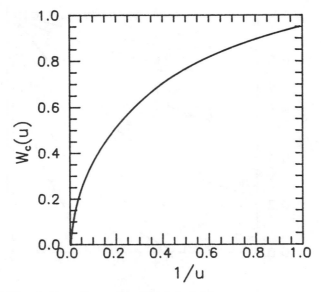

FIGURE 3.12   Arithmetic graph of Wc(u) vs 1/u.

$$\Delta m = \Delta m_a + \Delta m_c + \Delta m_i \tag{3.82}$$

$$\Delta m_t = \Delta m_a + \Delta m_{ct} + \Delta m_{it} \tag{3.83}$$

with

$$\Delta m_a = (\gamma_w/E_a)A_{ap} \tag{3.84}$$

$$\Delta m_c = (\gamma_w/E_{cc})A_{cp} \tag{3.85}$$

$$\Delta m_i = (\gamma_w/E_{ic})A_{ip} \tag{3.86}$$

$$\Delta m_{ct} = (\gamma_w/E_{cc})A_{cpt} \tag{3.87}$$

$$\Delta m_{it} = (\gamma_w/E_{ic})A_{ipt} \tag{3.88}$$

or

$$\Delta m_a = S_{sa}A_{ap} \tag{3.89}$$

$$\Delta m_{ct} = S_{scc} A_{cpt} \qquad (3.90)$$

$$\Delta m_{it} = S_{sic} A_{ipt} \qquad (3.91)$$

$$S_{sic} = \gamma_w / E_{ic} \qquad (3.92)$$

$$S_{sa} = \gamma_w / E_a \qquad (3.93)$$

$$S_{scc} = \gamma_w / E_{cc} \qquad (3.94)$$

where Δm is the cumulative ultimate aquifer, aquitard, and clay interbed consolidation (subsidence), Δm_a is the ultimate aquifer coarse-grained sediment consolidation, Δm_c is the ultimate aquitard consolidation, Δm_i is the ultimate clay interbed consolidation, γ_w is the unit weight of water, E_a is the compaction bulk modulus of elasticity of the aquifer coarse-grained sediment skeleton, E_{cc} is the compaction bulk modulus of elasticity of the aquitard skeleton, E_{ic} is the compaction bulk modulus of elasticity of the clay interbed sediment skeleton, A_{ap} is the area of the aquifer coarse-grained sediment two-dimensional depth-effective pressure head increase diagram, A_{cp} is the final area of the aquitard two-dimensional depth-effective pressure head increase diagram, A_{ip} is the final area of the clay interbed two-dimensional depth-effective pressure head increase diagram, Δm_t is the cumulative aquifer, aquitard, and clay interbed consolidation at time t, Δm_{ct} is the aquitard consolidation at time t, Δm_{it} is the clay interbed consolidation at time t, A_{cpt} is the area of the aquitard two-dimensional depth-effective pressure head increase diagram at time t, A_{ipt} is the area of the clay interbed two-dimensional depth-effective pressure head increase diagram at time t, S_{sic} is the clay interbed specific storativity, S_{sa} is the aquifer coarse-grained sediment specific storativity, and S_{scc} is the aquitard specific storativity.

With the multi-step artesian pressure head change shown in Figure 3.13, the equation governing the time-dependent consolidation (subsidence) is as follows (Domenico and Mifflin, 1965, pp. 563-576):

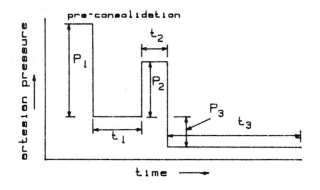

FIGURE 3.13   Conceptual graph of artesian pressure vs time.

$$\Delta m_t = \Delta m_{ap1c} - \Delta m_{ap2r} + \Delta m_{ap2c} + \Delta m_{ap3c} + \Delta m_{cp1-p2t1+t2+t3c}$$

$$+ \Delta m_{cp2t1c} - \Delta m_{cp2t2r} + \Delta m_{cp2t1cp} + \Delta m_{cp2t3-t1c}$$

$$+ \Delta m_{cp3t3c} + \Delta m_{ip1-p2t1+t2c} + \Delta m_{ip2t1c} - \Delta m_{ip2t2r}$$

$$+ \Delta m_{ip2t1cp} + \Delta m_{ip2t3-t1c} + \Delta m_{ip3t3c} \tag{3.95}$$

where Δm_{ap1c} is the ultimate aquifer coarse-grained sediment consolidation due to the artesian pressure decline p_1, Δm_{ap2r} is the ultimate aquifer coarse-grained sediment rebound due to the artesian
pressure increase p_2, Δm_{ap2c} is the ultimate aquifer coarse-grained
sediment consolidation due to the artesian pressure decline p_2, Δm_{ap3c}
is the ultimate aquifer coarse-grained sediment consolidation due to
the artesian pressure decline p_3, $\Delta m_{cp1-p2t1+t2+t3c}$ is the aquitard consolidation due to the artesian pressure decline $p_1 - p_2$ greater than the
maximum past effective pressure for a time period $t_1 + t_2 + t_3$, Δm_{cp2t1c}
is the aquitard consolidation due to the artesian pressure decline p_2
greater than the maximum past effective pressure $p_1 - p_2$ for a time
period t_1, Δm_{cp2t2r} is the aquitard rebound due to the artesian pressure
increase p_2 for a time period t_2, $\Delta m_{cp2t1cp}$ is the aquitard consolidation
due to the artesian pressure decline p_2 less than the maximum past
effective pressure p_1 for a time period t_1, $\Delta m_{cp2t3-t1c}$ is the aquitard
consolidation due to the artesian pressure decline p_2 greater than the
maximum past artesian pressure $p_1 - p_2$ for a time period $t_3 - t_1$,

Δm_{cp3t3c} is the aquitard consolidation due to the artesian pressure decline p_3 greater than the maximum past artesian pressure p_1 for a time period t_3, $\Delta m_{ip1-p2t1+t2+t3c}$ is the clay interbed consolidation due to the artesian pressure decline $p_1 - p_2$ greater than the maximum past effective pressure for a time period $t_1 + t_2 + t_3$, Δm_{ip2t1c} is the clay interbed consolidation due to the artesian pressure decline p_2 greater than the maximum past effective pressure $p_1 - p_2$ for a time period t_1, Δm_{ip2t2r} is the clay interbed rebound due to the artesian pressure increase p_2 for a time period t_1, $\Delta m_{ip2t1cp}$ is the clay interbed consolidation due to the artesian pressure decline p_2 less than the maximum past effective pressure p_1 for a time period t_1, $\Delta m_{ip2t3-t1c}$ is the clay interbed consolidation due to the artesian pressure decline p_2 greater than the maximum past effective pressure $p_1 - p_2$ for a time period $t_3 - t_1$, and Δm_{ip3t3c} is the clay interbed consolidation due to the artesian pressure decline p_3 greater than the maximum past artesian pressure p_1 for a time period t_3.

Equation 3.95 may be expanded and rewritten as:

$$\Delta m_t = (\gamma_w/E_{ac1})A_{ap1c} - (\gamma_w/E_{ar})A_{ap2r} + (\gamma_w/E_{ac2})A_{ap2c2}$$

$$+ (\gamma_w/E_{ac1})A_{ap3c} + (\gamma_w/E_{cc1})A_{cp1-p2t1+t2+t3c}$$

$$+ (\gamma_w/E_{cc1})A_{cp2t1c} - (\gamma_w/E_{cr})A_{cp2t2r} + (\gamma_w/E_{cc2})A_{cp2t1c}$$

$$+ (\gamma_w/E_{cc1})A_{cp2t3c} + (\gamma_w/E_{cc1})A_{cp3t3c}$$

$$+ (\gamma_w/E_{ic1})A_{ip1-p2t1+t2+t3c} + (\gamma_w/E_{ic1})A_{ip2t1c}$$

$$- (\gamma_w/E_{ir})A_{ip2t2r} + (\gamma_w/E_{ic2})A_{ip2t1c}$$

$$+ (\gamma_w/E_{ic1})A_{ip2t3-t1c} + (\gamma_w/E_{ic1})A_{ip3t3c} \qquad (3.96)$$

where E_{ac1} is the compaction bulk modulus of elasticity of the aquifer coarse-grained sediment skeleton with the artesian pressure below the maximum past artesian pressure, E_{ar} is the rebound bulk modulus of elasticity of the aquifer coarse-grained sediment skeleton, E_{ac2} is the compaction bulk modulus of elasticity of aquifer coarse-grained sediment skeleton with the artesian pressure above the maximum past artesian pressure, E_{cc1} is the compaction bulk modulus of elasticity of the aquitard with the artesian pressure below the maximum past artesian pressure, E_{cr} is the rebound bulk modulus of elasticity of the aquitard, E_{cc2} is the compaction bulk modulus of elasticity of the aquitard with the artesian pressure above the maximum past artesian pressure, E_{ic1} is the compaction bulk modulus of elasticity of the clay interbed with the artesian pressure below the maximum past artesian pressure, E_{ir} is the rebound bulk modulus of elasticity of the clay interbed, E_{ic2} is the compaction bulk modulus of elasticity of the clay interbed with artesian pressure above the maximum past artesian pressure, and $A_{..}$ is the area of the aquifer coarse-grained sediment, aquitard, or clay interbed two-dimensional depth-effective pressure change diagram for pressures and times indicated by subscripts.

Subsidence theory is discussed in detail by Bear and Corapcioglu (1981a, pp. 937-946; 1981b, pp. 947-958), Corapcioglu and Bear (1983, pp. 895-908), and Bear and Verruijt (1987, pp. 78-84, 105-113). Recent trends in subsidence conceptual models are towards developing a practical three-dimensional theory (Helm, 1982, pp. 103-139 and Helm, 1984, pp. 29-82). A FORTRAN program for estimating subsidence is listed by Narashimhan (1980).

4

Water Table Radial Flow

In the unsaturated zone, part of the soil pores are filled with air which physically obstructs water movement. Unsaturated aquifer hydraulic conductivity is less than saturated aquifer hydraulic conductivity. Unsaturated flow is complicated in that aquifer hydraulic conductivity is dependent on the water content which in turn is related to the negative pressure head in pores (Bouwer, 1978, pp. 234-244).

The capillary fringe in the unsaturated zone extends from the water table up to the limit of capillary rise of water. The thickness of the capillary fringe varies inversely with the pore size. Typical ranges of capillary fringe thickness for unconsolidated deposits (Lohman, 1972) are listed in Appendix B.

Water is discharged from the soil to the atmosphere by the process of evapotranspiration, a term combining evaporation from the capillary fringe above the water table and transpiration from plants whose roots penetrate to the capillary fringe or the water table. Evaporation is negligible unless the water table is within 3.3 to 4.9 ft (1 to 1.5 m) from the ground surface (Bear, 1979, p. 58) and is only 10 percent of pan evaporation at a depth of 3.3 ft (1 m). Root zone penetration is 1.6 ft (0.5 m) for shallow rooted crops such as certain grasses and vegetables, 3.3 to 6.6 ft (1 to 2 m) for most field crops, and 10 or more ft (3 or more m) for small to medium trees (Bouwer, 1978, p. 264). Very deep rooted plants may penetrate to depths of 49.2 to 65.6 ft (15 to 20 m) (Bear, 1979, p. 58).

Under water table conditions (Figure 4.1), water in the saturated

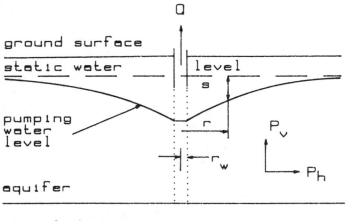

FIGURE 4.1   Cross section through production well in water table aquifer.

zone is released from aquifer storage in three stages. At early times during stage 1, water is released from storage primarily by the compaction of the aquifer skeleton and the expansion of the water itself. At intermediate times during the second stage, water is released from storage primarily by the drainage (delayed gravity yield) of the aquifer pores above the stage 1 cone of depression associated with an artesian storativity. As the second stage progresses, water released from storage above the stage 1 cone of depression is exhausted and water comes from storage at ever-increasing distances. At later times during the third stage, water is released from aquifer storage primarily by the drainage of the aquifer pores above an ever expanding cone of depression and delayed gravity yield impacts are negligible.

UNSATURATED FLOW

Unsaturated hydraulic conductivity is affected by hystersis. The relation between unsaturated hydraulic conductivity and negative pore pressure head depends on whether the head was reached by draining a wet deposit or wetting a dry deposit (Bouwer, 1978, pp.

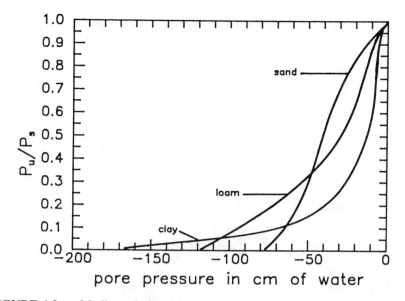

FIGURE 4.2   Median relationships between hydraulic conductivity and negative pore pressure in selected unsaturated deposits.

234-244). However, for rough estimates of unsaturated hydraulic conductivity, the relation is commonly considered unique and free from hystersis. Median relationships between the ratio of unsaturated and saturated hydraulic conductivities and negative pore pressure head, based on data presented by Bouwer (1964, pp. 121-144), are shown in Figure 4.2.

Values of the ratio of unsaturated and saturated hydraulic conductivities are estimated with the following approximate equation (Gardner, 1958, pp. 228-233; Bouwer, 1978, p. 239):

$$P_u/P_s = \{a/[(-h)^n + b]\}/P_s \qquad (4.1)$$

with (Bouwer, 1978, p. 239)

Deposit	a	b	n	$P_s = a/b$ (cm/day)
medium sands	5×10^9	10^7	5	500
fine sands, sandy loams	5×10^6	10^5	3	50
loams and clays	5×10^3	5×10^3	2	1

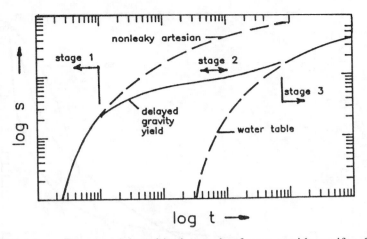

FIGURE 4.3   Conceptual logarithmic graph of s vs t with aquifer delayed gravity yield.

where P_u is the unsaturated hydraulic conductivity; P_s is the saturated hydraulic conductivity; a, b, and n are constants that decrease with decreasing particle size; and h is the pore pressure head.

DELAYED GRAVITY DRAINAGE

The time-drawdown curve has three segments with delayed gravity drainage coinciding with the three stages of release of water from aquifer storage described earlier (Figure 4.3). During the first segment, drawdown due to production well discharge is governed by the following equation (Theis, 1935, pp. 519-524):

$$s = QW(u_A)/(4\pi T) \qquad (4.2)$$

with

$$u_A = r^2S/(4Tt) \qquad (4.3)$$

Drawdown during the second segment enters a transitional period with a decreasing rate approaching a stable value and then an increasing rate approaching the third and final segment. During the third segment, drawdown is governed by the following equation (Neuman, 1972, pp. 1031-1045):

$$s = QW(u_B)/(4\pi T) \tag{4.4}$$

with

$$u_B = r^2 S_y/(4Tt) \tag{4.5}$$

where s is the drawdown at time t, Q is the constant production well discharge rate, t is the time after pumping started, $W(u_A)$ and $W(u_B)$ are the well function defined by Equation 2.2, T is the aquifer transmissivity, r is the distance between the production and observation wells, S is the aquifer artesian storativity, and S_y is the aquifer gravity yield.

The third segment starts at a time defined approximately by the following equation (Walton, 1962, p.6):

$$t = 5mS_y/P_v \tag{4.6}$$

where t is the time after pumping started beyond which delayed gravity yield impacts are negligible, m is the aquifer thickness, S_y is the aquifer specific yield, and P_v is the aquifer vertical hydraulic conductivity.

Drawdown during all three segments is governed by the following equation (Neuman, 1972, pp. 1031-1045):

$$s = QW(u_a,u_B,\beta,\sigma)/(4\pi T) \tag{4.7}$$

with

$$\beta = (r^2 P_v)/(m^2 P_h) \tag{4.8}$$

$$\sigma = S/S_y \tag{4.9}$$

$$u_B = \sigma u_A \tag{4.10}$$

where s is the drawdown at time t, Q is the constant production well discharge rate, t is the time after pumping started, T is the aquifer transmissivity, r is the distance between the production and observation wells, P_v is the aquifer vertical hydraulic conductivity, P_h is the aquifer horizontal hydraulic conductivity, m is the aquifer thickness, S is the aquifer storativity, and S_y is the aquifer specific yield.

Equations 4.7 to 4.10 are based on the following assumptions: the uniformly porous aquifer is homogeneous, anisotropic, infinite in areal extent, and constant in thickness throughout; drawdown is negligible in comparison to the initial aquifer thickness; principle horizontal and vertical hydraulic conductivities are oriented parallel to the coordinate axes; and wells fully penetrate the aquifer, have infinitesimal diameters and no wellbore storage.

Gravity drainage of aquifer pores under heavy pumping conditions can appreciably decrease the saturated aquifer thickness and therefore the aquifer transmissivity. In this case, the following equation (Jacob, 1944) may be used to adjust drawdowns calculated with Equations 4.7 to 4.10 for decreased saturated aquifer thickness:

$$s = s_o - s_o^2/(2m) \tag{4.11}$$

where s is the drawdown that occurs at time t with negligible decrease in saturated aquifer thickness as calculated with Equation 4.7, s_o is the drawdown that occurs at time t with appreciable decrease in saturated aquifer thickness, t is the time after pumping started, and m is the initial aquifer thickness.

Equation 4.11 is strictly applicable to third segment drawdown data and not to early first and second segment drawdown data (Neuman, 1975a, pp. 334-335). Also, Equation 4.11 cannot be used when the saturated aquifer thickness is severely decreased more than 50 percent.

Equations 2.21 to 2.27 and 4.7 to 4.11 may be combined to describe drawdown with partially penetrating wells. Neuman (1974, pp.303-312) and Boulton and Streltsova (1976, pp. 29-46) present equations describing drawdown in a water table aquifer with delayed gravity yield and partially penetrating wells.

Boulton and Streltsova (1978, pp. 527-532) present equations describing drawdown in a fractured rock water table aquifer with delayed gravity yield and fully penetrating wells. Hantush (1964, pp. 368-372) presents an equation describing drawdown distribution in sloping water table aquifers. A FORTRAN program for calculating drawdown with Equation 4.7 is listed by Neuman (1975b). This program evaluates a complex integral numerically using a self-adapting Simpson's Rule algorithm. A FORTRAN program for calculating drawdown in a water table aquifer with delayed gravity yield and wellbore storage is listed by Rushton and Redshaw (1979, pp. 242-244, 262). A BASIC version of that program is listed by Walton (1987, pp. 106-112).

Values of the well function $W(u_A,u_B,\beta,\sigma)$ presented by Neuman (1975a, 332-333), for the practical range of $1/u_A$, $1/u_B$, β, and a σ value of 10^{-9} are presented in Appendix G. Two families of logarithmic graphs of $W(u_A,\beta,\sigma)$ vs. $1/u_A$ and $W(u_B,\beta,\sigma)$ vs. $1/u_B$ for selected values of β are shown in Figure 4.4. The curves lying to the left of the β value labels are called type A curves and correspond to the top scale expressed in terms of $1/u_A$. The curves lying to the right of the β values labels are called type B curves and correspond to the bottom scale expressed as $1/u_B$. Type A curves apply to early time-drawdown data and type B curves apply to late drawdown data (see Neuman, 1975, p. 330).

Both type curves approach a set of long horizontal asymptotes. The type curves must be plotted on difference scales, one with respect to $1/u_A$ and the other with respect to $1/u_B$, because the two families of type curves are displaced from each other great distances. If S and S_y are known or can be estimated with reasonable accuracy thereby making it possible for σ to be calculated, a complete family of type curves for a particular field condition may be obtained with the following graphical technique (Neuman, 1975, p. 331). The family of type A curves are traced on logarithmic paper and the $1/u_A$ value (t_a) corresponding to the intersection of the $W(u_A,\beta,\sigma)$ curve and the $10^{-2}\ W(u_A,u_B,\beta,\sigma)$ horizontal line is noted. The family of type B curves are traced on the $1/u_A$ horizontal scale keeping in mind that the $1/u_B$ value corresponding to the intersection of the $W(u_B,\beta,\sigma)$ curve and the $10^{-2}\ W(u_A,u_B,\beta,\sigma)$ horizontal line must be equal to t_a multiplied by σ. The appropriate families of type A and

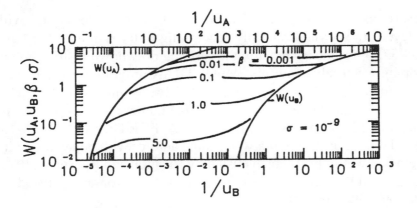

FIGURE 4.4   Logarithmic graphs of $W(u_A, u_B, \beta, \sigma)$ vs $1/u_A$ and $1/u_B$ for selected values of β and $\sigma = 10^{-9}$.

type B curves are joined by straight lines that are tangent to both curves and truncated portions of type curve traces are erased.

CONVERSION FROM ARTESIAN

When water levels decline below the top of a uniformly porous nonleaky artesian aquifer, a partial conversion from artesian to water table conditions occurs (Figure 4.5). The radial distance of conversion expands with time. Drawdown under conversion conditions due to production well discharge is governed by the following equations (Moench and Prickett, 1972, pp. 494-499):

when $r < R$

$$m - h_1 = QW(u_w, c)/(4\pi T) \qquad (4.12)$$

when $r > R$

$$h - h_2 = (h - m)W(u_a)/W(cS/S_y) \qquad (4.13)$$

with

$$u_a = r^2S/(4Tt) \qquad (4.14)$$

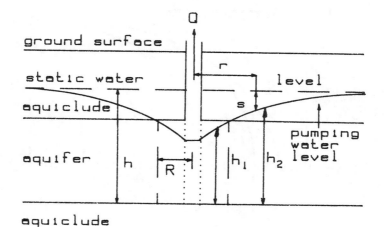

FIGURE 4.5   Cross section through production well in nonleaky artesian aquifer undergoing partial conversion to water table conditions.

$$u_w = r^2 S_y/(4Tt) \qquad (4.15)$$

$$c = R^2 S_y/(4Tt) \qquad (4.16)$$

where r is the distance between the production and observation wells, R is the distance between the production well and the point of conversion, m is the aquifer thickness, h_1 is the vertical distance from the aquifer base to the water table surface at time t, Q is the constant production well discharge rate, $W(u_w,c)$ is a well function, T is the aquifer transmissivity, h is the vertical distance from the aquifer base to the nonpumping potentiometric surface, h_2 is the vertical distance from the aquifer base to the cone of depression at time t when $r > R$, $W(u_a)$ and $W(cS/S_y)$ are the well function defined in equation 2.2, t is the time after pumping started, S is the aquifer artesian storativity, and S_y is the aquifer gravity yield.

Values of $W(u_w, c)$, for the practical range of $1/u_w$ and c, are presented in Appendix G. A family of logarithmic graphs of $W(u_w,c)$ vs. $1/u_w$ for selected values of c is shown in Figure 4.6.

The value of R may be determined by using the method of successive approximations. A first estimate of R is made and values of h_1 and h_2 are calculated with equations 4.12 or 4.13. A cone of

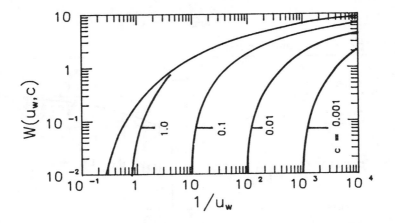

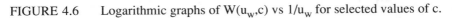

FIGURE 4.6   Logarithmic graphs of $W(u_w,c)$ vs $1/u_w$ for selected values of c.

depression profile is drawn and R is ascertained from the profile. This value of R is compared with the first estimate of R. If the comparison is favorable, the estimate of R is declared valid, otherwise a second estimate of R is made and the process is repeated until a favorable comparison is obtained.

Equations 4.12 to 4.16 are based on the following assumptions: the uniformly porous aquifer is homogeneous, isotropic, infinite in areal extent, and constant in thickness throughout; the aquifer is overlain and underlain by aquicludes; dewatering after conversion from artesian to water table conditions does not significantly reduce aquifer transmissivity; delayed gravity yield impacts are negligible; wells fully penetrate the aquifer; and the production well has an infinitesimal diameter and no wellbore storage.

MOUND BENEATH SPREADING AREA

Mounding of a water table aquifer can occur below certain types of waste disposal facilities. The impact of mounding on groundwater flow patterns and contaminant migration needs to be considered in evaluating the effectiveness of different remedial measures (Boutwell, et al., 1986, pp. 70-71, 101-106). The rise of the water table beneath circular spreading areas (Figure 4.7) is described by

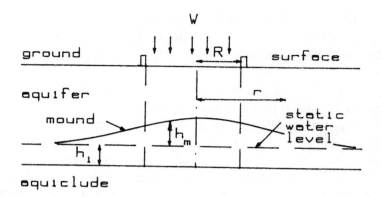

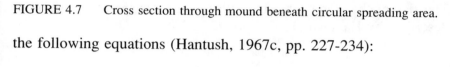

FIGURE 4.7   Cross section through mound beneath circular spreading area.

the following equations (Hantush, 1967c, pp. 227-234):

when $t \geq r^2 S_y/(2P_h m_a)$ and $r \leq R$

$$h_m = \{[(WR^2)/(2P_h)]\{W(u_o) - (r/R)^2 \exp(-u_o) + (1/u_o)$$
$$[1 - \exp(-u_o)]\} + h^2_i\}^{0.5} \qquad (4.17)$$

when $t \geq r^2 S_y/(2P_h m_a)$ and $r \geq R$

$$h_m = \{[(WR^2)/(2P_h)][W(u) + 0.5u_o \exp(-u)] + h^2_i\}^{0.5} \quad (4.18)$$

with

$$u = R^2 S_y/(4P_h m_a t) \qquad (4.19)$$

$$u_o = r^2 S_y/(4P_h m_a t) \qquad (4.20)$$

$$m_a = 0.5[h_i + h_m] \qquad (4.21)$$

where h_m is the height of the water table above the aquifer base with recharge from the circular spreading area at time t, W is the spreading area recharge rate, R is the radius of the circular spreading area, P_h is the aquifer horizontal hydraulic conductivity, $W(u)$ and $W(u_o)$ are the well function defined in Equation 2.2, r is the distance between the center of the circular spreading area and the observation well, h_i is the initial height of the water table above the aquifer base, t is the time after pumping started, and S_y is the aquifer specific yield.

Equations 4.17 to 4.21 are applicable if the rise of the water table relative to the initial depth of saturation does not exceed 50%. The uniformly porous aquifer underlain by an aquiclude is assumed to be homogeneous, isotropic, infinite in areal extent, and constant in thickness throughout. The water table is assumed to remain below the base of the spreading area. A BASIC program for defining mounds under circular spreading areas is listed by Walton (1984b). The rise of the water table beneath rectangular spreading areas is described by Hantush (1967c, pp. 227-234).

5

Boundaries And Flow Nets

Most analytical equations assume the aquifer system is infinite in areal extent, homogeneous, and isotropic. The existence of hydrogeologic boundaries serves to limit the continuity of most aquifers in one or more directions to distances ranging from a few hundred feet (hundred meters) or less to a few miles (5000 meters) or more. Commonly, aquifer systems are anisotropic and have nonhomogeneities. Under these conditions, the flow field is distorted and the image-well theory (Ferris, et al., 1962, pp. 144-166) is applied with analytical model equations to describe distorted cones of depression. The impact of anisotrophy and nonhomogeneities depends partly on the ratio of the area of influence of pumping and the area of anisotrophy or nonhomogeneity.

Hydrogeologic boundaries may be divided into two types: barrier and recharge. A barrier boundary is defined as a line (streamline) across which there is no flow and it may consist of folds, faults, or relatively impervious deposits (aquiclude) such as shale or clay. A recharge boundary is defined as a line (equipotential) along which there is no drawdown and it may consist of rivers, lakes, and other bodies of surface water hydraulically connected to the aquifer. A discontinuity is defined as a line separating two major areas of different hydraulic conductivities. Most hydrogeologic boundaries and discontinuities are not clear-cut straight-line features but are irregular in shape and extent. However, in practice complicated boundaries and discontinuities are simulated with straight-line effective demarcations.

IMAGE-WELL THEORY

The image-well theory may be stated as follows: the effect of a barrier boundary on the drawdown in a well, as a result of pumping from another well, is the same as though the aquifer were infinite in areal extent and a like discharging well were located across the real boundary on a perpendicular thereto and at the same distance from the boundary as the production well. The principle is the same for a recharge boundary except the image well is assumed to be recharging the aquifer system instead of pumping from it. Thus, the impacts of hydrogeologic boundaries on drawdown can be simulated by use of hypothetical image wells. Boundaries are replaced by imaginary wells which produce the same disturbing effects as the boundaries. Boundary problems are thereby simplified to consideration of an infinite aquifer in which real and image wells operate simultaneously (Ferris, et al., 1962, pp. 144-166). Total drawdown is the algebraic summation of real and image well drawdown and/or recovery components. If s_o is the total drawdown in an observation well at time t, s_p is the component of drawdown caused by a production well at time t, and s_i is the component of drawdown or buildup caused by an image well associated with a single boundary at time t, then

$$s_o = s_p \pm s_i \qquad (5.1)$$

Appropriate well functions are utilized to calculate s_p and s_i depending on existing aquifer system conditions. For example, with nonleaky artesian conditions:

$$s_o = QW(u)/(4\pi T) +- Q_iW(u_i)/(4\pi T) \qquad (5.2)$$

or

$$s_o = [Q/(4\pi T)]W_b(u) \qquad (5.3)$$

with

$$W_b(u) = W(u) +- W(u_i) \qquad (5.4)$$

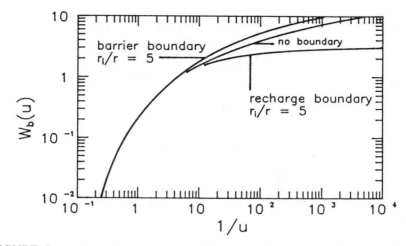

FIGURE 5.1   Logarithmic graphs of $W_b(u)$ vs $1/u$ for selected values of r_i/r.

$$u = r^2S/(4Tt) \qquad (5.5)$$

$$u_i = r_i^2S/(4Tt) \qquad (5.6)$$

where Q is the constant production well discharge rate; W(u) and $W(u_i)$ are well functions for the production and image wells defined in Equation 2.2; T is the aquifer transmissivity; Q_i is the constant image-well discharge or recharge rate which is equal to Q; $W_b(u)$ is the boundary summation well function; r is the distance between the production and observation wells; S is the aquifer storativity; t is the time after pumping started; and r_i is the distance between the observation and image well.

Boundary summation well functions for a single barrier or recharge boundary and a r_i/r ratio of 5 are shown in Figure 5.1. Boundary summation well functions for other r_i/r ratios are illustrated by Stallman (1963, pp. C45 -C47). The boundary summation well function concept is extended to cover multiple boundary situations by Kruseman and Ridder (1976, pp. 115-120) and Streltsova (1988, pp. 210-253).

BARRIER BOUNDARY

An aquifer system bounded on one side by an impervious forma-

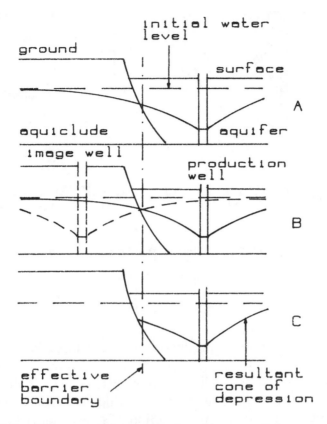

FIGURE 5.2 Diagrammatic representation of image-well theory as applied to a barrier boundary.

tion (barrier boundary) is illustrated in Figure 5.2. The impervious formation cannot contribute water to the production well. Water cannot flow across a line that defines the effective limit of the aquifer system. A hypothetical infinite hydraulic system that will satisfy the boundary conditions of the finite aquifer system is created with twin real and image wells (see Figure 5.2b).

Consider the cone of depression that would exist if the barrier boundary were not present (Figure 5.2a). The hydraulic gradient causes flow across the boundary. An imaginary discharging well placed across the boundary perpendicular to and equidistant from the boundary produces a hydraulic gradient from the boundary toward the image well equal to the hydraulic gradient from the

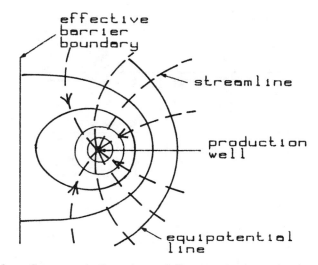

effective
barrier
boundary

streamline

production
well

equipotential
line

FIGURE 5.3   Conceptual plan view of flow net near production well and barrier boundary.

boundary toward the production well. A groundwater divide exists at the boundary (Figure 5.2b) and this is true everywhere along the boundary. The condition of no flow across the boundary line is fulfilled. Therefore, the imaginary hydraulic system of a production well and it's image counterpart in an infinite aquifer system satisfies the boundary conditions dictated by the finite barrier boundary situation. The resultant cone of depression is the summation of the components of both the production and image well cones of depression as shown in Figure 5.2c. The resultant profile of the cone of depression is flatter on the side of the production well toward the boundary and steeper on the opposite side away from the boundary than it would be if no boundary were present. A conceptual plan view of the flow net in the vicinity of a production well near a barrier boundary is presented in Figure 5.3.

With barrier boundary conditions, water levels in wells decline at an initial rate under the influence of the production well only. When the cone of depression of the image well appreciably impacts the production well, the time-rate of drawdown increases because the total rate of withdrawal from the aquifer system is then equal to that of the production well plus that of the discharging image well. Thus, the time-drawdown curve of the production well is deflected downward.

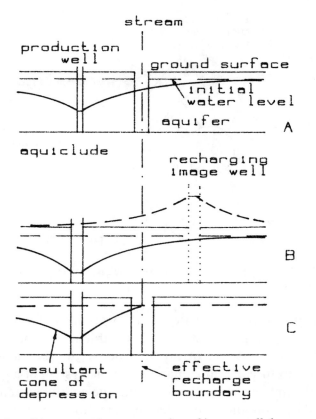

FIGURE 5.4   Diagrammatic representation of image-well theory as applied to a recharge boundary.

RECHARGE BOUNDARY

An aquifer system bounded on one side by a recharge boundary is shown in Figure 5.4a. The cone of depression cannot spread beyond the boundary. The condition is established that there is no drawdown along an effective line of recharge somewhere offshore. The imaginary hydraulic system of a production well and its image counterpart in an infinite aquifer system (Figure 5.4b) satisfies the foregoing boundary condition. An imaginary recharge well is placed directly opposite of and at the same distance from the boundary as the production well.

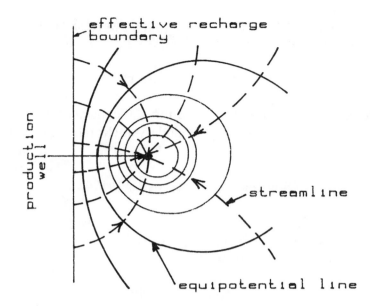

FIGURE 5.5   Conceptual plan view of flow net near production well and recharge boundary.

The resultant cone of depression is the algebraic summation of the components of the production well cone of depression and the image well cone of impression (Figure 5.4c). The resultant profile of the cone of depression is steeper on the boundary side of the production well and flatter on the other side of the boundary than it would be if no boundary were present. A conceptual plan view of the flow net in the vicinity of a production well near a recharge boundary is shown in Figure 5.5.

With recharge boundary conditions, water levels in wells decline at an initial rate under the influence of the production well only. When the cone of impression of the image well appreciably impacts the production well, the time-rate of drawdown changes and thereafter continuously decreases until equilibrium conditions prevail and recharge equals discharge.

Steady state equations for calculating drawdown with a single recharge boundary are presented by Kruseman and Ridder (1976, pp. 112-115). Theis (1963, pp. C101-C105) presents equations for calculating drawdown under steady state conditions in the vicinity of a finite straight-line recharge boundary.

Under steady state conditions, the drawdown near a recharge boundary is defined as (Rorabaugh, 1956, pp. 101-169):

$$s = [Q/(4\pi T)]\log(r_i/r) \qquad (5.7)$$

This equation may be rewritten in terms of the distance between the production well and the recharge boundary as (Rorabaugh, 1956, pp. 101-169):

$$s = Q\log[(4a^2 + r^2 - 4arcosB_r)^{0.5}/r]/(4\pi T) \qquad (5.8)$$

For the particular case where the observation well is on a line parallel to the recharge boundary, the following equation applies (Rorabaugh, 1956, pp. 101-169):

$$s = Q\log[(4a^2 + r^2)^{0.5}/r]/(4\pi T) \qquad (5.9)$$

For the particular case where the observation well is on a line perpendicular to the recharge boundary and on the boundary side, the following equation applies (Rorabaugh, 1956, pp. 101-169):

$$s = Q\log[(2a - r)/r]/(4\pi T) \qquad (5.10)$$

where s is the stabilized drawdown with the recharge boundary impacts, Q is the constant production well discharge rate, T is the aquifer transmissivity, r_i is the distance between the production and image wells, r is the distance between the production and observation wells, a is the distance between the production well and the effective line of recharge, and B_r is the angle between a line connecting the production and image wells and a line connecting the production and observation wells.

When a production well near a stream hydraulically connected to an aquifer (recharge boundary) is pumped, water is first withdrawn from storage within the aquifer in the immediate vicinity of the production well. The cone of depression then spreads, drawing water from aquifer storage within an increasing area of influence. Water levels in the vicinity of the stream are lowered, and more and

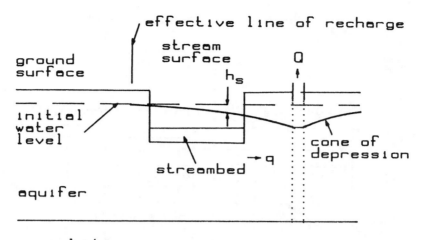

FIGURE 5.6   Cross section through production well with induced streambed infiltration.

more of the water which under natural conditions would have discharged into the stream as groundwater runoff is diverted toward the
production well. Water levels are lowered below the surface of the
stream in the immediate vicinity of the production well and the
aquifer is recharged by the influent seepage of surface water
(Hantush, 1959b, pp. 1921-1932; Mikels and Klaer, 1956, pp. 232-
242; Walton, 1963; and Hantush, 1965, pp. 2829-2838).

The cone of depression continues to grow until it intercepts sufficient area of the streambed and is deep enough beneath the streambed so that induced streambed infiltration balances discharge (see
Figure 5.6). If the hydraulic conductivity of the streambed is high,
the cone of depression may expand only partway across the streambed; if the hydraulic conductivity of the streambed is low, the
cone of depression may expand across and beyond the streambed.

Recharge by induced infiltration takes place over an area of the
streambed. However, to make the flow problem amenable to mathematical treatment, the area is replaced by a recharging image well.
The assumption is made that water levels in the aquifer will behave
the same way whether recharge occurs over an area or through a
recharging image well located at an effective distance from the
production well. It is further assumed that streambed partial pen-

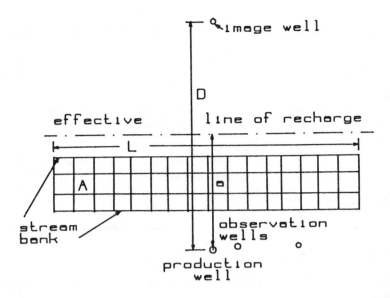

FIGURE 5.7   Plan view of production and observation wells near a stream with induced streambed infiltration.

etration and aquifer stratification impacts are integrated into that effective distance.

Aquifer transmissivity may be determined with steady state distance-drawdown data for observation wells on a line through and close to the production well and parallel to the streambed (see Figure 5.7). These observation wells are approximately equidistant from the recharging image well and the impacts of the induced streambed infiltration on water levels in these wells are approximately equal. Thus, the hydraulic gradient of the cone of depression near the production well and parallel to the stream is not distorted to any appreciable extent and it closely describes the hydraulic gradient that would exist if the stream did not exist. A plot of drawdown in the observation wells parallel to the streambed vs. the logarithm of the distances between the production and observation wells will yield a straight line graph. The slope of the straight line and the production well discharge rate are inserted into Equation 5.9 to calculate aquifer transmissivity.

Although the hydraulic gradient near the production well and parallel to the streambed is not distorted, the total values of drawdown in the observation wells are much less than they would be without recharge from the stream. Aquifer specific yield cannot be directly determined from the distance-drawdown graph for observation wells near the production well and parallel to the streambed. Instead, specific yield is determined by the method of successive approximations based on steady state drawdown data and the location of the recharging image well.

The distance between the production and recharging image wells is calculated with the following equation (Rorabaugh, 1956, pp. 101-169):

$$D = 2\{[r^2(10^b)^2 - r^2]/4\}^{0.5} \tag{5.11}$$

with

$$b = 4\pi Ts/Q \tag{5.12}$$

where D is the distance between the production and recharging image wells, r is the distance between the production and observation wells, T is the aquifer transmissivity, s is the drawdown in the observation well, and Q is the constant production well discharge rate.

Equation 5.11 is valid for observation wells on a line parallel to the streambed and the distance D is measured at a right angle to the streambed.

With the recharging image well located and aquifer transmissivity calculated, aquifer specific yield is determined with observation well steady state drawdown data. The production, observation, and image wells are drawn to scale on a map and the distances between the recharging image and observation wells are measured. Several values of aquifer specific yield are assumed and observation well drawdowns for each assumed value of specific yield are calculated with Equations 2.1 and 2.2. Calculated values of drawdown are compared with actual values of drawdown measured in the observation wells, and that specific yield used to calculate drawdowns equal to observed drawdowns is assigned to the aquifer. A BASIC program for determining specific yield under induced streambed infiltration conditions is listed by Walton (1987, pp. 131-133).

The rate of streamflow depletion (rate of recharge by induced streambed infiltration) is calculated with the following equation (Jenkins, 1968, p. 16):

$$q = Qerfc[a^2S/(4Tt)]^{0.5} \qquad (5.13)$$

where q is the rate of streamflow depletion at time t, Q is the constant production well discharge rate, a is the distance between the production well and the effective line of recharge, S is the aquifer storativity, t is the time after pumping started, and T is the aquifer transmissivity.

A BASIC program for calculating streamflow depletion is listed by Walton (1987, pp. 134-135). Equations for estimating depletion of streamflow in right-angle streambed bends due to pumping from a production well near a streambed are developed by Hantush (1967d, pp. 235-240).

The production well, recharging image well, and streambed are again drawn to scale on a map (Figure 5.7). A grid is superposed over the map. Several points at grid line intersections within the streambed up and down stream are selected for calculation of drawdown beneath the streambed. Drawdowns at these points are calculated as the algebraic summation of the drawdown due to the production well discharge and the buildup due to the recharging image well. The reach of the streambed L within the area of influence of the production well is ascertained by noting the points up and down stream where drawdown beneath the streambed is negligible (< 0.01 ft or 3.048 × 10⁻³m). The area of induced streambed infiltration is then equal to the product of L and the average distance between the shoreline and the effective line of recharge or the average width of the streambed depending upon the position of the effective line of recharge. A BASIC program for calculating values of drawdown beneath a streambed is listed by Walton (1987, pp. 136-140).

The induced streambed infiltration rate per unit area is calculated with the following equation (Walton, 1963):

$$I_a = q/A \qquad (5.14)$$

where I_a is the average induced streambed infiltration rate per unit area, q is the rate of streamflow depletion at time t, A is the area of induced streambed infiltration, and t is the time after pumping started.

The average head loss due to the vertical percolation of water through the streambed is determined from data for observation wells installed within the streambed area of induced infiltration at depths just below the streambed. In many cases, the installation of observation wells in the stream channel is impractical and the average head loss must be estimated with Equations 2.1 and 2.2. Drawdowns (head losses beneath the streambed) at grid line intersections within the reach of the streambed influenced by the production well on the map previously discussed are calculated as the algebraic summation of the drawdown due to the production well discharge and the buildup due to the recharging image well. These drawdowns are then averaged.

The average induced streambed infiltration rate per unit area per unit of head loss is calculated with the following equation (Walton, 1963):

$$I_h = I_a/h_s \qquad (5.15)$$

where I_h is the average induced streambed infiltration rate per unit of area per unit of head loss, I_a is the average induced streambed infiltration rate per unit of area, and h_s is the average head loss beneath the streambed.

Representative values of induced streambed infiltration rates are presented in Appendix B. The induced streambed infiltration rate per unit area per unit of head loss varies with the surface water temperature. A decline in the temperature of surface water of 1°F will decrease the rate about 1.5% (Rorabaugh, 1956, pp. 101-169) through the practical range of surface water temperature changes encountered in the field. The induced streambed infiltration rate for any particular surface water temperature is calculated with values of dynamic viscosity of water presented in Appendix B and the following equation (Walton, 1963):

$$I_t = I_h V_a / V_t \qquad (5.16)$$

where I_t is the average induced streambed infiltration rate per unit area per unit of head loss at temperature t, I_h is the calculated average induced streambed infiltration rate per unit area per unit of head loss at temperature a, V_a is the dynamic viscosity of the surface water at temperature a, and V_t is the dynamic viscosity of the surface water at temperature t.

The infiltration rate may not remain stable over a long period of time because of alternating sedimentation and scouring by the stream. Potential recharge by induced streambed infiltration is estimated from the calculated infiltration rate, streamflow records, and surface water temperature data with the following equation (Walton, 1963):

$$R = I_t h_s A \qquad (5.17)$$

where R is the rate of recharge from potential induced streambed infiltration (the induced streambed infiltration rate per unit area per unit of head loss at the potential average surface water temperature), h_s is the potential average head loss or depth of water in the stream whichever is largest, and A is the potential streambed area of induced infiltration.

The fact that the depth of water in the stream will decrease as the result of induced streambed infiltration must be considered in estimating R. Provided the water table remains below the streambed, the least amounts of induced streambed infiltration occur during extended dry periods when streamflow and the surface water temperature are low. Profiles of the stream channel are used to determine the average depth of water in the stream and the average width of the streambed for different stream stages during these periods.

Usually, available induced infiltration data is for a discharge rate which is a fraction of the potential capacity of a proposed well field near a stream. Where the streambed is narrow, the depth of water in the stream is small, and the induced streambed infiltration rate is low, drawdown data for the low discharge rate may indicate a certain position of the effective line of recharge which may not be valid for a higher potential higher discharge rate. At a higher discharge rate, water may be withdrawn at a rate in excess of the ability of the streambed to transmit it, and as a result the water table may decline below additional portions of the streambed. In such a case, the effective line of recharge moves away from the production well as maximum induced streambed infiltration occurs in the reach of the streambed in the immediate vicinity of the production well, and the cone of depression is forced to spread further up and down stream than it did during the low discharge rate period.

The effective line of recharge at a potentially higher discharge rate is determined by the method of successive approximations. Several positions of the effective line of recharge are assumed and potential drawdowns beneath the streambed and streambed induced infiltration areas are calculated. Values of R are calculated keeping in mind that h_s is either the average head loss beneath the streambed or the average depth of water in the stream depending on the relative elevations of the cone of depression and the streambed base. The position of the effective line of recharge which results in R balancing the desired higher discharge rate is judged to be valid. In these calculations, impacts associated with any seepage of water through unsaturated materials beneath the stream are assumed to be negligible (see Peterson, 1989, pp. 899-927).

Induced streambed infiltration can also occur due to flood stages in streams (Cooper and Rorabaugh, 1963).

DISCONTINUITY

Aquifer transmissivity may decrease beyond a line of discontinuity thereby creating a partial barrier boundary situation. In this case, the image well simulating the partial boundary is placed as described earlier and the image well strength is calculated with the following equation (Muskat, 1937):

$$Q_i = QD_t \qquad (5.18)$$

with

$$D_t = (T_p - T_d)/(T_p + T_d) \qquad (5.19)$$

where Q_i is the constant image-well strength, Q is the constant production well discharge rate, T_p is the aquifer transmissivity at the production well, and T_d is the aquifer transmissivity beyond the discontinuity.

MULTIPLE BOUNDARY SYSTEMS

Aquifers are often delimited by two or more boundaries. Two converging boundaries delimit a wedge-shaped aquifer, two parallel boundaries delimit an infinite-strip aquifer, two parallel boundaries intersected at right angles by a third boundary delimit a semi-infinite strip aquifer, and four boundaries intersecting at right angles delimit a rectangular aquifer. The image-well theory may be applied to such cases by taking into consideration successive image well reflections on the boundaries (Ferris, et al., 1962, pp. 151-161).

A number of image wells are associated with a pair of converging boundaries. A primary image well placed across each boundary balances the impacts of the production well at each boundary. However, each primary image well produces an unbalanced impact at the opposite boundary. Secondary image wells must be added at appropriate positions until the impacts of the production and primary image wells are balanced at both boundaries.

Although image-well systems can be devised regardless of the wedge angle involved, simple solutions of closed image-well systems are preferred. The actual aquifer wedge angle is approximated as equal to one of certain aliquot parts of 360°. These approximate angles were specified by Ferris, et al. (1962, p. 154) as follows: if the aquifer wedge boundaries are of like character, the approximate angle must be an aliquot part of 180°; if the aquifer wedge boundaries are not of like character, the approximate angle must be an aliquot part of 90°; and if the production well is on the bisector of the wedge angle and the aquifer wedge boundaries are like in character and both barrier, the approximate angle must be an odd aliquot part of 360°. Under these conditions, the exact number of required image wells is given by the following equation (Ferris, et al., p. 154):

$$N_i = (360°/W_e) - 1 \qquad (5.20)$$

where N_i is the number of image wells associated with an approximate aquifer boundary wedge angle W_e.

The locus of image well positions is a circle whose center is at the apex of the wedge and whose radius is equal to the distance from the production well to the apex.

The character of each image well is the same if the aquifer wedge boundaries are of like character. If the aquifer wedge boundaries are not of like character, the character of each image well, recharge or discharge, is ascertained by balancing the image well system considering each boundary separately with the following rules (Walton, 1962, pp. 20-21): a primary image well placed across a barrier boundary is discharging in character and a primary image well placed across a recharge boundary is recharging in character, a secondary image well placed across a barrier boundary has the same character as its parent image well, and a secondary image well placed across a recharge boundary has the opposite character of its parent image well. Image well systems for selected wedge-shaped aquifers are presented in Figure 5.8.

Two parallel boundaries require the use of an image well system extending to infinity (Ferris, et al. 1962, pp. 156-159). Each successively added secondary image well produces a residual impact at the opposite boundary. However, in practice it is only necessary to add pairs of image wells until the next pair has negligible influence (<0.01 ft or 0.003 m) on the sum of all image well impacts out to that point. The distance beyond which it is no longer necessary to add image wells is often estimated from Equation 2.3 by setting u = 10. Image well systems for selected parallel boundary situations are presented in Figure 5.9.

BASIC programs for calculating drawdown with hydrogeologic boundary impacts are listed by Clark (1987, pp. 10.1-10.13) and Walton (1988).

FLOW NET ANALYSIS

Two-dimensional flow and contaminant concentration graphical patterns are commonly analyzed to estimate aquifer system hydraulic properties or mass transport rates. Flow net analysis involves a study of potentiometric surface and water table contour maps or profiles and the law of the conservation of mass. A flow net (a graphical solution of a flow pattern) consists of families of streamlines and equipotential lines (Ferris, et al. , 1962, pp. 139-144; Strack, 1989, pp. 219-240; and Bear, 1979, pp. 165-169). Streamlines represent paths followed by a particle of water as it moves through the aquifer in the direction of decreasing head. The aquifer volume (domain) between two adjacent streamlines is called a flow channel (Figure 5.10). Usually, equipotential lines intersect streamlines at right angles and represent contours of equal head in the aquifer. Equipotential lines intersect barrier boundaries at right angles and are parallel to constant-head boundaries. Flow nets are constructed according to these concepts so that curvilinear rectangles are formed.

BASIC programs that generate and display streamlines are listed by Kinzelbach (1986, pp. 235-238), Rounds and Bonn (1989, pp. 329-350), and Bear and Verruijt (1987, pp. 302-308). Without dispersion, streamlines represent paths followed by contaminants particles through the aquifer in the direction of decreasing concentration. A program for producing flow nets, streamline plots, and capture zone maps is listed by Bonn and Rounds (1989). An interactive groundwater flow path analysis model is described by Shafer (1987).

The mathematical description of the law of the conservation of mass (mass balance equation) is as follows (Wood, et al. , 1984, pp. 36-48):

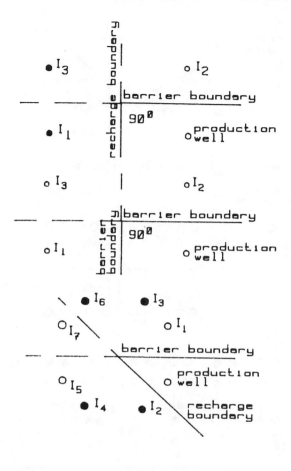

° discharging well
• recharging well

image wells are numbered
in the sequence in which
they were considered

FIGURE 5.8   Image-well systems for selected wedge-shaped aquifers.

$$q_c = q_i - q_o \pm q_{as} \qquad (5.21)$$

where q_c is the rate of change in storage of a mass in a volume with time, q_i is the amount of that mass flowing into the volume per unit of time, q_o is the amount of that mass flowing out of the volume per

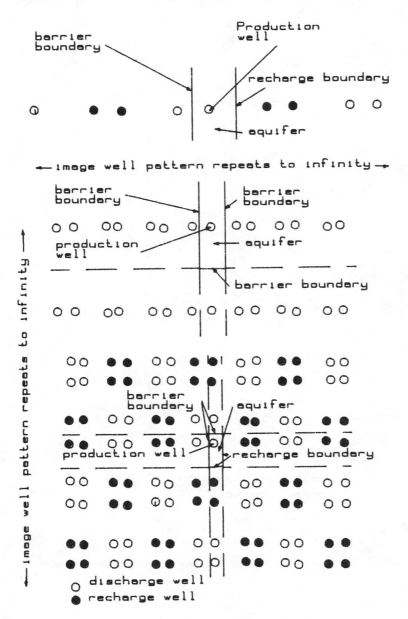

FIGURE 5.9   Image-well systems for selected parallel-boundary situations.

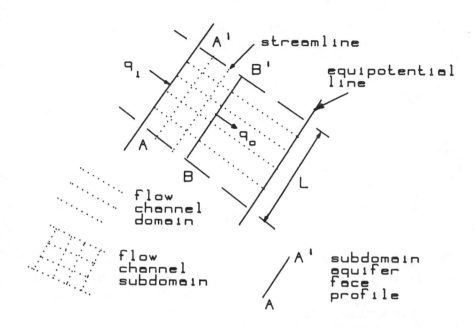

FIGURE 5.10   Plan view of flow net and flow channel subdomain.

unit of time, q_{as} is the amount of that mass added to or subtracted from the volume by sources or sinks per unit of time.

In flow net analysis, flow channels are divided into as many subdomains as can be supported by the data base. The more subdomains used, the higher the accuracy of the components of the domain mass balance equation. A flow channel subdomain delimited by two adjacent streamlines and two adjacent water level contours is shown in Figure 5.10. There is flow into the subdomain through the aquifer face A-A′ and flow out of the subdomain through the aquifer face B-B′. Streamlines are treated as barrier boundaries because there is no flow across streamlines. Under steady state conditions with no change in subdomain storage and no sinks or sources within the subdomain, $q_1 = q_o$. With a sink or source within the subdomain and no change in subdomain storage, $q_1 - q_o \pm q_{as} = 0$. Under nonsteady state conditions with a change in subdomain storage and no sinks or sources, $q_c = q_1 - q_o$. With a source within the subdomain, $q_c = q_1 - q_o + q_{as}$.

Flow of water and contaminant mass through aquifer faces A-A′

and B-B′ is calculated based on Darcy's law with the following equations (Ferris, et al., 1962, p. 141):

$$q_w = P_h I A \tag{5.22}$$

$$q_m = c q_w \tag{5.23}$$

with

$$I = i/d \tag{5.24}$$

$$d = A/L \tag{5.25}$$

where q_w is the rate of flow of water through a subdomain aquifer face, P_h is the aquifer horizontal hydraulic conductivity at a subdomain aquifer face, I is the hydraulic gradient of the water level map at a subdomain aquifer face, A is the plan view subdomain area, q_m is the rate of flow of contaminant mass through a subdomain aquifer face, c is the contaminant concentration at a subdomain aquifer face, i is the water level contour interval, and L is the width of a subdomain's aquifer face.

Although contaminant concentration is defined in units of mass per mass, for common dilute solutions it is commonly expressed in units of mass per volume (see Wood, et al. ,1984, p. 27). 1 ppm is equivalent to 1 mg/l and 1 ppb is equivalent to 1 μg/l.

If q_w, I, and A are known then P_h may be estimated with Equation 5.22. In cases where there are two concentric closed water level contours around a production well or a well field as shown in Figure 5.11, P_h can be determined with the following equation (Lohman, 1972):

$$P_h = Q/[m(L_1 + L_2)\Delta h_c/\Delta r] \tag{5.26}$$

where P_h is the aquifer horizontal conductivity, Q is the discharge rate of a production well or well field, m is the aquifer thickness, L_1 is the length of water level contour 1, L_2 is the length of water level contour 2, Δh_c is the water level contour interval, and Δr is the average distance between the two closed water level contours.

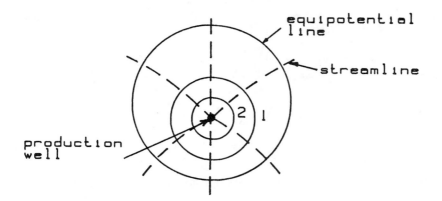

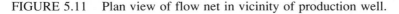

FIGURE 5.11   Plan view of flow net in vicinity of production well.

Aquifer storativity may be estimated with the following equation provided there are no other sources or sinks within a subdomain:

$$S = (v_{w2} - v_{w1})/(\Delta h_t A) \tag{5.27}$$

where S is the aquifer storativity, v_{w1} is the volume of flow of water through subdomain aquifer face A-A′, v_{w2} is the volume of flow of water through subdomain aquifer face B-B′, A is the plan view subdomain area, and Δh_t is the average rate of water level decline or rise within the subdomain.

If water levels within the subdomain are steady state and there are no sources or sinks except leakage through an aquitard into or from the subdomain then P' may be estimated with the following equation:

$$P' = (q_{w2} - q_{w1})m'/(h_a A) \tag{5.28}$$

where q_{w1} is the rate of flow of water through subdomain aquifer face A-A′, q_{w2} is the rate of flow of water through subdomain aquifer face B-B′, A is the plan view subdomain area, P' is the vertical hydraulic conductivity of the aquitard overlying the subdomain, m' is the aquitard thickness, and h_a is the difference between the head in the aquifer subdomain and that in the sourcebed above the aquitard.

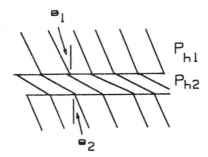

FIGURE 5.12   Profile of streamline refraction in layered aquifer.

In layered and heterogeneous aquifers, streamlines passing from a subdomain of one hydraulic conductivity to a subdomain of another hydraulic conductivity which is much greater than or less than that of the first subdomain refract (Figure 5.12) according to the following equation (Freeze and Cherry, 1979, pp. 172-173):

$$P_{h1}/P_{h2} = \tan\theta_1/\tan\theta_2 \qquad (5.29)$$

where P_{h1} is the aquifer hydraulic conductivity in subdomain 1, P_{h2} is the hydraulic conductivity in subdomain 2, θ_1 is the angle the streamline in subdomain 1 makes with the normal to the subdomain boundary, and θ_2 is the angle the streamline in subdomain 2 makes with the normal to the subdomain boundary.

In anisotropic aquifers, equipotential lines and streamlines are not orthogonal and flow nets must be constructed in transformed sections. The scale of the flow region is transformed such that coordinates XZ in the transformed region are related to those in the original region xz as follows (Freeze and Cherry, 1979, pp. 174-178):

$$X = x \qquad (5.30)$$

$$Z = zP_x^{0.5}/P_z^{0.5} \qquad (5.31)$$

where X, x, Z, and z are space coordinates; P_x is the aquifer hydraulic conductivity in the x-direction; and P_z is the aquifer hydraulic conductivity in the z-direction. The flow net is constructed in the transform section according to the rules for a homogeneous, isotropic aquifer. The flow net is then inverted to the original scale with Equations 5.30 and 5.31.

The rate at which contaminant mass is adsorbed or desorbed from the aquifer skeleton within a subdomain is estimated with the following equation provided there are no other contaminant mass sources or sinks and no contaminant mass storage changes within the subdomain:

$$q_s = q_{m1} - q_{m2} \qquad (5.32)$$

where q_s is the rate at which contaminant mass is adsorbed on the aquifer skeleton within the subdomain, q_{m1} is the contaminant mass flow rate at subdomain aquifer face A-A', and q_{m2} is the contaminant mass flow rate at subdomain aquifer face B-B'.

If aquifer faces A-A' and B-B' are closely spaced then contaminant mass storage changes will be negligible even through there is a minor time rate of change in contaminant concentration.

Contaminant source input rates or contaminant mass leakage through an aquitard into an aquifer is estimated with the following equation provided there are no other contaminant sources or sinks and no contaminant mass storage changes within the subdomain:

$$q_i = q_{m2} - q_{m1} \qquad (5.33)$$

where q_i is the contaminant source input rate.

Profile flow nets for homogeneous and isotropic aquifers with only horizontal flow have equipotential lines which are perpendicular to the aquifer base (Figure 5.13). In contrast, profile flow nets for aquifers with both horizontal and vertical components of flow have equipotential lines which are not perpendicular to the aquifer base (see Hubbert, 1940, pp. 785-944). As a result, the water level elevations in adjacent piezometers of different depth are not equal as they would be if flow were entirely horizontal. A water level contour map based on data for piezometers of one equal depth differ from a water level contour map based on data for piezometers of another

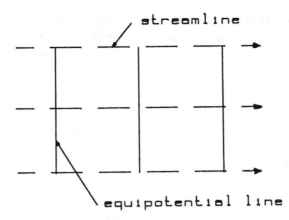

FIGURE 5.13   Profile of flow net with horizontal flow.

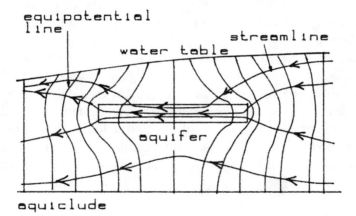

FIGURE 5.14   Profile of flow net with high conductivity lens in lower conductivity aquifer.

equal depth.

Regional flow nets can become quite complex because of the diversity of aquifer hydraulic characteristics, discontinuities, and boundaries; areas and magnitudes of aquifer recharge and discharge; and topography, stratigraphy, and aquifer anisotrophy (Toth, 1962, pp. 4375-4387; Toth, 1963, pp. 4795-4812; Freeze and Witherspoon, 1967, pp. 623-634). Typical regional flow patterns in layered aquifers are illustrated in Figure 5.14. Winter (1976, 1978,

1981, and 1983) describes flow net profiles in the vicinity of lakes and groundwater-lake interactions.

6

Aquifer and Tracer Tests

An aquifer test is defined as a field in situ study aimed at obtaining controlled aquifer system response data. A production well is pumped at a constant rate and water levels are measured at frequent intervals in the production well and nearby observation wells. Time-drawdown and distance-drawdown data are analyzed with type-curve matching, straight-line matching, or inflection-point selection techniques presented by Ferris, et al (1962); Walton (1962); Bentall (1963); Walton (1970); Stallman (1971); Lohman (1972); Neuman (1975a); Kruseman and De Ridder (1976); Reed (1980); Walton (1987); and Streltsova (1988).

A variety of aquifer system and facility conditions may influence aquifer test data including:

1. Production well discharge partly derived from water stored within the wellbore (wellbore storage)
2. Partially penetrating wells with aquifer stratification
3. Water table decline with delayed gravity yield

Erroneous conclusions may be reached if these conditions are ignored in aquifer test analysis. Impacts on time-drawdown and distance-drawdown data are site and facility dependent. In general, wellbore storage impacts tend to be negligible (see Equations 2.10 and 2.11) except for the first few minutes of the pumping period with moderate to high (>10000 gpd/ft or 124.2 m^2/day) transmissivities and small (<1 ft or .35 m) well radii. Appreciable wellbore storage impacts occur with low transmissivities and/or large well radii.

Partially penetrating well impact distances increase with increases in aquifer thickness and the ratio P_h/P_v from less than 10 feet (3.048 m) to more than 100 feet (30.48 m) from the production well (see Equation 2.12). Time durations when delayed gravity yield impacts are appreciable (see Equation 2.13) increase with aquifer thickness and decrease with aquifer vertical hydraulic conductivity from less than 100 minutes to more than 7 days.

Other important aquifer system and facility conditions which may influence aquifer test data include:

1. Nearby aquifer system boundaries or discontinuities
2. Aquitard storativity and source bed drawdown
3. Decreased transmissivity with water table decline
4. Interference from nearby production wells
5. Changes in barometric pressure or stream stages
6. Tidal fluctuations

AQUIFER TEST DESIGN

Aquifer test design is commonly guided by a pretest conceptual modeling effort (Walton, 1987, pp. 14-15) which involves estimating aquifer system hydraulic characteristics and boundary conditions with available hydrogeologic data and predicting aquifer system response to pumping with analytical models. Optimal depths, locations, and number of observation wells and pumping rate and duration are selected based on aquifer system behavior predicted with the pretest conceptual model.

In pretest conceptual analytical models, boundary and aquifer discontinuity impacts are simulated with image wells. Interference from nearby production wells is simulated with multiple-well system concepts. Whether or not boundary, discontinuity, or interference impacts are appreciable may be ascertained with the following equation (Ferris, et al, 1962; p. 93; set u = 5):

$$r_n = (20Tt/S_{aw})^{1/2} \qquad (6.1)$$

where r_n is the distance from the observation well beyond which boundary and discontinuity image well or nearby production well impacts are negligible, T is the aquifer transmissivity, t is the time after pumping started, and S_{aw} is the aquifer artesian or water table storativity (0.005 for leaky artesian conditions).

In general, boundary, discontinuity, or interference impacts with a pumping period of 1 day tend to be negligible when image or production wells are beyond distances of 1000, 5000, and 10000 feet (304.8, 1524, and 3048 m) under water table, leaky artesian, and nonleaky artesian conditions, respectively.

In developing a pretest conceptual model, initial aquifer system hydraulic characteristics, boundaries, and discontinuities and well construction features are selected based on available hydrogeologic data. Drawdown distribution with initial conceptual aquifer system hydraulic characteristic values and boundary conditions is estimated. The initial pretest conceptual model is modified if selected well construction features and the duration the test are not reasonable in light of the calculated drawdown distribution. The modified pretest conceptual model is used to recalculate drawdown distribution. The process is repeated until satisfactory results are obtained. Final design features are based on the pretest conceptual model results. A range of pretest conceptual models may be selected for analysis to bracket the range of possible aquifer response.

A BASIC program is listed by Walton (1987, pp. 106-112) for use with pretest conceptual models. It is a modification of a numerical aquifer test program developed in polar form by Rushton and Redshaw (1979, pp. 231-293). The program covers nonleaky artesian, leaky artesian, and water table conditions; wellbore storage; storativity conversion; delayed gravity yield; and decrease in transmissivity with water table decline.

The aquifer test production well discharge rate and diameter must be specified. The discharge rate is the specific capacity of the production well multiplied by the available drawdown usually defined as the distance between the initial water level and the aquifer top under artesian conditions or the top of the screen in the production well under water table conditions. Specific capacity may be estimated with Figures 8.1 or 8.2.

Criteria used in aquifer test site evaluation are presented by

Stallman (1971, pp. 6-8). Aquifer tests involve large investments in time and money and, to some degree, design is dictated by nonscientific considerations. Ease of data analysis and minimizing uncertainties are important design considerations. Data for partially penetrating production wells or wells with appreciable wellbore storage is much more difficult and uncertain to analyze than data for fully penetrating production wells with negligible wellbore storage. Data complicated by variable production well discharge rates and/or interference from nearby production wells is difficult to analyze with a reasonable degree of certainty.

In general, at least three observation wells at various distances from the production well are desirable (Walton, 1987, pp. 10-11). Observation well spacing is usually logarithmic and designed to provide at least one logarithmic cycle of distance-drawdown data. A typical spacing is 100, 400, and 1000 feet (30.48, 121.9, and 304.8 m). To avoid any partial penetration impacts, the distance from the closest observation well to the production well should be equal to or greater then $1.5m(P_h/P_v)^{0.5}$ where m is the aquifer thickness, P_h is the aquifer horizontal hydraulic conductivity, and P_v is the aquifer vertical hydraulic conductivity. If measurement of partial penetration impacts is important, an observation well should be closer to the production well at a distance equal to or less than the aquifer thickness.

With boundary conditions, observation wells are usually spaced along a line through the production well and parallel to the boundary to minimize the effects of the boundary on distance-drawdown data. It is desirable to space observation wells also on a line perpendicular to the boundary and at variable distances and directions from the image well associated with the boundary. Cone of depression deviations from symmetry may be ascertained by locating observation wells on two radial lines at a right angle with one another.

Under induced streambed infiltration conditions, the production well commonly is located at a distance equal to the aquifer thickness from the stream bank to minimize streambed partial penetration impacts. At least two observation wells are located one logarithmic cycle apart on a line through the production well and parallel to the streambed. Distances from the production well to these two observation wells usually are about 50 and 500 feet (15.24 and 152.4 m), respectively. Two additional observation wells are located on a line through the production well and at a right angle to the streambed. One observation well is located at the near stream bank. The other observation well is located across the stream and at a distance equal to the aquifer thickness from the far stream bank [if the streambed width exceeds 500 feet (152.4 m) then this observation well is deleted].

The open (screened) portions of both production and observation wells are usually spaced vertically opposite the same aquifer zones (Walton, 1987, pp. 11-12). Fully penetrating wells are commonly desirable. With partial penetration conditions, at least one nest of observation wells is desirable near the production well. One observation well of the nest is open opposite the aquifer zone open to the production well and the other observation well of the nest is open near the aquifer top or base.

With leaky artesian conditions, at least one nest of aquifer and aquitard observation wells is desirable. The nest is located about twice the aquifer thickness from the production well. The aquifer observation well is open opposite the aquifer zone open to the production well and the aquitard observation well is open opposite a lower portion of the aquitard at least 20% of the aquitard thickness above the top of the aquifer. Aquifer and aquitard observation well diameters usually exceed 1 inch (25.4 mm) and are often 4 to 6 inches (101.6 to 152.4 mm) when float-operated recorders are required.

An understanding and appreciation of the extent of the aquifer system volume affected by pumping facilitates aquifer test design. Without boundaries, that volume is a cylinder with a radius equal to the distance from the production well beyond which drawdown is negligible. Under nonleaky artesian and water table conditions, the aquifer system volume cylinder height affected by pumping is the aquifer saturated thickness. Under leaky artesian conditions, the cylinder height is the aquifer thickness plus overlying and underlying portions of aquitards and source beds within the vertical extent of the cone of depression. The vertical hydraulic conductivity of any window through an aquitard is integrated with the vertical hydraulic conductivity of the aquitard outside any window within the extent of the cone of depression.

Hydraulic characteristic values estimated with aquifer test data are means of hydraulic characteristic values at many sites within a cylindrical volume. A point specific hydraulic characteristic value based on a well sample may deviate significantly from aquifer test results. For this reason, correlation between aquifer test and well sample hydraulic characteristic values should proceed with due caution.

Under water table conditions, the storativity and vertical hydraulic conductivity values calculated with aquifer test data are mean values for the portion of the aquifer system volume between the initial water table and the cone of depression, not the entire aquifer thickness. The aquitard vertical hydraulic conductivity value calculated with aquifer test data is the mean value for the portion of the aquitard affected by pumping, not necessarily the entire aquitard thickness. With partially penetrating wells, the aquifer vertical hydraulic conductivity value calculated with aquifer test data is the mean value for the portion of the aquifer system volume having a radius equal to $1.5m(P_h/P_v)^{0.5}$.

Aquifer tests are commonly 8 hours to 5 days in duration. With a tight aquitard, an aquifer test without aquitard observation wells may have a duration of several months. Aquitard observation wells respond to pumping much less rapidly than do aquifer observation wells. Commonly, aquitard observation wells require a pumping period of at least 1 day before drawdown in the aquitard observation well is appreciable. This factor should be considered in the design of an aquifer test. If a boundary occurs within the aquifer system volume likely to be affected by pumping, the duration of the aquifer test should be long enough so that drawdown deviations due to the boundary are clear.

Time intervals for observation well water level measurements vary from short at the start of the aquifer test when water level declines are rapid to long at the end of the aquifer test when the time rate of drawdown is small. A typical range of time intervals for observation well water level measurements is as follows (Walton, 1987, p. 14):

Time after pumping started (min)	Time interval
1—2	10 seconds
2—5	30 seconds
5—15	1 minute
15—50	5 minutes
50—100	10 minutes
100—500	30 minutes
500—1000	1 hour
1000—5000	4 hours
5000—end	1 day

A typical aquifer test schedule for an artesian aquifer system is as follows:

Day 1. Water levels measurements to establish antecedent trend
Day 2. 1-hour trial test to adjust equipment followed by a 1-hour recovery period. 3-hour step-drawdown test to determine production well loss coefficient followed by a 20-hour recovery period
Day 3. 24-hour constant rate test to determine aquifer system hydraulic characteristics and boundary conditions
Day 4. 24-hour recovery test to verify aquifer system hydraulic characteristics and boundary conditions

For details concerning aquifer test data collection see Stallman (1971, p. 11) and Driscoll (1986, pp. 534-552).

AQUIFER TEST ANALYSIS

To determine drawdown or recovery, the water level trend before pumping started or after pumping stopped (antecedent trend) is extrapolated through the pumping or recovery period, and differences between extrapolated and observed water levels are calculated. The accuracy of drawdown or recovery calculations is equally dependent upon extrapolated trends and measured water levels.

Recovery is a mirror image of drawdown provided aquifer system hydraulic characteristic and boundary conditions do not change during the aquifer test and extrapolations are correct. The analysis of drawdown and recovery data should yield the same results. Drawdown or recovery is the difference between where water levels are and where water levels would be if the aquifer test was not conducted. Extrapolation is often facilitated by obtaining water level data in an observation well immediately outside the area of influence of the aquifer test. Erroneous aquifer test results will be obtained if antecedent trends are ignored.

Before water levels data are analyzed, they are adjusted for any pumping rate changes in the test production well or in nearby production wells and/or barometric pressure changes. Time rate of drawdown changes due to interference and/or barometric pressure trend changes may be erroneously interpreted as boundary impacts.

Under artesian conditions, barometric pressure change adjustments are made by selecting a time interval during which water levels are not affected by pumping rate changes. Barometric readings are inverted and plotted on plain coordinate paper, together with water level data for the production and observation wells. Prominent barometric changes expressed in feet of water (1 inch of mercury = 1.13 feet of water) are compared to corresponding water level changes.

Barometric scales may be graduated in inches (in.), millimeters (mm), or millibars (mb). Conversion from one scale to another is facilitated by using the following equation: 1 in. = 25.4 mm = 33.86 mb.

The amount of rise in water level as a result of a decrease in barometric pressure and the amount of decline in water level as a result of an increase in barometric pressure are calculated. The barometric efficiency is then calculated with the following equation (Ferris et al, 1962; p. 85):

$$BE = (W/B)100 \tag{6.2}$$

where BE is the barometric efficiency, W is the change in water level, and B is the change in barometric pressure.

Barometric changes may exceed 1 inch of mercury (1.13 ft of water) and barometric efficiency under artesian conditions commonly exceeds 50%. Barometric efficiency usually is negligible under water table conditions.

Drawdown data are adjusted for barometric pressure changes occurring during an aquifer test with records of barometric pressure changes and the following equation:

$$W = (BE\ B)/100 \qquad (6.3)$$

Antecedent barometric trends are considered in estimating changes in barometric pressure.

The tidal or river efficiency may be calculated and drawdown data may be adjusted for any surface water stage changes occurring during an aquifer test by obtaining a record of surface water stage fluctuations prior to and during the aquifer test and using the following equations (Ferris et al, 1962; p. 85):

$$TE = (WS/H)100 \qquad (6.4)$$

$$RE = (WS/H)100 \qquad (6.5)$$

$$WS = (TE\ H)/100 \qquad (6.6)$$

$$WS = (RE\ H)/100 \qquad (6.7)$$

where TE is the tidal efficiency, RE is the river efficiency, WS is the change in water level, and H is the change in surface water stage.

The application of heavy loads in the vicinity of some artesian wells causes changes in water levels. Fluctuations in water levels sometimes occur when railroad trains or trucks pass aquifer test sites. Drawdown data are adjusted for these changes in aquifer loading before they are used to determine aquifer system hydraulic characteristic values. Earth tides and earthquakes affect water levels and are given due attention (Ferris et al, 1962; pp. 86-87 and Bredehoeft, 1967, pp. 3075-3087).

Under water table conditions, gravity drainage of interstices due to pumping may appreciably decrease the saturated aquifer thickness and, therefore, transmissivity. Equations for analyzing aquifer test data are based on the assumption that drawdown is negligible in comparison to the initial saturated thickness of the aquifer. Thus, drawdown data must be adjusted for the effects of dewatering before they are used to determine aquifer system hydraulic characteristic values.

Data manipulation including drawdown adjustment may be facilitated with a microcomputer general purpose database management system such as REFLEX (trademark of Borland International,Inc.). This software allows users to view data in several ways: in single record form, in lists, in graphs, and in cross tabulations. REFLEX provides an unusually large area to design a form on the screen and offers convenient editing commands including the ability to move the cursor left or right five characters at a time. Records may be statistically analyzed, sorted, and filtered. For example, fields (water level data and adjustments) may be added or subtracted to determine record (time) drawdown totals (Ericson and Moskol, 1986). REFLEX sorts up to five fields at once and provides a large number of arithmetic, date, and logic functions. The software groups and subtotals information and displays a graph next to a database list. REFLEX permits 250 fields, allows 254 characters per field and supports an IBM PC, XT, AT, or compatible, a variety of dot matrix printers, and plotters.

In the type curve matching technique, adjusted water level data are matched to the well function type curve which best suits the specific aquifer system and facility conditions encountered in the field. Match-point coordinates are inserted in analytical model equations to quantify aquifer system hydraulic characteristics (Ferris, et al, 1962, p. 94). The pretest conceptual analytical model serves as a frame of reference for selecting the appropriate type curve and guides corroboration of the reliability of aquifer test results.

Reed (1980, pp. 57-106) and Neuman (1975b) list FORTRAN programs for calculating artesian and water table system well function values, respectively. These programs are useful in pretest conceptual analytical model calculations. Clark (1987, pp. 1.1-1.16), Kinzelbach (1986, pp. 224-226), and Walton (1989a, pp. 9-18, 59-74) list BASIC programs for calculating artesian system well function values.

Graphical microcomputer-assisted well function curve matching techniques are available for aquifer test analysis (Dansby, 1987, pp. 1523-1534). Drawdown or recovery data are plotted on the microcomputer screen and an appropriate well function type curve is selected, overlain, and matched to the aquifer test data directly on the screen. The hydraulic characteristics of the aquifer system are automatically calculated with curve match data. Well functions $W(u)$, $W(u,S,\rho)$, $W(u,r/B)$, and $W(u_a,u_b,\beta)$ are supported in software called "Graphical Well Analysis Package" distributed by Groundwater Graphics, San Diego, California through the National Water Well Association.

Type curve analysis tends to be nonunique because field measurements are usually limited in accuracy, type curves seldom completely simulate reality, and several combinations of aquifer system hydraulic characteristic and boundary conditions may satisfy type curve equations. Analysis accuracies of 15% for aquifer hydraulic conductivity and 30% for aquifer storativity are commonly acceptable.

Aquifer system hydraulic characteristics cannot be directly determined with flow equations and aquifer test data because aquifer transmissivity occurs in the flow equation well function argument and again as a divisor of the well function. However, Theis (Wenzel, 1942; p. 88) devised a convenient type curve graphical method of superposition to analyze aquifer test data. Briefly, that method involves matching logarithmic time-drawdown or distance-drawdown graphs to theoretical logarithmic well function curves (type curves) for appropriate analytical models. Thus, type curves such as $W(u)$ vs. $1/u$ or $W(u)$ vs. u are matched to aquifer test data plots of drawdown s vs. time t, s vs. distance squared r^2, or s vs. r^2/t, respectively (Figure 6.1). With fully penetrating wells, one set of type curves may be applied to all observation wells. With partially penetrating wells, a special set of type curves are generated for each observation well.

To illustrate the type curve matching technique, consider the single $W(u)$ vs. $1/u$ type curve. Values of $W(u)$ are plotted against values of $1/u$ on logarithmic paper to describe a $W(u)$ function type curve. Values of adjusted drawdown s are plotted on logarithmic paper of the same scale as used in preparing the type curve against

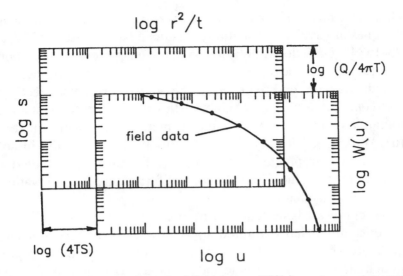

FIGURE 6.1   Conceptual match of field data and W(u) vs u type curve graphs.

values of time after pumping started t to describe a time-drawdown curve. W(u) is related to 1/u in the same manner that s is related to t, thus, the time-drawdown curve is analogous to the type curve. The type curve is superposed over the time-drawdown curve keeping the W(u) axis parallel to the s axis and the 1/u axis parallel to the t axis. In the matched position, a common match point for the two curves is chosen, and the four coordinates W(u), 1/u, s, and t are recorded. For convenience, the match point may be chosen at the intersection of the major axes of the type curve. Match point coordinates are substituted into Equations 2.1 and 2.3 to calculate aquifer hydraulic characteristics. Transmissivity is calculated first and then storativity.

Interpretations of aquifer test data based solely on time-drawdown data are weak. Distance-drawdown data compliment time-drawdown data and they are analyzed whenever possible to strengthen interpretations. W(u) is related to u in the same manner as s is related to the square of the distances from the production well to the observation wells r^2 for a selected time. Thus, the s vs. r^2 distance-drawdown curve is analogous to the W(u) vs. u type curve. The distance-drawdown curve is matched to the type curve and aquifer hydraulic characteristics are calculated with match point coordinates. Plots of s vs. t/r^2 may be matched to the $W(u)$ vs. $1/u$ type curve or plots of s vs. r^2/t may be matched to the $W(u)$ vs. u type curve when several observation wells are spaced at unequal distances from the production well (Ferris, et al, 1962, p. 94). However, scattered data may mask minor hydrogeologic boundary impacts. A type curve matching technique for analyzing aquifer test data with variable discharge rates is presented by Stallman (1962, pp. 118-121).

The type curve matching technique may be extended to cover a nonleaky artesian aquifer with two barrier boundaries simulated with two discharging image wells. The production and image wells operate simultaneously and at the same discharge rate. The $W(u)$ vs. $1/u$ type curve is matched to the early portion of the time-drawdown curve for an observation well unaffected by the image well and hydraulic characteristics are calculated as described earlier. The type curve is moved up and to the right and matched to later data affected by the first image well. The correctness of the match position is judged by noting that the s value underlying a selected $W(u)$ value in the early data unaffected by the image well is 1/2 the s value underlying that same selected $W(u)$ value in later data affected by the image well. The difference between the first type curve extrapolated trace and the second type curve trace s_{i1} is determined for a selected time t_j. The aquifer hydraulic characteristics calculated with early data, the production well discharge rate, and values of s_{i1} and t_j are substituted into Equations 2.1 and 2.3 to determine the distance from the observation well to the image well r_{i1}.

The type curve is moved further up and to the right and matched to late data affected by both image wells. The correctness of the match position is judged by noting that the s value underlying a selected $W(u)$ value in the early data unaffected by both image wells is 1/3 the s value underlying that same selected $W(u)$ value in late data affected by both image wells. The difference between the second type curve extrapolated trace and the third type curve trace s_{i2} is determined for a selected time t_j. The aquifer hydraulic characteristics calculated with early data, the production well discharge, and values of s_{i2} and t_j are substituted into equations 2.1 and

2.3 to determine the distance from the observation well to the second image well r_{i2}. In the case of partially penetrating wells, observed drawdowns are adjusted for partial penetration impacts to obtain drawdowns that would occur under fully penetrating conditions before multiple image well analysis is performed.

To illustrate the family of type curve technique, consider the family of W(u,b) vs. 1/u type curves. Values of W(u,b) are plotted against values of 1/u on logarithmic paper for various values of b. Values of s plotted on logarithmic paper of the same scale as the type curve scale against values of t describe a time-drawdown curve that is analogous to one of the family of type curves. The family of type curves is superposed on the time-drawdown curve, keeping the W(u,b) axis parallel with the s axis and the 1/u axis parallel to the t axis. A particular type curve is selected as being analogous to the time-drawdown curve. In the matched position a common match point is selected and coordinates W(u,b), 1/u, s, and t are recorded. Coordinates and Equations 3.1 and 3.2 are used to determine the aquifer transmissivity and storativity. The value of b used to prepare the particular type curve found to be analogous to the time-drawdown curve is substituted into Equation 3.3 to determine the aquitard vertical hydraulic conductivity. Hund-Der Yeh and Hund-Yuang Han (1989, pp. 655-663) present a numerical method for analyzing leaky artesian aquifer test data using the nonlinear least-squares and finite-difference Newton's techniques.

Analysis of aquifer test data under water table conditions is complicated because radial flow equations under water table conditions describe two asymptotic families of type curves labeled $W(u_A,\beta,\sigma)$ vs. $1/u_A$ (type A) and $W(u_B,\beta,\sigma)$ vs. $1/u_B$ (type B) curves (Neuman, 1975a; pp. 330-331). Both families of type curves approach a set of horizontal asymptotes the lengths of which depend upon the ratio $\sigma = S/S_y$ (Neuman, 1975a; pp. 330-331). The type A family of curves is superposed on early time-drawdown data, keeping the $W(u_A,\beta,\sigma)$ axis parallel to the s axis and the $1/u_A$ axis parallel to the t axis. A particular type curve with a β value is selected as analogous to early and intermediate time-drawdown data. A common match point is selected and coordinates are recorded. Coordinates are substituted into Equations 4.2 and 4.3 and the aquifer transmissivity and artesian storativity are calculated. The value of β is substituted into equation 4.8 to determine the P_v/P_h ratio.

The type B curve with a value of β previously selected is matched to intermediate and late time-drawdown data. A common match point is selected and coordinates are recorded. Coordinates are substituted into Equations 4.4 and 4.5 to calculate aquifer transmissivity and water table storativity (specific yield). Since the value of Beta used in matching both type A and type B curves is the same, calculation of the ratio P_v/P_h need not be repeated. The calculated values of aquifer transmissivities for early and late time-drawdown data are usually the same. Another way to analyze time-drawdown data is to superpose both type A and type B curves on the time-drawdown curve at the same time (Neuman, 1975a; pp. 330-331).

With limited and imperfect field data, the point of beginning in type curve matching is to assume that the earliest few time-drawdown values are not affected by aquitard leakage or delayed gravity yield. Analysis of intermediate and late time-drawdown data may dictate that this initial assumption be revised.

Type curve, time-drawdown, and distance-drawdown graphs may be rapidly displayed with dot-matrix printers or plotters utilizing general purpose chart or graphic microcomputer programs such as GRAPHER (trademark of Golden Software, Inc.) that support linear-linear, log-linear, linear-log, and log-log graphs; superimposition of graphs; and linear, logarithmic, exponential, power, and cubic spline curve fitting.

Aquifer test data may be analyzed with the following method of successive approximations in cases where the complexities of aquifer system conditions greatly exceed that assumed in type curve equations. A subjective set of hydraulic characteristics and boundary conditions are used as the data base for an analytical or a numerical model microcomputer program simulating appropriate radial flow. Usually, the data base reflects available geological information and past experience with aquifer system response to pumping. Drawdowns calculated with the numerical model are compared with measured values of drawdown.

The procedure is repeated for selected data bases until calculated and observed values of drawdown match and the data base is declared valid. The uniqueness of the solution is defended based on the reasonableness of the data base in light of hydrogeologic information and sensitivity analyses. Time-drawdown curves generated with numerical models are analogous to complex well function type curves.

A type curve method for analyzing aquifer test data under three-dimensional homogeneous and anisotropic conditions was developed by Way and McKee (1982, pp. 594-603). A type curve method for analyzing aquifer test data with dike and fracture conditions is described by Boonstra and Boehmer (1989, pp. 171-180).

The commonly used straight line matching technique is based on Equations 2.1-2.3 (Cooper and Jacob, 1946; pp. 526-534) and the fact that graphs of W(u) vs. the logarithm of u and W(u) vs. logarithm of 1/u describe straight lines when u ≤ 0.02. The technique may be applied in cases where wellbore storage, well partial penetration, and delayed gravity yield impacts are negligible. The technique is particularly applicable to data for a production well and data collected under induced streambed infiltration conditions because u becomes small usually after a few minutes of pumping in the case of a production well and steady state conditions prevail after a relatively short pumping period in the case of recharge from a nearby stream. An anthology of 18 calculator and microcomputer program listings related to the straight line matching technique is available from the National Water Well Association.

Values of drawdown are plotted against the logarithms of time after pumping started. A straight line may be fitted to portions of the time-drawdown semilogarithmic graph where u ≤ 0.02 utilizing the BASIC program listed by Walton (1987, pp. 128-130). That program utilizes the method of least squares, finds the best-fit straight line to the data points, and calculates the slope and zero-drawdown intercept of the line. Aquifer transmissivity and storativity are then calculated with time-drawdown slope and zero-drawdown intercept equations.

Values of drawdown for a specified aquifer test period in two or more observation wells at difference distances from the production well may be plotted against the logarithms of the respective distances. A straight line is fitted to portions of the distance-drawdown graph where u ≤ 0.02. Aquifer transmissivity and storativity are calculated with distance-drawdown slope and zero-drawdown intercept equations.

Scattered drawdown data are sometimes interpreted as describing a straight line when actually they plot as a gentle curve. The time that must elapse before a semilogarithmic time-drawdown or distance-drawdown graph will yield a straight line may vary from several minutes under artesian conditions to more than one day under water table conditions. After tentative values of transmissivity and storativity have been calculated, the segment of the data where u ≤ 0.02 should be determined and compared with the segment of data through which the straight line was drawn. The time that must elapse before the straight line matching technique can be properly applied to aquifer test data is as follows (Walton, 1962; p. 9):

$$t = r^2 S/(0.08T) \tag{6.8}$$

where t is the time that must elapse before a semilogarithmic time-drawdown or distance-drawdown graph will yield a straight line, r is the distance between the production and observation wells, S is the aquifer storativity, and T is the aquifer transmissivity.

Aquifer storativity cannot be determined with any reasonable degree of accuracy from data for the production well because the effective radius of the production well is seldom known and drawdowns in the production well are often affected by well losses which cannot be determined precisely.

Calculation of aquifer storativity by the straight line matching method may involve appreciable error. The zero-drawdown intercept is poorly defined where the slope is small. Intercepts often occur at points where the values of time are very small and minor deviations in extrapolating the straight line will result in large variations in calculated values of storativity.

Time-drawdown slope and zero-drawdown intercept equations are as follows (see Cooper and Jacob, 1946; pp. 526-534):

$$T = 2.3Q/(4\pi\Delta s) \tag{6.9}$$

$$S = 2.25Tt_0/r^2 \qquad\qquad (6.10)$$

Distance-drawdown slope and zero-drawdown intercept equations are as follows (see Cooper and Jacob, 1946; pp. 526-534):

$$T = 2.3Q/(2\pi\Delta s) \qquad\qquad (6.11)$$

$$S = 2.25Tt/r_0^2 \qquad\qquad (6.12)$$

where T is the aquifer transmissivity, Q is the constant production well discharge rate, Δs is the drawdown per logarithmic cycle of time or distance, t is the time after pumping started, t_0 is the time intercept of the straight line with the zero-drawdown axis, r is the distance between the production and observation wells, and r_0 is the distance intercept of the straight line with the zero-drawdown axis.

TIGHT FORMATIONS

Tight formations (shales, compacted clays, rock matrix, etc.) with typical vertical hydraulic conductivities ranging from 10^{-9} to 10^{-16} m/s are often considered as cap rocks to contain hazardous wastes. The ability of tight formations to restrict the migration of contaminants is governed by the vertical hydraulic conductivity and thickness of the tight formation and the vertical hydraulic gradient. An aquifer test using observation wells in an aquifer adjacent to a tight formation would have to be conducted for impractical long periods of time to reflect the impact of leakage through the tight formation on water levels in the aquifer and they generally are not feasible. Slug tests in tight formations can be too long for practical applications. However, there are several single- and multiple-well field tests for evaluating the hydraulic characteristics of tight formations which can be conducted in a reasonably short period of time.

Burns (1969, pp. 743-752), Prats (1970, pp. 637-643), Hirasaki (1974, pp. 75-90), and Bredehoeft and Papadopulos (1980, pp. 233-

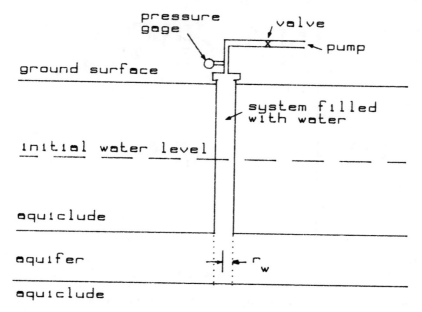

FIGURE 6.2   Pressurized test facilities in tight formation.

238) developed methods for determining the hydraulic characteristics of tight formations from aquifer test data collected in a single well in a tight formation. The Burns, Prats and Hirasaki methods require the use of sophisticated downhole equipment including packers and pressure gauges. The commonly used Bredehoeft and Papadopulos method involves a pressurized test.

In the Bredehoeft and Papadopulos method, the well is filled with water to the surface and suddenly pressurized with an additional amount of water (Figure 6.2). It is assumed that the water level decline which follows the filling of the system with water is at such a slow rate that it is either negligible for the duration of the pressurization period or it can be projected to the end of this period without significant errors. The well is shut-in and the decay of pressure or head is observed (Figure 6.3). The hydraulic characteristics of the tight formation are calculated with pressure or head decay data and the following equation (Bredehoeft and Papadopulos, 1980, p. 234):

$$H/H_0 = W(u_i, \rho_i)$$
(6.13)

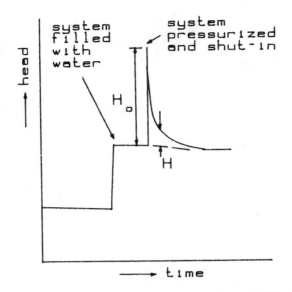

FIGURE 6.3   Variation of head during pressurized test in tight formation.

with

$$u_i = \pi r_w^2 S/(V_w C_w \rho_w g) \qquad (6.14)$$

$$\rho_i = \pi T t/(V_w C_w \rho_w g) \qquad (6.15)$$

where H is the head change at any time t after pressurization, H_o is the maximum head change due to the pressurization, $W(u_i, \rho_i)$ is a well function, r_w is the well radius, S is the tight formation storativity, T is the transmissivity of the tight formation tested interval, V_w is the volume of water within the pressurized section of the system, C_w is the compressibility of water, ρ_w is the density of water, t is the time after pressurization, and g is the gravitational acceleration.

Neuzil (1982, pp. 439-441) discusses the limitation of the Bredehoeft and Papadopulos method and proposes some modifications in the setup and procedure for performing the test.

Values of $W(u_i, \rho_i)$, in terms of the practical range of u_i and ρ_i, are listed in Appendix G. These numerical values are plotted as a family

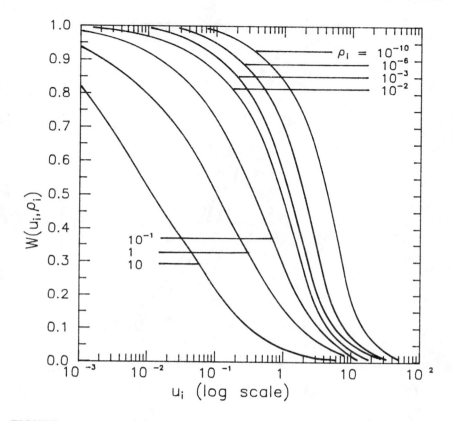

FIGURE 6.4   Selected slug test type curves of $W(u_i,\rho_i)$ vs u_i for selected values of ρ_i.

of semilogarithmic type curves with values of $W(u_i,\rho_i)$, for selected values of ρ_i, on the arithmetic y-axis and values of u_i on the logarithmic x-axis in Figure 6.4.

Values of H/H_0 plotted on semilogarithmic paper of the same scale as the type curve scale against values of the logarithm of time t describe a field data curve that is analogous to one of the family of $W(u_i,\rho_i)$ vs. u_i type curves. The family of type curves is superposed on the field data curve, keeping the $W(u_i,\rho_i)$ axis parallel with the H/H_0 axis and the u_i axis parallel to the t axis. The family of type curves is moved horizontally until a match or interpolated fit is made. A match point with u_i, ρ_i, and t coordinates is selected.

Transmissivity is calculated with these coordinates from $T = \rho_i V_w C_w \rho_w g/(\pi t)$ and storativity is calculated from $S =$

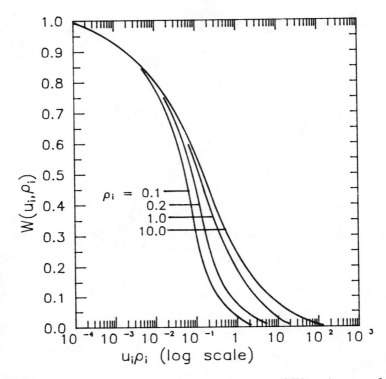

FIGURE 6.5   Selected pressurized test type curves of $W(u_i,\rho_i)$ vs $u_i\rho_i$ for selected values of ρ_i.

$u_i V_w C_w \rho_w g/(\pi r_w^2)$. The determination of storativity has questionable reliability because of the similar shape of the type curves (Bredehoeft and Papadopulos, 1980, p. 236). If $u_i > 0.1$, only the product TS may be calculated by matching the field data curve with a family of type curves of $W(u_i,\rho_i)$ vs. the product $u_i\rho_i$ (Figure 6.5).

Witherspoon and Neuman (1967, pp. 949-955) developed a multiple-well test to determine the hydraulic characteristics of a tight formation based on the ratio of the drawdown in a tight formation to that measured in the aquifer at the same time and the same radial distance from an aquifer production well (Figure 6.6). The ratio test requires a nest of observation wells and consists of an aquifer production well, an aquifer observation well, and a tight formation observation well. The nested observation wells must be within a few hundred feet of the production well (Neuman and Witherspoon, 1972, p. 1288). It is assumed that the effect of

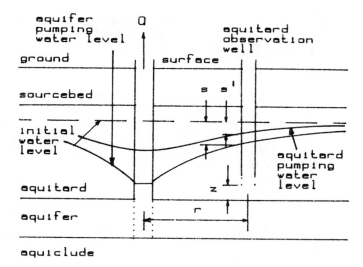

FIGURE 6.6   Multiple-well test facilities with tight aquitard.

pumping the aquifer does not reach the tight formation top and t $\leq$ 0.1S'm'/P' where t is the time after pumping started, S' is the tight formation storativity, m' is the tight formation thickness, and P' is the tight formation vertical hydraulic conductivity (Neuman and Witherspoon, 1972, pp. 1287-1288).

In the ratio method, aquifer transmissivity and storativity are calculated with aquifer observation well time-drawdown data and Equations 3.54 and 3.55. Several arbitrary values of time are selected and corresponding values of drawdowns in the aquifer and tight formation are plotted on logarithmic paper, smooth curves are drawn through the data, and representative values of s and s' are selected based on the smooth curves. Values of u and s'/s are calculated for each selected value of time based on the calculated aquifer transmissivity and storativity. Values of u' = z^2S'/(4P'm't) corresponding to calculated values of s'/s and associated values of u are interpolated from data in Appendix G or Figure 6.7 tabulated by Witherspoon and Neuman (1967). z is the distance between the aquifer top and the tight formation observation well base, S' is the tight formation storativity, P' is the tight formation vertical hydraulic conductivity, m' is the tight formation thickness, and t is the time after pumping started.

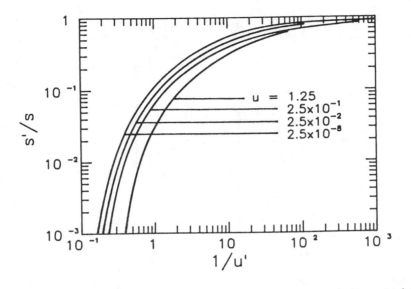

FIGURE 6.7   Multiple-well tight aquitard test type curves of s′/s vs 1/u′ for selected values of u.

Values of P′ are calculated from P′ = z²S′/(4u′m′t) based on interpolated values of u′ and values of S′ determined from standard laboratory consolidated tests on core samples or estimated by correlating published results on similar sediments. The average value of P′ calculated for selected time periods is assigned to the tight formation between the aquifer top and the bottom of the tight formation observation well.

The approximate drawdown distribution in the aquifer and aquitard are given by the following equations (Sauveplane, 1984, p. 215):

$$s = QK_0(a^{0.5})/(2\pi T) \tag{6.16}$$

$$s' = QK_0(a^{0.5})\sinh(d)/[\pi T\sinh(c)] \tag{6.17}$$

with

$$a = u/2 + [4\tau/(2/u)^{0.5}]\coth(c) \tag{6.18}$$

$$u = r^2S/(Tt) \tag{6.19}$$

$$c = (4\tau/B^2)[1/(2u)^{0.5}] \tag{6.20}$$

$$\tau = (r/4)[(S'P')/(TSm')]^{0.5} \tag{6.21}$$

$$B = r[P'/(Tm')]^{0.5} \tag{6.22}$$

$$d = \{[4\tau(1 - z/m')]/B^2\}[1/(2/u)^{0.5}] \tag{6.23}$$

$$\coth(c) = \cosh(c)/\sinh(c) \tag{6.24}$$

$$\cosh(c) = [\exp(c) + \exp(-c)]/2 \tag{6.25}$$

$$\sinh(c) = [\exp(c) - \exp(-c)]/2 \tag{6.26}$$

where s is the drawdown in the aquifer at time t, s' is the drawdown in the tight formation at time t, Q is the constant production well discharge rate, T is the aquifer transmissivity, r is the distance between the production and observation wells, S is the aquifer storativity, t is the time after pumping started, S' is the tight formation storativity, P' is the tight formation vertical hydraulic conductivity, m' is the tight formation thickness, and z is the vertical distance from the aquifer top to the bottom of the tight formation observation well.

SLUG TEST

The single-well slug test is a primary method for evaluating the hydraulic characteristics of aquifers with low to medium hydraulic conductivity particularly those beneath hazardous waste sites. In a slug test, a well is installed and developed. The initial water level is measured. A small volume of water is instantaneously injected or withdrawn from the well. Alternately, a solid piece of metal (slug) may be rapidly lowered into the well and submerged below the initial water level. This is equivalent to adding a volume of water equal to the volume of the slug.

The water level immediately after the well is perturbated and the subsequent rate of water level rise or decline in the well are observed. Recovery to the initial water level in moderately permeable aquifers may require only 30 seconds or less. A pressure transducer along with automatic electronic signal recording equipment is often used to measure and record the water level every second. Recovery to the initial water level in low hydraulic conductivity aquifers may require hours to days and water levels can be measured with a steel tape.

Slug test data may be adversely affected by the following factors: bridging of packer seals, leaky casing joints, entrapped air, boundary conditions, damaged or enhanced zone (skin) around the well, partial penetration of well, anisotrophy of formation, wellbore storage, and presence of fractures. Estimates of hydraulic characteristics based on slug test data may be misleading (Faust and Mercer, 1984, pp. 504-506) and result in unrealistic and inappropriate estimates of contaminant migration if these factors are not considered in slug test analysis. Proper construction and development of the well and pre-test examination to determine if the well is open to the aquifer is an important aspect of slug test planning. Clogged boreholes, piezometers, or monitor wells are inappropriate for slug tests.

The extensive development of slug test theory is summarized by Sageev (1986, pp. 1323-1324). Available model equations simulate wellbore storage, full or partial penetration of the well, skin, and fractures (double-porosity) under artesian and water table conditions. Complex numerical inversion computer procedures are required in solving equations simulating complicated combinations of factors.

Many slug tests are analyzed with the type curve method developed by Cooper, et al (1967, pp. 263-269). The method assumes a fully penetrating finite diameter well with wellbore storage and no skin in a uniformly porous aquifer infinite in areal extent, uniform in thickness, and overlain and underlain by aquicludes (Figure 6.8).

The equation governing the response of the water level in a well under these conditions to an instantaneous slug injection or withdrawal of water is as follows (Cooper, et al, pp. 263-269):

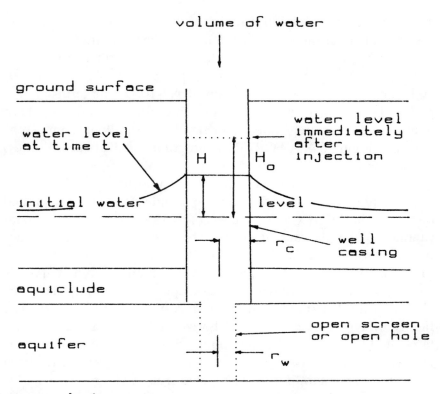

FIGURE 6.8   Cross section through Cooper slug test model.

$$H/H_0 = W(u_i, \rho_i) \qquad (6.27)$$

with

$$u_i = Tt/r_c^2 \qquad (6.28)$$

$$\rho_i = r_w^2 S/r_c^2 \qquad (6.29)$$

where H is the height of the water level in the well above or below the initial water level immediately after the slug injection or withdrawal, H_0 is the height of the water level above or below the initial water level at a time t after the slug injection or withdrawal, $W(u_i, \rho_i)$ is a well function, T is aquifer transmissivity, S is the aquifer storativity, r_w is the effective radius of the well screen or open hole, t is the time after the slug injection, and r_c is the radius of the well casing in the interval over which the water level fluctuates.

Values of $W(u_i,\rho_i)$ presented by Cooper, et al (1967, pp. 263-269) and Papadopulos, et al (1973, pp. 1087-1089), in terms of the practical range of u_i and ρ_i, are presented in Appendix G. These numerical values are plotted as a family of semilogarithmic type curves in Figure 6.4. A FORTRAN program that evaluates $W(u_i,\rho_i)$ is listed by Reed (1980, pp. 99-102).

In the type curve method, values of $W(u_i,\rho_i)$ for selected values of ρ_i are plotted on the arithmetic y-axis and values of u_i are plotted on the logarithmic x-axis. Values of H/H_0 plotted on semilogarithmic paper of the same scale as the type curve scale against values of the logarithm of time t describe a field data curve that is analogous to one of the family of type curves. The family of type curves is superposed on the field data curve, keeping the $W(u_i,\rho_i)$ axis parallel with the H/H_0 axis and the u_i axis parallel to the t axis. The family of type curves is moved horizontally until a match or interpolated fit is made. A match point with u_i, ρ_i, and t coordinates is selected. Transmissivity is calculated with these coordinates from $T = u_i r_c^2/t$ and storativity is calculated from $S = \rho_i r_c^2/r_w^2$.

The determination of S has questionable reliability because of the similar shape of the type curves (Cooper, et al, 1967, p. 267). The radius of investigation of a slug test, defined as the radius from the well beyond which water level response is negligible (<0.01), may be about 1000 multiplied by the piezometer or monitor well radius or less under typical artesian conditions (Sageev, 1986, p. 1332).

Hvorslev (1951) developed a quick and simple method for obtaining order-of-magnitude estimates of horizontal hydraulic conductivity of uniformly porous formations from data collected during a slug test with a partially penetrating piezometer or auger hole with no skin (Figure 6.9). The method has been expanded to simulate a variety of field conditions (NAVFAC, 1971; Cedergren, 1977; Chapuis, 1989, pp. 647-654). The most commonly used equation in the method is as follows (Hvorslev, 1951):

$$P_h = [r_w^2 \ln(L_s/r_c)]/(2L_s t_o) \tag{6.30}$$

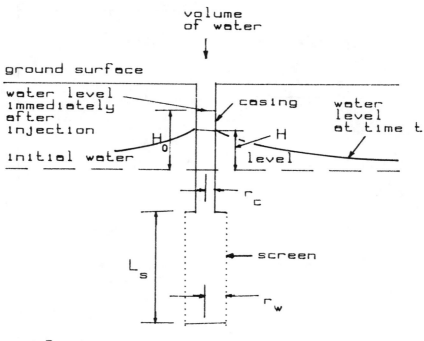

FIGURE 6.9   Cross section through Hvorslev method slug test model.

where P_h is the horizontal hydraulic conductivity, r_w is the radius of the piezometer or auger hole casing, r_c is the radius of the piezometer or auger hole screen or the distance from the center of the piezometer or auger hole to the outside edge of the gravel pack, L_s is the length of the piezometer or auger hole screen or gravel pack, and t_o is the time it takes for the water level to rise or fall to 37 percent of the maximum water level change due to slug injection or withdrawal.

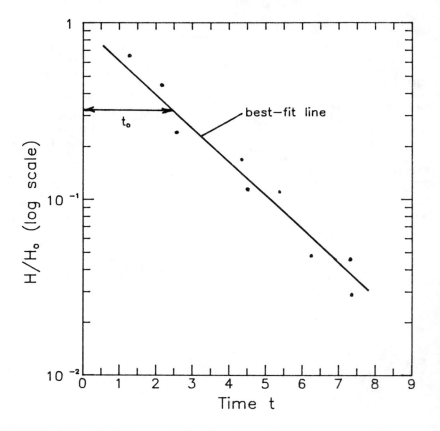

FIGURE 6.10   Field data graph for Hvorslev method slug test analysis.

It is assumed that the length of the piezometer or auger hole is more than 8 times the radius of the piezometer or auger hole screen.

The log of the head ratio H/H_0 is plotted against time on semilogarithmic paper (see Figure 6.10). H_0 is the height of the water level immediately upon slug injection or withdrawal and H is the height of the water level at a time t after slug injection or withdrawal. A line is fitted to the data and t_0 is determined from the line. Thompson (1987, pp. 212-218) presents a table of equations for analyzing slug test data with differing piezometer or auger hole geometry and formation conditions and lists a BASIC program for statistically fitting a line to data and solving equations of the Hvorslev method.

Nguyen and Pinder (1984, pp. 222-240) develop analytical equations and procedures for analyzing slug test data for uniformly porous aquifers with fully or partially penetrating wells having no skin. Wellbore storage is considered and it is assumed that the effects of slow gravity drainage under water table conditions or leakage from an aquitard during the short test period are negligible. Bouwer and Rice (1976, pp. 423-428) develop an analytical procedure for analyzing slug test data for a uniformly porous water table aquifer with fully or partially penetrating wells having wellbore storage and no skin.

Sageev (1986, pp. 1323-1333) developed equations for analyzing slug test data for uniformly porous aquifers with fully penetrating wells having wellbore storage and skin. A numerical inversion procedure involving the Stehfest (1970, pp. 47-49) algorithm is used to solve these equations and generate type curves. Dougherty and Babu (1984, pp. 1116-1122) develop equations for analyzing slug test data for a fractured rock aquifer with fully or partially penetrating wells having wellbore storage and skin. Equation solution is obtained by numerical inversion of the Laplace transform using the Stehfest (1970, pp. 47-49) algorithm.

WELL PRODUCTION TEST

In a well production test, the pump in the production well is operated during successive periods of 1 hour in duration at three or more constant fractions of full capacity and water levels in the production well are measured at frequent intervals. Well loss and the well loss constant are often estimated with step-drawdown well production test data (Rorabaugh, 1953, pp. 362-1 to 362-23; Lennox, 1966, pp. 25-48; and Bierschenk, 1964, pp. 493-506) and the following equations (Jacob, 1947):

$$s_w = CQ^2 \qquad (6.31)$$

$$C = (\Delta s_i/\Delta Q_i - \Delta s_{i-1}/\Delta Q_{i-1})/(\Delta Q_{i-1} + \Delta Q_i) \qquad (6.32)$$

where s_w is the component of drawdown in the production well due

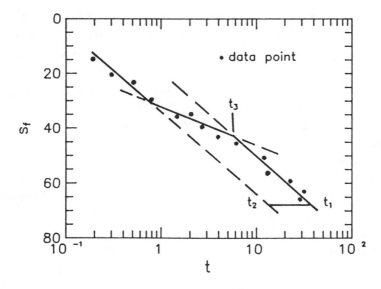

FIGURE 6.11   Time-drawdown semilogarithmic graph for fractured rock aquifer.

to well loss, C is the well loss constant, Q is the production well discharge rate, and Δs_i is the increment of drawdown at the end of pumping period i due to the increment of discharge ΔQ_i during pumping period i.

The drawdown increments are chosen for the same pumping period duration. Calculation of the well loss constant is impractical with discharge rates of a few gpm or less. Equation 6.32 assumes that drawdown under water table conditions is inappreciable in comparison to the initial aquifer saturated thickness. Because water levels in the production well during pump operation tend to be unstable, the precision of estimated well loss and the well loss constant is not high.

For steps 1 and 2:

$$C_1 \text{ and } 2 = (\Delta s_2/\Delta Q_2 - \Delta s_1/\Delta Q_1)/(\Delta Q_1 + \Delta Q_2) \quad (6.33)$$

For steps 2 and 3:

$$C_2 \text{ and } 3 = (\Delta s_3/\Delta Q_3 - \Delta s_2/\Delta Q_2)/(\Delta Q_2 + \Delta Q_3) \quad (6.34)$$

A BASIC program for calculating values of C is listed by Walton (1987, pp. 141-142). Equation 6.32 assumes that the production well is stable and that the value of C does not change during the well production test. However, newly completed and old wells are sometimes unstable and the value of C is affected by changes in the discharge rate. The value of C calculated for steps 1 and 2 may be greater or less than that calculate for steps 2 and 3. Sand and gravel often shift outside the production well during discharge periods under the influence of high discharge rates resulting in either the development or clogging of the pores of the well face. If the value of C for steps 2 and 3 is considerably less than the value of C for steps 1 and 2, it is probable that development has occurred during the well production test. A large increase in the value of C with higher discharge rates indicates clogging has occurred during the well production well test. If the production well is unstable, C may be calculated with Equation 6.34 and data for steps 1+2 and 3 or 2+3 and 1.

Clogging due to incomplete well development or well deterioration is generally negligible when C is <5 sec^2/ft^5. Values of C between 5 and 10 sec^2/ft^5 indicate mild clogging or well deterioration, and clogging or well deterioration is severe when C is >40 sec^2/ft^5 (Walton, 1962, p. 27). Deteriorated wells may be returned to near original yields by one of the several rehabilitation methods described by Driscoll (1986, pp. 630-669). The success of the rehabilitation can be appraised with the results of well production tests conducted prior to and after rehabilitation.

Skin refers to an alteration in near-wellbore hydraulic conductivity resulting from well completion, mud filtrate during drilling, production, acidizing, hydraulic fracturing, or formation erosion. The relation between skin and well loss is developed by Ramey (1982, pp. 265-271).

FRACTURED ROCK

The single-well constant discharge straight-line matching technique described by Streltsova (1988, pp. 385-388) is commonly used to analyze time-drawdown data for a production well in a fractured rock aquifer when u_f is ≤ 0.02 (see Equation 2.71). Values of drawdown are plotted against the logarithm of time after pumping started as shown in Figure 6.11. Three linear segments in the time-drawdown curve are evident. The transmissivity of the fractured rock aquifer is determined from the slopes of the first or third segments which are equal and the following equation (Streltsova, 1988, p. 385):

$$T = Q/(5.457\Delta s) \tag{6.35}$$

where T is the fractured rock transmissivity, Q is the constant production well discharge rate, and s is the drawdown difference per log time cycle.

Transmissivity is also be determined from the slope of the second segment which is half of the slopes of the first or third segments with the following equation:

$$T = Q/(10.914\Delta s) \tag{6.36}$$

Calculation of the fissure and/or block storativity is not possible with any reasonable degree of accuracy because the effective radius of the production well is seldom precisely known and drawdown is affected by well losses which cannot be determined precisely.

The block-to-fissure storativity ratio is determined from the horizontal displacement of the first and third segments (Figure 6.11) with the following equation (Streltsova, 1988, p. 386):

$$S_b/S_f = (t_2/t_1) - 1 \tag{6.37}$$

If the half thickness of the average block unit and the storativity of the block are known, the block vertical hydraulic conductivity is determined with the time value at the intersection of the second and third segments (Figure 6.11) and the following equation (Streltsova, 1988, p. 383):

$$P_b = 0.73m_b{}^2 S_b/(t_3m) \tag{6.38}$$

where P_b is the block vertical hydraulic conductivity, m_b is the half thickness of the average block unit, S_b is the block storativity, S_f is the fissure storativity, t_1 is the time coordinate of the intersection of a selected value of s and the third segment of the time-drawdown graph, t_2 is the time coordinate of the intersection of the selected value of s and the extension of the first segment of the time-drawdown graph, t_3 is the time coordinate of the intersection of the second and third segments of the time-drawdown graph, and m is the fractured rock aquifer thickness.

The multi-well constant discharge type curve matching technique described by Streltsova (1988, pp. 388-391) may be applied to early and intermediate time-drawdown data for an observation well. The time-drawdown field data curve is matched to the appropriate τ type curve in Figure 2.22 and match point coordinates $W(u_f,\tau)_{mp}$, $(1/u_f)_{mp}$, $(s_f)_{mp}$, and t_{mp} are used to calculate fractured rock transmissivity with $T_f = QW(u_f,\tau)_{mp}/[4\pi(s_f)_{mp}]$ and then fissure storativity is calculated with $S_f = 4T_f t_{mp}/[r^2(1/u_f)_{mp}]$. Knowing the block storativity and the half thickness of the average block unit, the selected value of τ_{mp} and the calculated values of T_f and S_f are used to calculate the vertical hydraulic conductivity of the block with $P_b = P_f\{\tau_{mp}m_b/[0.25r(S_b/S_f)^{1/2}]\}^2$.

FIELD TRACER TEST

Field tracer tests are conducted to determine aquifer dispersivity and effective porosity and solute distribution coefficients (Sauty, 1977, pp. 82-90; Sauty, 1980, pp. 145-158; Sauty, 1988, pp. 33-56; Pickens, et al. 1981, pp. 529-544; and Sauty and Kinzelbach, 1988, pp. 33-56). In these tests, tracer-labelled water is injected into a well and groundwater is sampled for tracer analysis at monitoring well sites usually 3.281 to 16.41 ft (1 to 5 m) from the injection well. Tracers (Davis, et al., 1985, pp. 61-148) are used whose concentrations can be accurately measured and which can be easily and safely introduced into and withdrawn from the aquifer. Practical aspects of tracer test planning are discussed by Davis, et al. (1985, pp. 21-60).

Tracer tests are conducted under several combinations of flow patterns (linear or radial) and injection processes (continuous or instantaneous), however, tracer tests with continuous tracer injection and radial flow are the most economical because they can use aquifer test facilities. Conservative (nonreactive) tracer test data is commonly analyzed with type curves expressing the relation between tracer radial movement and dimensionless concentration, dimensionless time, and Peclet number defined by Sauty (1977, pp. 82-90) as:

$$c_r = c/c_m \tag{6.39}$$

$$t_r = Qt/(n_e \pi r^2 m) \tag{6.40}$$

$$P_e = r/\alpha_L \tag{6.41}$$

where c_r is the dimensionless concentration at time t, t is the time after injection started, c is the aquifer concentration at time t, c_m is the maximum aquifer concentration, t_r is the dimensionless time, Q is the constant injection rate, n_e is the aquifer effective porosity, r is the radial distance between the injection and monitor wells, m is the aquifer thickness, P_e is the peclet number, and α_L is the aquifer longitudinal dispersivity.

A field data graph of dimensionless concentration and real time is matched to one of the family of type curves generated with the numerical computer program listed by Javandel, et al. (1984, pp. 167-170) as shown in Figure 6.12. Equations for use with this family of type curves are:

$$\alpha_L = r/P_e \tag{6.42}$$

$$n_e = Qt_c/(\pi r^2 m) \tag{6.43}$$

where Q is the constant injection rate, n_e is the aquifer effective porosity, r is the radial distance between the injection and monitor wells, m is the aquifer thickness, P_e is the Peclet number of the matched type curve, t_c is the real time corresponding to $t_r = 1$ in the matched position, and α_L is the aquifer longitudinal dispersivity.

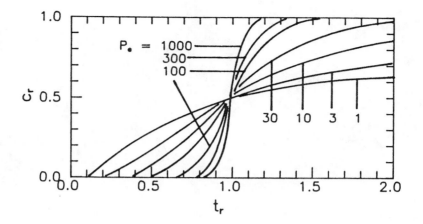

FIGURE 6.12   Arithmetic graphs of c_r vs t_r for selected values of P_e.

Conservative tracer test data with diverging radial flow conditions are also commonly analyzed with the following equations (Pickens, et al., 1981, p. 530):

$$n_e = t_a(Q/m)/(r^2\pi) \tag{6.44}$$

$$\alpha_L = [3r/(16\pi)](\Delta t/t_a)^2 \tag{6.45}$$

where n_e is the aquifer effective porosity, t_a is the time at which relative concentration $c/c_i = 0.5$, c is the aquifer concentration at time t, t is the time after injection started, c_i is the injected solute concentration, Q is the constant injection rate, m is the aquifer thickness, r is the radial distance between the injection and monitor wells, α_L is the aquifer longitudinal dispersivity, and Δt is the time increment between the intercepts of the breakthrough curve tangent line at 0.5 c_i with the $c/c_i = 0.0$ and 1.0 lines.

7

Contaminant Migration

Numerous analytical model equations are available for predicting conservative contaminant migration with advection (transport in accordance with the water level hydraulic gradient and the average pore velocity) and hydrodynamic dispersion in uniformly porous aquifers (Javandel, et al, 1984, pp. 9-68; Baetsle, 1967, pp. 576-588; Baetsle, 1969; Codell and Schreiber, 1979, pp. 1193-1212; Sauty, 1980; and Walton 1984, pp. 289-307). Most of these equations simulate one- and two-dimensional mass transport with steady one-directional groundwater flow in homogeneous infinite, semi-infinite, or finite uniformly porous aquifers having uniform flow and mass transport characteristics. Point-sources or planar-sources are either slug (momentary) or continuous inputs. Usually, it is assumed that source input rates are negligible in relation to regional flow rates, differences in density and viscosity between source inputs and native groundwater are negligible, and there is no vertical leakage of contaminants into or out of the aquifer other than from the source.

Semi-analytical model equations are available for predicting two-dimensional contaminant migration with complex source geometries and assumptions concerning uniform flow (Hunt, 1973, pp. 13-21; Lenau, 1973, pp. 1247-1263; Lenau, 1972, pp. 331-344; Hoopes and Harleman, 1967; Cleary, 1978). FORTRAN programs utilizing many of these model equations are listed by Cleary (1978). Complex integrals are solved numerically in many of these programs. Several three-dimensional mass transport equations, based on a rather restrictive assumption of infinite aquifer thickness, are available (Shen, 1976, pp. 707-716; Hunt, 1978, pp. 75-85; Domenico and Palciauska, 1982, pp. 303-311; Sager, 1982, pp. 47-62; and Domenico and Robbins, 1985, pp. 476-485).

Yeh (1981) and Prakash (1984, pp. 1642-1658) present analytical model equations describing three-dimensional contaminant migration with finite aquifer thickness. A generalized FORTRAN program for predicting contaminant migration with a wide variety of three-dimensional situations is listed by Yeh (1981, pp. 62-79). A three-dimensional analytical model for predicting contaminant migration from a partially penetrating strip source in a finite thickness aquifer is presented by Huyakorn, et al. (1987, pp. 588-598).

BASIC and FORTRAN programs for generating contaminant migration paths in aquifers are listed by Bear and Verruijt (1987, pp. 286-298) and Javandel, et al. (1984, pp. 175-204). These programs are based on semi-analytical advective mass transport equations. A BASIC program for estimating dispersion-free contaminant migration paths and travel times with semi-analytical equations is listed by Kinzelbach (1986, pp. 235-238).

HYDRODYNAMIC DISPERSION

Hydrodynamic dispersion has two components: molecular diffusion and mechanical dispersion (Bear, 1972, pp. 579-582). At low groundwater velocities molecular diffusion is the dominant component and at other groundwater velocities mechanical dispersion is the dominant component. Commonly, the mechanical dispersion coefficient is one or more orders of magnitude larger than the molecular diffusion coefficient and the effects of molecular diffusion can be neglected. The hydrodynamic dispersion coefficient is expressed as (Bear, 1979, pp. 233, 235):

$$D_h = D + D^* \qquad (7.1)$$

with

$$D^* = D_0\tau_t \tag{7.2}$$

where D_h is the hydrodynamic dispersion coefficient, D is the mechanical dispersion coefficient, D^* is the molecular diffusion coefficient for porous medium, D_O is the molecular diffusion coefficient for solutes in water, and τ_t is a dimensionless parameter (tortuosity) relating actual flowpath length to straight-line flowpath length.

Tortuosity in unconsolidated materials is generally assumed to be 0.67 (Bear, 1972, p. 112). Alternative expressions for the relationship between D_* and D_O are given by Domenico (1977).

Molecular diffusion is the process whereby ionic and molecular constituents migrate under the influence and in the direction of their concentration gradient (Freeze and Cherry, 1979. p. 103). Through molecular diffusion, contaminants migrate from zones of high concentration to zones of low concentration even in the absence of groundwater flow.

Mechanical dispersion is a mixing process whereby contaminants spread over a greater volume of aquifer than would be predicted solely from an analysis of advective transport (Figure 7.1) with the average pore velocity of flow (Keely, 1987, p. 11 and Gillham and Cherry, 1982, pp. 31-59). Deviation of groundwater velocities from the average pore velocity is caused by irregularly shaped pores, tortuous microscopic flowpaths, and macroscopic aquifer heterogeneities (Fried, 1975, pp. 6-8).

Mechanical dispersion causes contaminants to arrive at a sink (production well or surface water body) prior to the arrival time calculated with the average pore velocity (Freeze and Cherry, pp. 388-391). With mechanical dispersion, the solute-native groundwater interface does not remain an abrupt one as with advection alone, but, instead a transition zone is created across which the concentration varies from that of the solute to that of the native groundwater. The time-concentration curve is S-shaped with mechanical dispersion instead of rectangular in shape as with advection alone (Gillham and Cherry, pp. 31-59).

The mechanical dispersion coefficient is related to the average pore velocity as follows (Marsily, p. 238):

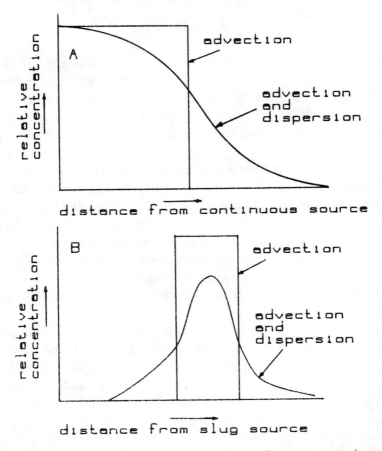

FIGURE 7.1   Influence of dispersion on contaminant concentrations down gradient from continuous (A) and slug (B) sources.

$$D = \alpha v \qquad (7.3)$$

with

$$v = V/n_e \qquad (7.4)$$

where D is the aquifer mechanical longitudinal or transverse dispersion coefficient, α is the aquifer longitudinal or transverse dispersivity, v is the average pore velocity, V is the Darcy velocity, and n_e is the aquifer effective porosity.

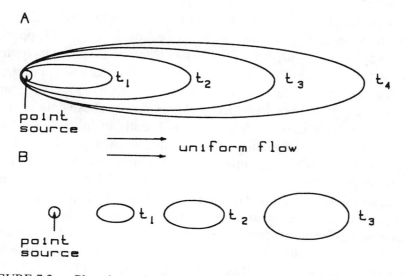

FIGURE 7.2   Plan views of contaminant plumes from continuous (A) and slug (B) sources.

Mechanical dispersion is anisotropic (directionally dependent) and occurs in the horizontal direction both parallel to (x-direction) and normal to (y-direction) the flowpath and in the vertical direction parallel to (z-direction) the flowpath. Mechanical dispersion in the x-direction is called longitudinal dispersion and dispersion in the y- and z-directions is usually called transverse dispersion. Mechanical dispersion is stronger in the x-direction of flow than it is in the y-direction normal to the direction of flow and in the z-direction of flow (Gillham and Cherry, 1982, pp. 31-59). As a result, solute plumes develop an elliptical 2-D plan view shape (Figure 7.2) and a half of a football 3-D shape even though the aquifer is isotropic.

In general, dispersivity is directionally dependent on aquifer vertical and horizontal hydraulic conductivity variations, and increases with the degree of aquifer heterogeneity and anisotrophy (Mercado, 1967, pp. 23-26). Dispersivity is influenced by the degree of and variations in stratifications, varies from one aquifer layer to another, and depends on whether the flow is through uniformly porous aquifers or fractured rock aquifers. Commonly, horizontal transverse dispersivity is 1/5 to 1/10 of horizontal longitudinal dispersivity and vertical transverse dispersivity is 1/50 to 1/100 of horizontal longitudinal dispersivity (Marsily, 1986. p. 238).

Dispersivity may be determined in the field with tracer tests and type curve matching techniques described by Sauty (1977, pp. 82-90) and Sauty (1980, pp. 145-158). Values of dispersivity determined from single well field tests, two well field tests, and areal model calibration studies as compiled by Anderson (1979) and Borg, et al. (1976) are presented in Appendix B. These data suggest that dispersivity is scale dependent and tends to increase and approach some maximum asymptotic value with expansions in the volume of the aquifer occupied by the contaminant. An asymptotic scale dependency equation which can be used to estimate dispersivity was developed by Pickens and Grisak (1981a, pp. 1191-1212). For mean contaminant migration distances less than the maximum asymptotic dispersivity, longitudinal dispersivity in uniformly porous aquifers is usually estimated with the following equation (Pickens and Grisak, 1981b, pp. 1701-1711):

$$\alpha_L = 0.1 \ L \tag{7.5}$$

where α_L is the aquifer longitudinal dispersivity and L is the mean contaminant migration distance.

Although maximum asymptotic dispersivities up to 1640 ft (500 m) have been reported, maximum dispersivity is commonly assumed to be 328 ft (100 m).

Graphs showing the relation between longitudinal dispersivity and mean contaminant migration distances in uniformly porous aquifers are presented by Anderson (1984, p. 41), Kinzelbach (1987, p. 201), and Gelhar, et al. (1985). The graph presented in Figure 7.3 is commonly used in groundwater studies. The complexities and uncertainties surrounding the estimation of dispersivity are discussed by Anderson (1984, pp. 37-45).

If the contaminant migration model could simulate all of the variations in aquifer heterogeneities and associated velocities, dispersive migration would not have to be considered and contaminant migration would be defined in detail by the advective process (Reilly, et al., 1987, pp. 1-44). However, field data on heterogeneities and velocities is never available in sufficient detail and completeness for contaminant migration to be accurately simulated

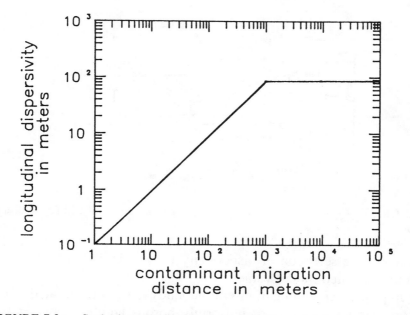

FIGURE 7.3   Scale dependency of longitudinal dispersivity.

by the advection process alone. Even if the data base was adequate, unreasonable computational effort would be required to simulate contaminant migration with the advection process alone.

Equations most commonly used to simulate contaminant migration in groundwater are based on the Fickian advection-dispersion equation whose validity is being debated (Anderson, 1984, pp. 37-45). It is argued that dispersion is non-Fickian on the order of 10's to 100's meters from sources and generally becomes Fickian at large times or contaminant migration distances when constant dispersivity values is achieved. A Fickian model with constant dispersivity often does not describe observed field conditions. However, it has been found that Fickian models based on calibrated dispersivities which increase with time or contaminant migration distance increases and approach a maximum asymptotic value simulate known contaminant plumes with reasonable accuracy.

Dispersivities are scale dependent with respect to numerical model grids and depend on how explicitly field velocity distribution and actual geometry of aquifer heterogeneity are simulated in numerical models (Davis, 1986). Predictions of future plume migration

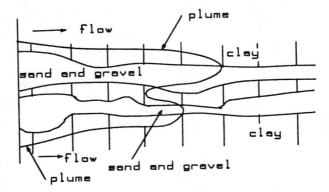

FIGURE 7.4   Contaminant plume profile with influence of aquifer stratification.

may require selection of dispersivity values significantly different from calibrated model values.

In profile, contaminant migration tends to be concentrated in the more permeable zones of uniformly porous aquifers causing fingering and irregularity in plumes. However, diffusion from permeable zones into adjacent less permeable zones tends to create more regular plumes (Figure 7.4) as concentrations are reduced in main zones of advection and dispersion and concentrations are increased in the zones of lesser flow (Gillham and Cherry, 1982. pp. 31-59). In unfractured silty or clayey deposits, diffusion generally controls contaminant migration. In fractured rock aquifers, rapid migration along fissures occurs with dispersion due to mixing at fissure intersections and variations in fissure aperture width and diffusion from fissures into the matrix block.

Molecular diffusion coefficients depend on the solute, solute concentration, and temperature. Molecular diffusion coefficients for electrolytes in aqueous solutions are well known for major ions in water and have values ranging from 1×10^{-9} to 2×10^{-9} m^2/s at 25°C (Freeze and Cherry, 1979, pp. 103). Molecular diffusion coefficient values of 1×10^{-10} to 1×10^{-11} m^2/s are typical for chemical species in claylike materials (Freeze and Cherry, 1979, p 393). Values for coarse grained unconsolidated materials may be higher than 1×10^{-10} m^2/s but less than the coefficients for the chemical species in water (2×10^{-9} m^2/s). Dispersion in the z-direction is commonly assumed to consist only of molecular diffusion in worst case scenarios.

SORPTION

Some contaminants (nonconservative) react with the aquifer skeleton and the groundwater, undergo biological decay, or undergo radioactive decay. Chemical and biological reactions that affect contaminant migration include (Freeze and Cherry, 1979, p. 402): sorption (adsorption and desorption), ion-exchange, precipitation-dissolution, oxidation-reduction, complexation, co-solvation (enhanced solubility due to presence of another contaminant), immiscible phase partitioning, radionuclide decay, microbial population dynamics, biotransformation, and co-metabolism. Chemical reactions cause transfer of contaminant mass between the liquid and solid phases or conversion of dissolved species from one form to another.

In many groundwater studies, sorption is the dominating reaction. As a result of sorption, solute migration is retarded with respect to migration by advection and dispersion without sorption (Keely, 1987, p. 11 and Figure 7.5). Adsorption reduces the size and concentration of the solute plume to some fraction of that which can be attributed to advection and dispersion without adsorption. Adsorption reduces the solute migration velocity relative to the average pore velocity.

In general, relative adsorption by geological materials may be expressed by the series: gravel < sand < silt < clays. Many anions are not adsorbed and most cations undergo some adsorption. Chloride is either weakly or not at all adsorbed; potassium, ammonia, magnesium, silicon, and iron are moderately adsorbed; and lead, cadmium, mercury, and zinc can be strongly adsorbed. Heavy metals, transitional metals, metalloids, radionuclides, and other inorganic species can be mobile or immobile depending on hydrogeochemical conditions represented by the pH, redox condition, ionic strength, mineralogy, solid-phase surface area, and complexing capacity. Solubilities for selected inorganic compounds are listed in Appendix H (ORD, 1981).

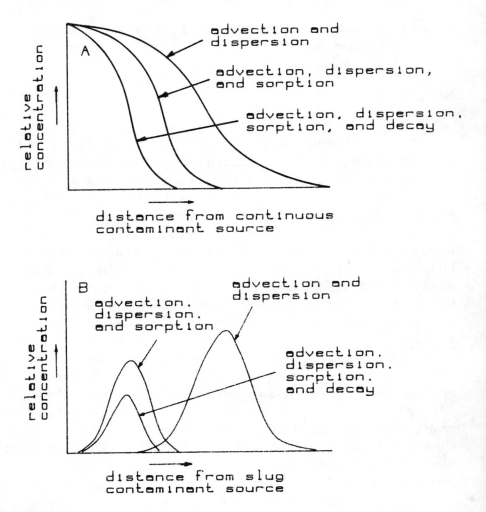

FIGURE 7.5   Influence of sorption and radioactive or biological decay on contaminant concentrations down gradient from continuous (A) and slug (B) sources.

Sorption is usually simulated by assuming that the contaminant concentration in the solution phase is a function of the contaminant concentration in the solid phase, equilibrium conditions exist, and the sorption reaction is fast in relation to the average pore velocity. Under these conditions, the relation between solution- and solid-phase concentrations is expressed as (Cherry, et al. ,1984, p. 49):

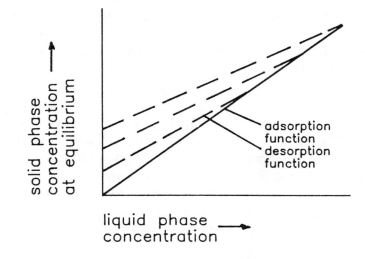

FIGURE 7.6   Conceptual linear adsorption-desorption isotherms.

$$c = bc^{*a} \qquad\qquad (7.6)$$

where c is the contaminant concentration in the solution phase, b is the distribution function, c* is the contaminant concentration in the solid phase, and a is an empirical coefficient.

The relation between c and c* is linear if a is 1. Under linear conditions, $b = K_d$ where K_d is the distribution coefficient. The concentration-distance profile in the flow direction is narrow and the contaminant migration rate is less than it would be with linear adsorption if $a > 1$ (Cherry, et al., 1984, p. 49). The concentration profile in the flow direction is broad and the contaminant migration rate is more rapid than it would be with linear adsorption if $a < 1$.

Desorption must be considered in the evaluation of remedial measures. Commonly it is assumed that adsorption and desorption are coincident (sorption process is completely reversible). However, desorption may display hystersis relative to the corresponding adsorption reactions (Marsily, 1986, pp. 256-257). Possible conceptual linear adsorption-desorption isotherms are shown in Figure 7.6.

The distribution function is normally determined in the laboratory by means of a batch test conducted under equilibrium conditions. During this test, a known mass of geological material is immersed in a solution of leachate or groundwater containing a specified contaminant concentration. The liquid-solid mixture is agitated for a period of hours or days and the contaminant concentration in solution and concentration adsorbed on the solids are determined. The test is repeated several times using different contaminant concentrations in solution and the relationship between c and c*, known as the adsorption isotherm, is determined.

It is not feasible to transfer geological material samples collected in the field to the laboratory without in some way altering the geochemical characteristics of the sample. Even if the contaminant solution used in the batch test has the same composition as the leachate, it is possible that the exchange properties of the geological material may change from that observed during the batch test as a result of precipitation, dissolution, oxidation, or reduction with field contaminant migration. For these reasons, predictions of contaminant concentration generated with laboratory distribution coefficient values are uncertain (Cherry, et al., p. 49).

Uncertainties can be reduced if adsorption is linear by using the results of a field tracer test involving a continuous solute injection well and a monitor well. Data on the relative migration rates of conservative and nonconservative tracers are commonly used with the following equation to determine K_d (Pickens, et al.,1981, pp. 529-544):

$$K_d = (n_e/\rho_s)[(t_c/t_{nc}) - 1] \qquad (7.7)$$

where K_d is the aquifer distribution coefficient, n_e is the aquifer effective porosity, t_c is the time at which $c/c_o = 0.5$ for the conservative tracer, t_{nc} is the time at which $c/c_o = 0.5$ for the nonconservative tracer, ρ_s is the bulk mass density of the dry aquifer skeleton, c is the contaminant concentration at time t_c or t_{nc}, and c_o is the initial contaminant concentration.

There are many functional forms of adsorption isotherms (Smith, 1970), however, the Freudlich linear isotherm (Helfferich, 1962) with $a = 1$ and $b = K_d$ is most commonly fitted to batch test data for groundwater contaminants (Freeze and Cherry, 1979, p. 403). K_d can be calculated theoretically using the chemical mass-action equilibrium equation (Cherry et al., 1979, p. 49). If the contaminant source contains multiple solutes, each will have its own K_d and there will be a number of solute fronts in the complex plume.

It is commonly assumed that migration of each contaminant can be calculated independently of its neighbors. This assumption is valid with very low contaminant concentrations. The limitations and uncertainties inherent in simulating sorption with linear equilibrium parameters are discussed in detail by Cherry, et, al. (1984, pp. 46-64). Typically, the contaminant mass in the solution phase is underestimated and contaminant retardation is over-estimated. Accurate simulation of sorption may require the incorporation of kinetic parameters and/or a nonlinear isotherm relationship to define contaminant migration. Nonequilibrium (kinetic) models of sorption are described by Voss (1984, pp. 43-46).

Measured values of K_d are generally reported as ml/g. Distribution coefficients for solutes commonly encountered range from values near zero to 10^3 ml/g or greater (Freeze and Cherry, 1979, p. 405). Solutes are considered to be immobile with K_d values orders of magnitude greater than 1. A K_d of 10 ml/g is considered to be modest.

Representative distribution coefficient ranges for selected radio-nuclides and rocks are listed in Appendix H (Borg, et al., 1976 and Haji-Djafari, 1981, p. 234). Data for determining the K_d for agricultural chemicals is presented by Dean, et al. (1984).

With adsorption, the velocity of the advancing contaminant plume front becomes the effective contaminant velocity v/R_d and the dispersion coefficient becomes the effective dispersion coefficient D/R_d. In addition, the injected mass is divided by R_d to reflect the fact that part of the injected contaminant mass is adsorbed on the matrix and does not contribute to dissolved concentration.

The ratio of the average pore velocity without adsorption and the rate of advance of a contaminant plume front with adsorption is calculated with the following equation (Gillham and Cherry, 1982, p. 50):

$$v/v^* = 1 + (\rho_s/n_e)K_d \tag{7.8}$$

where v is the average pore velocity without adsorption, v^* is the velocity of the $c/c_0 = 0.5$ point on the concentration profile of the plume with adsorption, ρ_s is the bulk mass density of the dry aquifer skeleton, n_e is the aquifer effective porosity, c is the contaminant concentration at time t, and c_0 is the initial contaminant concentration.

A retardation factor R_d, commonly used in contaminant migration studies, is calculated with the following equation (Gillham and Cherry, 1982, p. 50):

$$R_d = 1 + (\rho_s/n_e)K_d \qquad (7.9)$$

Common ranges of distribution coefficients, retardation factors, and mobilities for uniformly porous aquifers are shown in Figure 7.7. Mobility is defined as the relative rate of contaminant migration compared to the average pore velocity. The total mass in an aquifer (mass in solution plus mass adsorbed onto the aquifer matrix) is the product of the retardation factor and the dissolved mass.

A R_d of 1 is usually assumed when modeling the worst case migration scenario, whereas, a $R_d > 1$ is usually assumed when modeling the worst case remediation scenario. In preliminary studies, the following retardation factors for inorganics are commonly used: heavy metals — 50, cations — 5, and anions — 1. Data on the sorption of selected inorganic constituents are presented by Rai, et al. (1984). Retardation factors of 1 and 2 are found for many synthetic organic chemicals.

In fractured rock aquifers, migrating contaminants contact only the mineral surfaces exposed on the fissure walls and the amorphous geochemical weathering or alteration products that exist on these surfaces (Cherry, et al. ,1984, p. 61). Some contaminants diffuse out of the fissure and into the matrix.

The distribution coefficient for fractured rock aquifers is defined in terms of the effective surface area of reaction in the fissures. The retardation factor for fractured rocks R_{dr} is defined as follows (Marsily, 1986, pp. 266-267):

$$R_{dr} = 1 + b_f K_{df}/n_e \qquad (7.10)$$

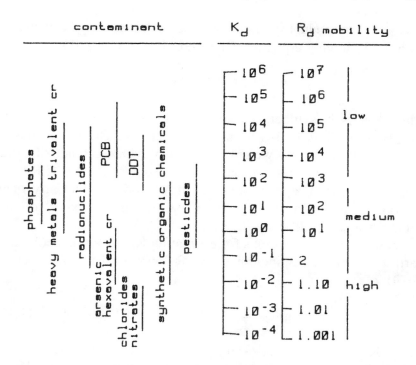

FIGURE 7.7   Common ranges of distribution coefficients, retardation factors, and mobilities in uniformly porous aquifers.

where b_f = area of fissure planes/volume of medium, R_{dr} is the retardation factor for the fractured rock, K_{df} is the distribution coefficient for the fractured rock, and b_f is defined by counting two planes for the walls of each fissure.

Synthetic organic chemicals in solution can be adsorbed by the organic carbon on the aquifer skeleton. The adsorption rate is largely proportional to the organic carbon content of the sediments and the degree of hydrophobicity of the solute as measured by the soil-water partition coefficient. K_d's for many chemicals are listed by Dawson, et al. (1980).

Synthetic organic chemical sorption is commonly greater on silt-clay-sized sediments than on sand-sized sediments. Adsorption of synthetic organic chemicals in permeable sand and gravel aquifers is frequently negligible. Sorption by the mineral surfaces of the aquifer skeleton may approach or exceed that by the organic carbon on the aquifer skeleton (Mackay, 1985, p. 385).

K_d's for individual synthetic organic chemicals may be estimated with the following equation (Cherry, et al. ,1984, p. 55):

$$K_{do} = K_{oc}OC \qquad (7.11)$$

where K_{do} is the synthetic organic chemical distribution coefficient, K_{oc} is the soil-water partition coefficient, and OC is the dimensionless fraction of dry weight of sediment which is made of solid-phase organic carbon and usually varies from 0.001 to 0.1. Ranges of OC for selected deposits are listed in Appendix H. McCarty, et al. (1981) developed an equation for estimating OC based on the octanol/water partition coefficient.

There are many equations for estimating K_{oc} from octanol/water partition coefficients including the following (Karickhoff, et al. 1979):

$$logK_{oc} = 1.00logK_{ow} - 0.21 \qquad (7.12)$$

where K_{ow} is the octanol/water partition coefficient.

Values of K_{oc} and solubility for many synthetic organic chemicals are listed in Appendix H (Fetter, 1988, pp. 403-405 and Roy and Griffin, 1985, pp. 241-247). The following physical and chemical properties of 137 organic compounds are described by Montgomery and Welkom (1989): appearance, odor, boiling point, dissociation constant, Henry's Law constant, ionization potential, log K_{oc}, $logK_{ow}$, melting point, solubility in organics, solubility in water, specific density, transformation productions, and vapor pressure.

Water solubility is affected by temperature, salinity, dissolved organic matter, and pH. Synthetic organic compounds are only rarely found in groundwater at solubility limit concentrations. Concentrations usually are 1/10 of the solubility limit or less (Mackay, et al., 1985, p. 387).

Partitioning of synthetic organic chemicals between the organic and aqueous phases may occur thereby decreasing or increasing the mobility of the chemical in relation to the mobility in the absence of the organic phase (Keely, 1987, p.17). This partitioning process increases the total volume of groundwater affected by the con-

taminant and may interfere with biological transformation processes.

Pathogenic bacteria and viruses can survive for long periods: *E.
coli* > 100 days, *Salmonella typhi* > 100 days, *Salmonella
typhimurium* ≤ 230 days, other Salmonellae ≤ 70 days, Yersinia sp.
≤ 200 days, and Poliovirus ≥ 250 days (Matthess, G. and A.
Pekdeger, 1985). Retardation factors between 1 and 2 are found for
bacteria (*E. coli* and *Serratia marcescens*) and retardation factors up
to 500 are found for polioviruses.

RADIOACTIVE DECAY

The change in concentration of radioactive materials in an aquifer
due to radioactive decay is expressed by the following equation
(Marsily, 1986, p. 265):

$$c_t = c_o e^{-dt} \tag{7.13}$$

where c_t is the radionuclide concentration in the aquifer at time t, c_o
is the initial radionuclide concentration in the aquifer, d is the
radioactive decay constant, and t is time after radioactive decay
starts.

The time for the concentration to decrease to one-half the initial
concentration is defined as the half-life of the radionuclide. The
relation between d and the half-life H_L is expressed by the following
equation (Marsily, 1986, p. 265):

$$d = 0.693/H_L \tag{7.14}$$

Half-lifes for selected radionuclides presented by Milnes (1985,
p.16-17) are listed in Appendix H.

If a substance c_i disappears through radioactive decay it means
that it gives birth to a daughter product (a different substance). In the
transport equation of the new substance c_j the source term will then
represent a creation of matter. The migration equation for c_j must be
solved after the one for c_i as the distribution in time and space of the

source term for element c_j is given by the solution of the migration of c_i (Marsily, 1986, pp. 265-266). Three-member decay chain analytical solutions have been developed (Harada, et al., 1980 and Pigford, et al., 1980).

Daughter products have chemical properties different from the parent and may be adsorbed to a greater or less extent than the parent. Therefore, the daughter migrates at a velocity faster or slower than the parent and the daughter is partitioned between the soil and water differently from the parent.

The degradation of organic contaminants occurs as a result of chemical and biological reactions including oxidation, hydrolysis, and reduction. Biological reactions are usually enzymatic reactions induced by bacteria. Significant microbial activity occurs in the saturated zone. Biodegradation of a broad range of organic compounds including pesticides, halogenated hydrocarbons, aromatic hydrocarbons, amines, and alcohols occurs in anaerobic environments (Cherry, et al., 1984, p. 57). Equations 7.13 and 7.14 can be applied to biological decay with H_L being the mean biological half-life of the substance. Compilations of measured rate constants for both chemical and biological degradation reactions are available (Callahan, et al. (1979); Mills, et al., 1982; Lyman, et al., 1982; and Dawson, et al, 1980). A few bioreclamation degradation rates for selected waste constituents are presented by Boutwell, et al. (1986, p. 307).

IMMISCIBLE CONTAMINANTS

Contamination of groundwater by petroleum products from leaky tanks, pipelines, or spills is a complex problem because oils and gasoline are less dense than water and are immiscible in water (Freeze and Cherry, 1979, pp. 444-447). Many petroleum products have low water solubility and migrate as nonaqueous phase liquids (NAPL). There is gas migration as the result of volatilization and dissolved organics migration. A multiphase system consisting of three separate phases, air, water, and NAPL exists. Analysis of such systems is difficult because separate equations for each of the fluids flowing simultaneously through the system and immiscible properties such as phase interactions, interfacial tension, and capillary forces must be considered.

Initially, petroleum product as a distinct phase migrates downward by gravity through the unsaturated zone. Capillary effects may cause lateral spreading. Some residual product is retained in the soil pore system due to capillary tension if the product volume is finite. A zone of volatilized product may develop and extend beyond the main zone of vertical migration. If the product volume is small and the unsaturated zone is thick, the entire volume of oil may become immobile as residual saturation prior to reaching the water table. If the product volume is sufficiently large, some product will eventually reach the water table. NAPL spreads on top of the water table in the capillary fringe and the product slightly depresses the water table. Due to the contact between the production and the water phase, soluble components may dissolve and migrate with groundwater flow.

Sophisticated or rigorous models for simulating immiscible contaminant migration have been developed (Corapcioglu and Baehr, 1987, pp. 191-200; Pinder and Abriola, 1986, pp. 109s-119s; Kuppusamy, et al. 1987, pp. 625-631; Parker, et al., 1989, pp. 301-312). These models have a system of highly nonlinear partial differential equations with a large number of hard-to-define parameters. El-Kadi (1989, pp. 453-471) developed a semi-analytical approach to simulating immiscible contaminant migration.

Schwille (1988) discusses chemical processes for dense solvent spills, how spill volumes translate into magnitudes of contamination zones, how to remedy existing spills, and how to predict spill behavior.

ONE-DIMENSIONAL MASS TRANSPORT

The continuous source equation governing the one-dimensional contaminant migration in a uniformly porous aquifer from a vertical planar surface source (Figure 7.8) when $x/(\alpha_L) > 10$ is as follows (Kinzelbach, 1986, p. 213):

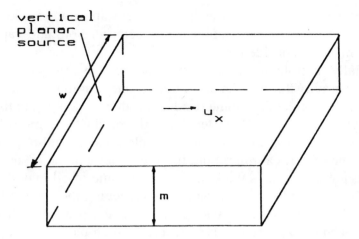

FIGURE 7.8   Three-dimensional view of vertical planar contaminant source.

$$c = M \, \text{erfc}[(x - u_x t)/(2d)] \tag{7.15}$$

with

$$M = c_0 Q/(2w n_e m u_x) \tag{7.16}$$

$$d = (\alpha_L u_x t)^{0.5} \tag{7.17}$$

where c is the aquifer contaminant concentration at time t, M is the source mass input rate, x is the x-coordinate of the monitoring point, α_L is the aquifer longitudinal dispersivity, u_x is the average pore velocity in the x-direction, w is the vertical planar surface source width, m is the aquifer thickness, n_e is the aquifer effective porosity, c_0 is the constant source input concentration, t is the time after the continuous source started, and Q is the constant source input rate.

Field situations which can be simulated with Equations 7.15 to 7.17 include contaminant migration from a stream and a line of closely spaced point sources. A BASIC program for solving Equations 7.15 to 7.17 is listed by Kinzelbach (1986, pp. 224-226).

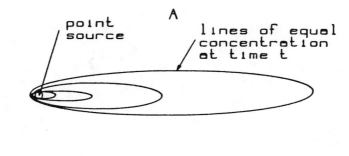

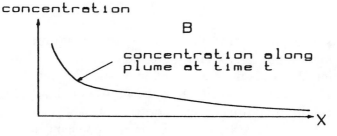

FIGURE 7.9   Plan view (A) and cross section view (B) of two-dimensional continuous contaminant source plume.

TWO-DIMENSIONAL MASS TRANSPORT

The most widely used continuous point source equation governing two-dimensional contaminant migration in a uniformly porous aquifer (Figure 7.9) when $x/B<50$ is (Wilson and Miller, 1978, pp. 504-505; Hunt, 1978, pp. 77-78; Hunt, 1983, pp. 136-140):

$$c = M\exp(x/B)W(a,b)/[4\pi n_e m(D_L D_T)^{0.5}] \qquad (7.18)$$

with

$$M = c_o Q \qquad (7.19)$$

$$B = 2D_L/u_x \qquad (7.20)$$

$$b = R/B \tag{7.21}$$

$$R^2 = x^2 + y^2(D_L/D_T) \tag{7.22}$$

$$a = R^2/(4D_L t) \tag{7.23}$$

$$D_L = \alpha_L u_x \tag{7.24}$$

$$D_T = \alpha_T u_x \tag{7.25}$$

where c is the aquifer contaminant concentration at time t, M is the constant mass input rate, x and y are the coordinates of the monitoring point, $W(a,b)$ is the well function defined in Equation 3.5, n_e is the aquifer effective porosity, m is the aquifer thickness, D_L is the aquifer longitudinal dispersion coefficient, D_T is the aquifer transverse dispersion coefficient, c_0 is the constant source input concentration, t is the time after the continuous source started, Q is the constant source input rate, u_x is the average pore velocity in the x-direction, α_L is the aquifer longitudinal dispersivity, and α_T is the aquifer transverse dispersivity.

Several methods are available for estimating contaminant mass input rates (Boutwell, et al., 1986, pp. 106-108). Mass input rates for landfills having ponded surfaces are often estimated with contaminant concentration data and Darcy's Law. The liquid input rate is equal to the vertical hydraulic conductivity of the landfill bed multiplied by the product of the vertical hydraulic gradient and landfill bed horizontal area. The mass input rate is equal to the liquid input rate multiplied by the contaminant concentration. A unit vertical hydraulic gradient is commonly assumed. This method is applicable where the soil beneath the ponded surface is much more permeable than the landfill bed and the depth of liquid is small compared with the landfill bed. Fenn, et al. (1975) developed a water balance method for estimating mass input rates without ponded surfaces.

$W(a,b)$ may be approximated as follows (Wilson and Miller, 1978, p. 505):

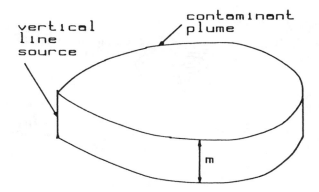

FIGURE 7.10   Three-dimensional view of vertical line contaminant source.

$$W(a,b) = [\pi/(2b)]^{0.5}\exp(-b)\mathrm{erf}\{-[(b - 2a)/(2a^{0.5})]\} \quad (7.26)$$

where $\mathrm{erf}(x)$ is defined in Equations 3.23 and 3.24.

Values of $\mathrm{erf}(x)$ and $\exp(x)$, for the practical range of x, are listed in Appendix G. The approximation in Equation 7.26 is reasonably accurate (within 10%) for $b > 1$ and very accurate (within 1%) for $b > 10$. Values of $W(a,b)$ for the practical range of a and b are listed in Appendix G. BASIC programs for solving Equations 7.18 to 7.25 are listed by Kinzelbach (1986, pp. 224-226) and Walton 1989a, pp. 39-44, 95-110).

Equations 7.18 to 7.25 assume the point source is a vertical line fully penetrating the aquifer (Figure 7.10).

Contaminant migration from a source area may also be simulated with an array of closely spaced point sources. The source area and contaminant load are divided into equal parts and assigned to a number of point sources. The cumulative effects of individual point sources are then calculated using the principle of superposition (Bear,1979, pp. 152-159). The simulation is reasonably accurate at distances beyond the source area where there is complete vertical mixing (concentration becomes essentially uniform with depth) as expressed by the following equation (Kinzelbach, 1986, p. 220):

$$x_m = 0.5m^2u_x/D_z \quad (7.27)$$

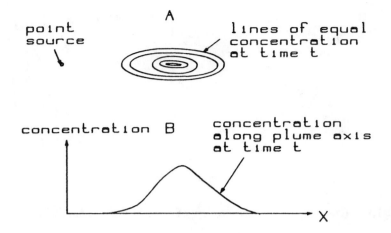

FIGURE 7.11 Plan view (A) and cross section view (B) of two-dimensional slug contaminant source plume.

with

$$D_z = \alpha_z u_z \tag{7.28}$$

where x_m is the distance beyond a partially penetrating point source where there is complete vertical mixing, m is the aquifer thickness, u_x is the average pore velocity in the x-direction, u_z is the average pore velocity in the z-direction, D_z is the aquifer vertical dispersion coefficient, and α_z is the aquifer vertical dispersivity.

A variable source input rate may be simulated by dividing the variable rate into step increments. These step increments may in turn be divided into a large number of short term (slug) input rate sub-step increments which are placed on top of one another at a selected site.

The slug point source equation governing two-dimensional contaminant migration in a uniformly porous aquifer (Figure 7.11) is (Wilson and Miller, 1978, p. 504; Hunt, 1978, p. 1977):

$$c = M \exp\{-[(x - u_x t)^2/(4D_L t)] - [y^2/(4D_T t)]\}/$$

$$[4\pi n_e m(D_L D_T)^{0.5} t] \tag{7.29}$$

with

$$M = c_0V_S \tag{7.30}$$

$$D_L = \alpha_L u_x \tag{7.31}$$

$$D_T = \alpha_T u_x \tag{7.32}$$

where c is the aquifer contaminant concentration at time t, M is the momentary mass load, n_e is the aquifer effective porosity, m is the aquifer thickness, x and y are the coordinates of the monitoring point, u_x is the average pore velocity in the x-direction, D_L is the aquifer longitudinal dispersion coefficient, D_T is the aquifer transverse dispersion coefficient, c_0 is the concentration of the source input, t is the time after the contaminant slug was injected into the aquifer, V_S is the volume of the source input, α_L is the aquifer longitudinal dispersivity, and α_T is the aquifer transverse dispersivity.

The slug source equation governing the contaminant migration in a uniformly porous aquifer from a vertical planar surface is as follows (Kinzelbach, 1986, p. 209):

$$c = M \exp[-(x - u_xt)^2/(4\alpha_Lu_xt)]/$$

$$[2wn_em(\pi\alpha_Lu_xt)^{0.5}] \tag{7.33}$$

with

$$M = c_0V_S \tag{7.34}$$

where c is the aquifer contaminant concentration at time t, M is the momentary mass load, x and y are the coordinates of the monitoring point, u_x is the average pore velocity in the x-direction, α_L is the aquifer longitudinal dispersivity, w is the length of the source vertical planar surface, m is the aquifer thickness, t is the time after the contaminant slug was injected into the aquifer, n_e is the aquifer

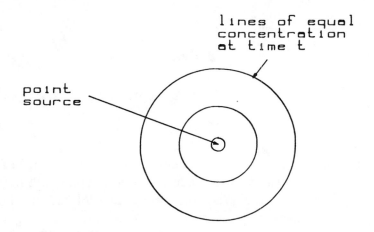

FIGURE 7.12   Plan view of radial contaminant point source plume.

effective porosity, c_O is the momentary source input concentration, and V_S is the volume of the source input.

RADIAL MASS TRANSPORT

Advection and dispersion of contaminants in continuous radial flow in a uniformly porous aquifer from a point source (Figure 7.12) is governed approximately by the following equation (Hoopes and Harleman, 1967, pp. 3595-3607):

$$c = 0.5c_0\mathrm{erfc}\{[(a^2/2) - b]/(4a^3/3)^{0.5}\} \qquad (7.35)$$

with

$$a = r/\alpha_L \qquad (7.36)$$

$$b = Qt/(2\pi m\alpha_L{}^2 n_e) \qquad (7.37)$$

where c is the aquifer contaminant concentration at time t, c_O is the constant source input concentration, r is the distance between the point source and the monitor well, α_L is the aquifer longitudinal dispersivity, Q is the constant source input rate, t is the time after the constant source started, m is the aquifer thickness, and n_e is the aquifer effective porosity.

A similar equation is presented by Konikow and Bredehoeft (1978, p. 27). The average pore velocity at any distance from the point source is equal to $Q/(2\pi r n_e m)$.

Moench and Ogata (1981, pp. 250-251) developed an exact solution in the Laplace transform domain using Airy functions for the radial dispersion from a point source in a uniformly porous aquifer. The numerical inversion of the Laplace transform (Stehfest, 1970) in this solution is incorporated in the FORTRAN program for describing radial mass transport listed by Javandel, et al. (1984, pp. 167-170).

An approximate equation describing the advection and dispersion of contaminants in continuous radial flow towards a production well in a uniformly porous aquifer is presented by Sauty and Kinzelbach (1988, p. 40).

ADVECTION FROM STREAM TO WELL

The equation governing the time-rate of change in the concentration of the water discharged from a production well in a uniformly porous aquifer due to the abrupt change in concentration of the water in a nearby stream (Figure 7.13) without dispersion is as follows (Kirkham and Affleck, 1977, p. 239):

$$t/b = \{1/[\sin^2 \pi c_p/c_o]\}[1 - \pi(c_p/c_o)\cot \pi(c_p/c_o)] \qquad (7.38)$$

with

$$b = n_e a^2 \ln(2a/r_w)/[Ph(h_e - h_w)] \qquad (7.39)$$

where t is the time after the abrupt change in the concentration of the water in the stream, c_p is the change in the production well concentration at time t, c_o is the abrupt change in the concentration of the water in the stream, n_e is the aquifer effective porosity, a is the

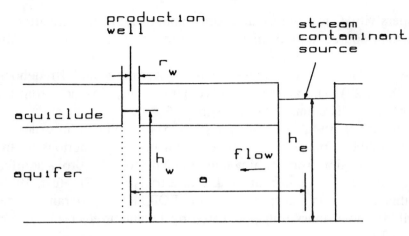

FIGURE 7.13 Cross section through production well and nearby stream contaminant source.

distance between the production well and the stream effective line of recharge, r_w is the production well effective radius, P_h is the aquifer horizontal hydraulic conductivity, h_e is the height of the stream surface above the aquifer base, and h_w is the height of the production well water level above the aquifer base.

Equations 7.38 and 7.39 are based on the following assumptions: the constant production well discharge is balanced by recharge from the stream, groundwater flow is steady state, the density and viscosity of the water in the stream and in the aquifer are essential the same, and dispersion is negligible. An arithmetic graph of values of the ratio $c_p(t)/c_o$ vs. the ratio $\log(t/b)$ is shown in Figure 7.14.

UPCONING BELOW WELL

Figure 7.15 shows a partially penetrating production well in a uniformly porous nonleaky artesian aquifer overlain and underlain by aquicludes. The aquifer is homogeneous, isotropic, infinite in areal extent, and constant in thickness throughout. Fresh water overlies salt water, the two fluids are separated by an abrupt inter-

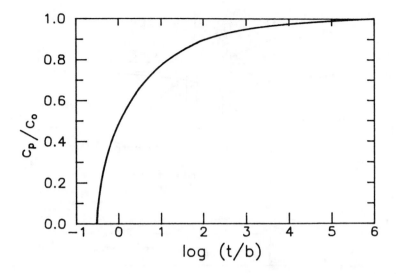

FIGURE 7.14   Arithmetic graph of ratio c_p/c_o vs log(t/b).

face, and there is negligible dispersion. The following equation describes the upconing of the fresh-salt water interface due to pumping (Schmorak and Mercado, 1969, pp. 1292-1293):

$$z = Q\{[1/(1 + R^2)^{0.5}] - \{1/[(1 + a)^2 + R^2]^{0.5}\}\}/$$

$$[2\pi(\Delta\rho/\rho_f)P_h d] \tag{7.40}$$

with

$$R = (r/d)(P_v/P_h)^{0.5} \tag{7.41}$$

$$a = (\Delta/\rho_f)P_v t/(2n_e d) \tag{7.42}$$

$$\Delta\rho = \rho_s - \rho_f \tag{7.43}$$

where z is the rise of the fresh-salt water interface above its initial position at time t, Q is the constant production well discharge rate, ρ_f is the fresh water density, ρ_s is the salt water density, r is the distance between the production and observation wells, P_v is the aquifer vertical hydraulic conductivity, P_h is the aquifer horizontal

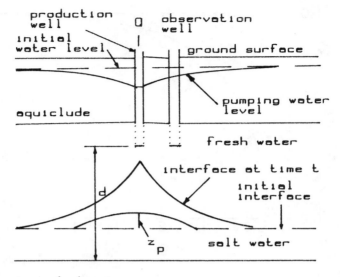

FIGURE 7.15 Cross section through salt water upcone beneath production well.

hydraulic conductivity, d is the height of the production well base above the aquifer base, t is the time after pumping started, and n_e is the aquifer effective porosity.

Beneath the production well Equation 7.40 becomes (Schmorak and Mercado, 1969, p. 1293):

$$z_p = Q\{1 - [1/(1 + a)]\}/[2\pi(\Delta\rho/\rho_f)P_hd] \qquad (7.44)$$

where z_p is the rise of the interface above its initial position just below the production well at time t.

The linear relation between z_p and Q is limited to a certain critical rise $z_c = n_ed$ above which the rate of rise of salt water reaches the base of the production well with a sudden jump. Assuming that salinization of the production well occurs when $z_p > z_c$, the maximum discharge rate with salt free water is expressed by the following equation (Schmorak and Mercado, 1969, p. 1293):

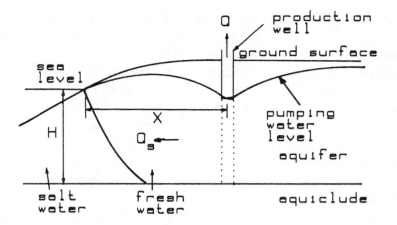

FIGURE 7.16   Cross section through production well and nearby saltwater wedge.

$$Q_m \leq 2\pi n_e d^2 (\Delta\rho/\rho_f) P_h \qquad (7.45)$$

where Q_m is the maximum production well discharge with salt free water.

PRODUCTION WELL NEAR SALTWATER WEDGE

Strack (1976, pp. 1165-1174) developed the following equation for estimating the maximum constant production well discharge near a coastal salt-water wedge (Figure 7.16) without salt water intrusion into the production well:

$$a = 2[1 - (b/\pi)]^{0.5} + (b/\pi)\ln\{[1 - (1 - b/\pi)^{0.5}]/$$

$$[1 + (1 - b/\pi)^{0.5}]\} \qquad (7.46)$$

with

$$a = P_h H^2 \rho_s (\rho_s - \rho_f)/(Q_s x \rho_f^2) \qquad (7.47)$$

$$b = Q/(Q_s x) \qquad (7.48)$$

where P_h is the aquifer horizontal hydraulic conductivity, H is the distance between the aquifer base and sea level, ρ_s is the salt water density, ρ_f is the fresh water density, Q_s is the seaward flow of fresh water per unit aquifer width, x is the distance from the sea shore to the production well, and Q is the maximum constant production well discharge rate without salt water intrusion into the production well.

Equation 7.46 assumes a sharp interface between the salt and fresh waters and negligible dispersion. The production well fully penetrates a uniformly porous aquifer which is homogeneous, isotropic, semi-infinite in areal extent, and constant in thickness throughout.

Q is estimated with the method of successive approximations. First a is calculated with the data base. Then an initial estimate of Q is substituted into Equation 7.48 and b is calculated. The calculated value of b is substituted into Equation 7.46 and a associated with the initial Q estimate is calculated. This value of a is compared with the value of a calculated with the data base. If the comparison is favorable, the solution is declared valid. Otherwise, the procedure is repeated.

Hantush (1968, pp. 40-60) developed equations for describing the growth of a lens resulting from the injection of fresh water at a constant rate through a partially penetrating well in a uniformly porous infinite water table saline aquifer, semi-infinite coastal water table saline aquifer, and semi-infinite water table saline aquifer with a barrier boundary.

FRESH-SALT WATER INTERFACE

Under natural conditions in coastal aquifers, a hydraulic gradient exists toward the sea that causes discharge of fresh water to the sea. There is a body of salt water, usually in the shape of a wedge, underneath fresh water as illustrated in Figure 7.17. The fresh-salt water interface is a transitional zone caused by dispersion, however, often the width of this zone is small compared to the aquifer thickness and it is appropriate to assume an abrupt interface ap-

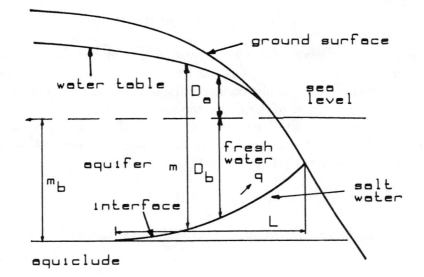

FIGURE 7.17   Cross section through saltwater wedge with seaward fresh water flow.

proximation. The hydrostatic balance between fresh and salt water assuming an abrupt interface and an infinitesimal outflow face width is expressed by the Ghyben-Herzberg principle as follows (Bear, 1979, p. 385):

$$h_s = ah_f \tag{7.49}$$

with

$$a = \rho_w/(\rho_s - \rho_w) \tag{7.50}$$

where h_s is the elevation of the fresh-salt water interface below mean sea level, ρ_w is the fresh water density, ρ_s is the salt water density, and h_f is the elevation of the water table or piezometric surface above mean sea level. Typically, $\rho_f = 1.000$ g/cm^3, $\rho_s = 1.025$ g/cm^3, and $h_s = 40h_f$.

The steady-state seaward fresh water flow rate for the nonleaky artesian aquifer shown in Figure 7.17 is calculated with the following equation (Bear, 1979, p. 395):

$$q = [(\rho_s - \rho_f)/\rho_f]P_h m^2/(2L) \qquad (7.51)$$

where q is the seaward fresh water flow rate per unit of sea front, ρ_s is the salt water density, ρ_f is the fresh water density, P_h is the aquifer horizontal hydraulic conductivity, m is the aquifer thickness, and L is the distance to the toe of the salt water wedge.

Fresh-salt water interface depths (Figure 7.17) assuming a finite outflow face width are described approximately by the following equations (Kashef, 1986, pp. 418-477):

$$D^2_b = 2q(x/a) + 0.5(q/a)^2 \qquad (7.52)$$

$$D_b/m_b = \{[D_a/(bm_b)] + 0.1375[1 - (x/L)]/d^2\}^{0.5} \qquad (7.53)$$

with

$$a = bP_h \qquad (7.54)$$

$$b = (\rho_s - \rho_f)/\rho_f \qquad (7.55)$$

$$d = L/m_b \qquad (7.56)$$

where D_b is the depth below sea level to the fresh salt water interface at a distance x from the sea front, q is the seaward fresh water flow rate per unit of sea front, x is the distance from the sea front to the observation point, D_a is the height of the water table above sea level at a distance x from the sea front, L is the distance to the toe of the salt water wedge from the sea front, P_h is the aquifer horizontal hydraulic conductivity, m_b is the aquifer thickness below sea level, ρ_s is the salt water density, and ρ_f is the fresh water density.

Equations 7.51 to 7.56 are based on the following assumptions: the uniformly porous aquifer is homogeneous, isotropic, semi-infinite in areal extent, and of the same thickness throughout; an aquiclude underlies the aquifer; dispersion is negligible and there is a sharp fresh-salt water interface; and the coastal outflow surface is vertical.

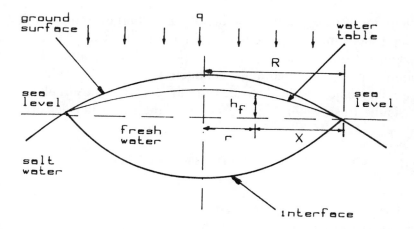

FIGURE 7.18   Cross section through fresh water lens underlying circular island.

ISLAND FRESH WATER LENS

Salt water underlies and surrounds oceanic islands as illustrated in Figure 7.18. The water table elevation at any distance from the shoreline of an infinite strip island receiving recharge from precipitation is calculated with the following equation (Fetter, 1988, p. 155):

$$h_f = q\{(w/2)^2 - [(w/2) - x]^2\}/[Ph(1 + a)] \qquad (7.57)$$

The water table elevation at any distance from the center of a circular island receiving recharge from precipitation is defined by the following equation (Fetter, 1988, p. 155):

$$h_f = q(R^2 - r^2)/[2Ph(1 + a)] \qquad (7.58)$$

with

$$a = \rho_f/(\rho_s - \rho_f) \qquad (7.59)$$

where h_f is the elevation of the water table above mean sea level at the point of observation, q is the rate of recharge from precipitation (depth per time), w is the width of the infinite strip island, x is the distance from the shoreline to the point of observation, P_h is the aquifer horizontal hydraulic conductivity, R is the radius of the circular island, r is the distance from the center of the circular island, ρ_f is the fresh water density, and ρ_s is the salt water density.

HEAT CONDUCTION AND CONVECTION

Heat storage in aquifer systems is a promising alternative for seasonal storage of low-grade thermal energy but may be a contaminant source. Aquifers are physically well suited to thermal energy storage because of their low heat conductivities, large volumetric capacities, ability to contain water under pressure, and widespread availability (Tsang and Hopkins, 1982, pp. 427-441). Theoretical aspects of groundwater heat-pump well systems are described by Schaetzle, et al. (1980) and Kazmann and Whitehead (1980, pp. 28-31).

Geothermal resources have received worldwide attention as alternative or supplemental sources of energy. Heat transport models have been developed for the exploitation of geothermal energy (Faust and Mercer, 1979, pp. 23-30; Faust and Mercer, 1979, pp. 31-46). Considerable attention has been directed toward developing analytical and numerical solutions for temperature distribution in the area of low-grade thermal energy injection (Chen and Reddell, 1983; Gringarten and Sauty, 1975; Tsang, et al., 1981; Voss, 1984).

Thermal distribution in aquifer systems is influenced primarily by heat convection and conduction. Convection is the migration of heated water along with, and in the direction of, groundwater flow. Conduction is the process whereby one part of the aquifer system is heated by direct contact with a source of heat, and neighboring parts become heated successively. Thermal conductivity is analogous to hydraulic conductivity in groundwater flow. The flow of heat is proportional to the difference in temperature between two points (Carslaw and Jaeger, 1959). Heat convection is similar to contaminant advection and heat conduction in the two phases, solid plus liquid, is similar to mechanical dispersion and molecular diffusion in solute transport (Marsily, 1986, p. 277).

Representative thermal properties and densities of selected dry rocks are presented in Appendix B. The specific heat of pure water is 1.00 cal/g °C, the density of pure water at 4°C is 1.0 g/cm^3, and the thermal conductivity of pure water is 1.43×10^{-3} cal/cm sec °C. The densities and specific heats of saturated and dry rocks are related as follows (Bear, 1972, p. 648):

$$c_a\rho_a = nc_w\rho_w + (1 - n)c_s\rho_s \qquad (7.60)$$

where c_a is the saturated rock specific heat, c_w is the water specific heat, c_s is the dry rock specific heat, ρ_a is the saturated rock density, ρ_w is the water density, n is the aquifer porosity, and ρ_s is the dry rock density.

The thermal conductivities of wet and dry rocks are related as follows (Bear, 1972, p. 648):

$$K_a = nK_w + (1 - n)K_s \qquad (7.61)$$

where K_a is the wet rock thermal conductivity, K_w is the water thermal conductivity, n is the aquifer porosity, and K_s is the dry rock thermal conductivity.

The affects of thermal dispersion can be simulated by adjusting the actual aquifer thermal conduction thereby creating an equivalent thermal conductivity as defined by the following equation (Marsily, 1986, p. 278):

$$K_e = K_a + \beta\rho_w c_w v \qquad (7.62)$$

where K_e is the equivalent saturated aquifer thermal conductivity, K_a is the actual saturated aquifer thermal conductivity, β is the aquifer thermal dispersivity, ρ_w is the water density, c_w is the water specific heat, and v is the Darcy groundwater flow velocity.

The thermal dispersivity is comparable to the contaminant migration dispersivity; the asymptotic value of the contaminant migration dispersivity can be five times stronger than the asymptotic value of the thermal dispersivity (Marsily, 1986, pp. 279-281). An advancing contaminant front arrives at a monitoring point before a thermal front as described by the following equations (Marsily, 1986, p. 279):

for advancing contaminant front

$$v_a = v/n_e \qquad (7.63)$$

for advancing thermal front

$$v_a = -\rho_w c_w v/(\rho_a c_a) \qquad (7.64)$$

where v_a is the average pore groundwater velocity, v is the Darcy groundwater flow velocity, n_e is the aquifer effective porosity, ρ_w is the water density, c_w is the water specific heat, ρ_a is the saturated aquifer density, and c_a is the saturated aquifer specific heat.

Most analytical heat migration equations are based on the following assumptions: all physical dimensions, hydraulic, and thermal characteristics of the aquifer system are constant in time and space; the aquifer is uniformly porous; the aquifer is uniform in thickness; aquicludes overlie and underlie the aquifer and have infinite thickness with respect to heat conduction; thermal losses through the aquicludes are governed only by vertical thermal conduction, aquiclude horizontal thermal conduction is negligible; heat migration in the aquifer is governed by horizontal thermal convection and conduction with convection dominating; no vertical temperature distribution exists in the aquifer; water throughout the aquifer system is at the same temperature prior to heated water injection; the heated water injection rate and temperature are constant; the heated water injection well or line source fully penetrates the aquifer; groundwater flow in the aquifer is steady state and the thermal regime is unsteady; heat dispersion is negligible; and there are negligible differences between the density and viscosity of the injected low-grade heated water and the native groundwater.

Lauwerier (1955, p. 149) developed the following equation defining the linear one-dimensional temperature distribution from a line source of low-grade heated water with these assumptions:

$$T_a = T_i W_h(a,b) \tag{7.65}$$

with

$$W_h(a,b) = erfc\{a/[2(b-a)^{0.5}]\}U(\tau - \varepsilon) \tag{7.66}$$

$$U(\tau - \varepsilon) = 0 \text{ when } \tau - \varepsilon \le 0 \tag{7.67}$$

$$U(\tau - \varepsilon) = 1 \text{ when } \tau - \varepsilon > 0 \tag{7.68}$$

$$a = \varepsilon/\theta \tag{7.69}$$

$$b = \tau/\theta \tag{7.70}$$

$$\varepsilon = 4K_c x/(m^2 \rho_w c_w v) \tag{7.71}$$

$$\tau = 4K_c t/(m^2 \rho_a c_a) \tag{7.72}$$

$$\theta = \rho_a c_a/(\rho_c c_c) \tag{7.73}$$

where T_a is the groundwater temperature change at time t, T_i is the constant difference between the heated water line source temperature and the initial groundwater temperature, $W_h(a,b)$ is a heated water line source or well function, erfc(x) is the complementary error function of the argument x which is defined in Equation 3.21, $U(\tau - \varepsilon)$ is a unit step function, K_c is the effective saturated aquiclude thermal conductivity, x is the distance from the heated water line source, m is the aquifer thickness, ρ_w is the water density, c_w is the water specific heat, v is the heated water injection velocity, ρ_a is the saturated aquifer density, c_a is the saturated aquifer specific heat, ρ_c is the saturated aquiclude density, and c_c is the saturated aquiclude specific heat.

Equation 7.65 is applicable to an injection well source of low-grade heated water (see Figure 7.19) with dominant radial flow, negligible regional groundwater flow, and no dispersion when ε is redefined as (Spillette, 1972, p. 22):

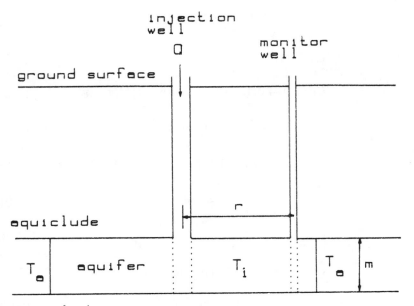

FIGURE 7.19   Cross section through heated water injection well.

$$\varepsilon = 4K_c\pi r^2/(m\rho_w c_w Q) \tag{7.74}$$

where r is the distance from the heated water injection well and Q is the heated water constant injection rate.

Values of $W_h(a,b)$, for the practical range of a and b, based on a graph presented by Spillette (1972, p. 22), are listed in Appendix G. A family of arithmetic graphs of $W_h(a,b)$ vs. a for selected values of b is shown in Figure 7.20. A BASIC program for calculating temperature distribution from a heated water injection well is listed by Walton (1984a, pp. 357-359).

The fraction of the injected heat that remains in the uniformly porous aquifer in Figure 7.19 is calculated with the following equation (Spillette, 1972, p. 22):

$$H_f = (1/b)[2(b\pi)^{0.5} - 1 + e^b erfc(b)^{0.5}] \tag{7.75}$$

where H_f is the fraction of injected heat that remains in the aquifer at time t and b is defined in Equation 7.70.

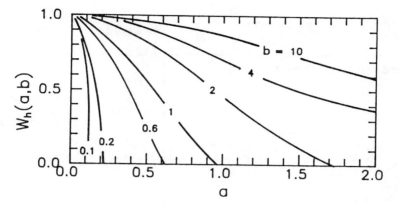

FIGURE 7.20   Arithmetic graphs of $W_h(a,b)$ vs a for selected values of b.

Voss (1984, pp. 186-187) developed the following approximate equation simulating the propagation of the temperature front as it moves radially away from a low-grade heated water injection well fully penetrating a uniformly porous aquifer with negligible regional groundwater flow, negligible heat conduction and convection through overlying and underlying aquicludes, and appreciable heat dispersion in the aquifer:

$$T_a = 0.5 T_i \text{erfc}\{(r^2 - a^2)/\{2[(4\alpha_L/3)a^3 + (b/c)a^4]^{0.5}\}\}$$
(7.76)

with

$$a = (2ct)^{0.5}$$
(7.77)

$$b = nK_W + (1 - n)K_S$$
(7.78)

$$c = (n\rho_W c_W/f)d$$
(7.79)

$$d = Q/(2\pi nm\rho_W)$$
(7.80)

$$f = n\rho_W c_W + (1 - n)\rho_S c_S$$
(7.81)

where T_a is the groundwater temperature change at time t, T_i is the constant difference between the heated water injection well tem-

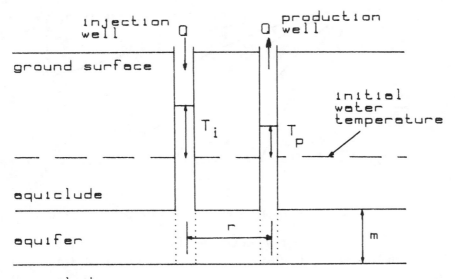

FIGURE 7.21   Cross section through heated water injection and production wells.

perature and the initial groundwater temperature, erfc(x) is the complementary error function of the argument x which is defined in equation 3.21, K_S is the dry rock thermal conductivity, K_W is the water thermal conductivity, m is the aquifer thickness, ρ_W is the water density, c_W is the water specific heat, ρ_S is the dry rock density, c_S is the dry rock specific heat, n is the aquifer porosity, α_L is the aquifer longitudinal heat dispersivity, r is the distance from the heated water injection well, t is the time after heated water injection started, and Q is the constant heated water injection rate.

With a heated water injection well and a heated water production well as shown in Figure 7.21 having equal and constant injection and withdrawal rates, the equation governing the temperature change in the heated water production well with negligible heat dispersion is as follows (Gringarten and Sauty, 1975, pp. 4956-4962):

$$T_p = T_i W_h(\alpha, \lambda) \tag{7.82}$$

with

$$\lambda = [\rho_w c_w \rho_a c_a/(K_c \rho_c c_c)](Qm/r^2) \qquad (7.83)$$

$$\alpha = (\rho_w c_w/\rho_a c_a)(Qt/r^2 m) \qquad (7.84)$$

where T_p is the temperature change of water discharged from the heated water production well due to heated water injection at time t, T_i is the difference between the temperature of the injected heated water and the initial temperature of the groundwater, $W_h(\alpha,\lambda)$ is a well function, ρ_w is the water density, c_w is the water specific heat, ρ_a is the saturated aquifer density, c_a is the saturated aquifer specific heat, K_c is the effective saturated aquiclude thermal conductivity, ρ_c is the saturated aquiclude density, c_c is the saturated aquiclude specific heat, Q is the constant injection rate which is equal to the withdrawal rate, r is the distance between the heated water injection and production wells, t is the time after injection and production started, and m is the aquifer thickness.

Values of $W_h(\alpha,\lambda)$, for the practical range of α and λ, based on a graph presented by Gringarden and Sauty (1975, p. 4960) are listed in Appendix G. A semilogarithmic family of $W_h(\alpha,\lambda)$ vs. α for selected values of λ is shown in Figure 7.22.

Gringarden and Sauty (1975, p. 4962) developed the following equation to determine the spacing between heated water injection and production wells if there is to be negligible change in the temperature of the pumped water for a period equal to the useful life of the wells and heat dispersion is negligible:

$$D = \{2Q\Delta t/\{[n + (1 - n)\rho_c c_c/(\rho_w c_w)]m +\{[n + (1 - n)\rho_c c_c/$$

$$(\rho_w c_w)]^2 m^2 + 2K_c \rho_c c_c \Delta t/(\rho_w c_w)^2\}^{0.5}\}\}^{0.5} \qquad (7.85)$$

where D is the required spacing between heated water injection and production wells, Q is the heated water constant injection rate which is equal to the withdrawal rate, Δt is the useful life of the heated water injection and production wells, n is the aquifer porosity, ρ_c is the saturated aquiclude density, c_c is the saturated aquiclude specific heat, ρ_w is the water density, c_w is the water specific heat, m is the

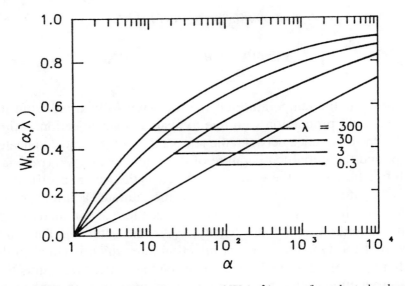

FIGURE 7.22   Semilogarithmic graphs of $W_h(\alpha,\lambda)$ vs α for selected values of λ.

aquifer thickness, and K_C is the effective saturated aquiclude thermal conductivity.

Bredehoeft and Papadopulos (1965, pp. 325-328) developed equations describing vertical steady flow of groundwater and heat through an aquitard. Using these equations, vertical groundwater velocities may be estimated with temperature data at several depths within the aquitard.

8

Production And Drainage Facilities

Production and drainage facilities include production wells, collector wells, galleries, construction dewatering well-point and deep well systems, drains, and mines.

Production well design procedures described herein pertain mainly to wells in unconsolidated deposits and involve choosing the optimum well casing diameter, screen slot size, screen length, well completion method, and well spacing. Procedures are based on a study of drilling cutting samples (grain-size distribution curves) and hydrogeologic data.

The specific capacity of a production well is defined as its yield per unit of drawdown. Well efficiency is specified for a particular pumping period duration and is defined as the actual production well specific capacity divided by the theoretical production well specific capacity. The theoretical specific capacity is based on calculated drawdown affected by well geometry, aquifer hydraulic characteristics, degree of well penetration, well loss, and any aquifer boundaries or discontinuities.

A collector well is constructed by lowering a large-diameter cylindrical concrete caisson into unconsolidated deposits, sealing the bottom with a concrete plug, and jacking perforated pipes horizontally into the aquifer. The construction features of collector wells are described by Spiridonoff (1964). Collector wells are particularly effective in shallow sand and gravel aquifers near streams.

Galleries are horizontal groundwater supply facilities consisting of open trenches, ditches, buried conduits, buried drains, or tunnels (see Driscoll, 1986, pp. 761-768). Analytical equations describing steady state and unsteady state flow under nonleaky artesian, leaky artesian, and water table conditions to fully or partially penetrating and infinite length or finite length galleries are developed by Huisman (1972, pp. 13-76).

Dewatering well-point and/or deep well systems are utilized to lower the water table or piezometric surface for the purpose of facilitating safe and dry construction of trenches, sewers, water mains, and excavations (Anon, 1977, pp. 351-370; Driscoll, 1986, pp. 734-760; Powers, 1981). Well-point systems are closely spaced small diameter wells usually connected to a header pipe or manifold and pumped by suction lift. Deep well systems are closely spaced water supply wells equipped with turbine pumps.

High water tables which reduce crop yields and interfere with farming operations are commonly lowered with underground drains to optimal positions usually ranging between 2 and 3 ft (0.6 and 0.9 m) in sandy soils and between 3.3 and 4.9 ft (1 and 1.5 m) in clayey soils (Bouwer, 1978, pp. 294-299). Drains generally consist of perforated plastic tube, clay tile, or concrete tile .16 to .66 ft (5 to 20 cm) in diameter and commonly discharge into ditches. Parallel drains usually are spaced from 26 to 164 ft (8 to 50 m) (Bouwer, 1978, pp. 294-299).

Drains of finite length are frequently used to control contaminant plume movement, divert groundwater flow, and depress water levels in remedial actions at uncontrolled hazardous waste sites (Boutwell, et al., 1986). They can range from fully penetrating vertical trenches to partially penetrating ditches or perforated pipes. Cohen and Miller (1983, pp. 86-91) compiled a large number of available analytical equations for simulating drains. Sandberg, et al. (1981) list calculator programs for simulating steady-state drawdown around finite line sinks, steady-state flow to finite line sinks with drawdown as given, and finite line sinks for nonsteady-state conditions.

One of the most commonly encountered mine situations is that created by an advancing linear mine box cut or pit incising an aquifer. The hydrologic balance is significantly impacted by mine discharge (Wilson and Hamilton, 1978, pp. 1213-1223 and Van

Voast and Hedges, 1975; Cook, 1982. pp. 397-405).

PRODUCTION WELLS

Production well specific capacity is commonly expressed as the discharge rate divided by the drawdown and varies with the duration of pumping, radius of the well, and aquifer hydraulic characteristics as indicated in the following equation (Cooper and Jacob, 1946):

$$Q/s = 4\pi T/\{-0.5772 - \ln[r^2_w S/(4Tt)]\} \qquad (8.1)$$

where Q/s is the specific capacity of the production well at time t, t is the time after pumping started, Q is the constant production well discharge rate, s is the drawdown in the production well, T is the aquifer transmissivity, r_w is the production well effective radius, and S is the aquifer storativity.

Equation 8.1 assumes that the production well fully penetrates a uniformly porous nonleaky artesian aquifer; well loss (drawdown due to turbulent flow of water through the screen or well face and inside the casing to the pump intake) is negligible; wellbore storage is negligible; and the aquifer is homogeneous, isotropic, infinite in areal extend, and of the same thickness throughout. Specific capacity varies with the logarithm of $1/r^2_w$. Large increases in the production well effective radius result in comparatively small changes in specific capacity.

Equation 8.1 may be used to determine specific capacity under water table conditions provided drawdown is negligible in comparison with the original aquifer thickness by substituting S_y for S. Specific capacity decreases with the logarithm of the pumping period duration because the cone of depression declines with time. For this reason, it is important to state the pumping period duration for which a particular value of specific capacity is calculated. If well loss is negligible, specific capacity is independent of the discharge rate.

Logarithmic graphs of production well specific capacity vs. aquifer transmissivity for selected pumping period durations, artesian (S

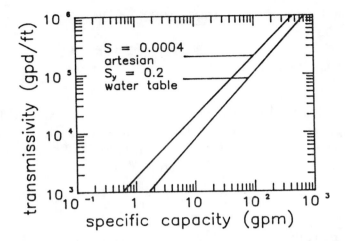

FIGURE 8.1   Logarithmic graph of aquifer transmissivity vs production well specific capacity for pumping period of 1 hour.

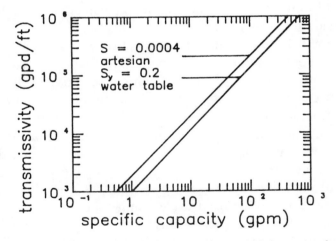

FIGURE 8.2   Logarithmic graph of aquifer transmissivity vs production well specific capacity for pumping period of 1 day.

= 0.0004) conditions, and water table (S = 0.2) conditions calculated with equation 8.1 (Walton,1962, pp. 12-13) are shown in Figures 8.1 and 8.2. Because specific capacity varies with the logarithm of 1/S, large changes in aquifer storativity result in comparatively small changes in the specific capacity-aquifer transmissivity relation. The graphs may be used to estimate aquifer transmissivity with specific capacity data.

If well loss is appreciable, specific capacity decreases as the discharge rate increases and specific capacity is less than that calculated with Equation 8.1. If the production well partially penetrates the aquifer and/or if the production well is near a barrier boundary, specific capacity is less than that calculated with Equation 8.1. Specific capacity is more than that calculated with Equation 8.1 if the production well is near a recharge boundary. Under water table conditions, specific capacity is less than that calculated with Equation 8.1 because aquifer thickness and therefore T is diminished as a result of aquifer dewatering within the cone of depression. All of these factors must be considered in the use of specific capacity data.

Production well completion must provide for ready entrance of groundwater into the well. In consolidated formations where the material of the well wall is stable, water enters directly into the uncased borehole. In unconsolidated deposits, a screen or perforated casing is required to hold back sediments of the well wall and allow water to flow into the production well without excessive head loss or the passage of sediments during pumping. Detailed information concerning production well design in unconsolidated and consolidated formations, well drilling methods, installation of screens, well development, and well specifications is provided by Driscoll (1986, pp. 268-701). A well cost analysis for municipal, industrial, and domestic wells and pumps is provided by Campbell and Lehr (1973, pp. 377-400).

Sediment size distribution (see Driscoll, 1986, pp. 405-412), expressed by the grain-size distribution curve, guides the design of production wells. The slope of the major part of the grain-size distribution curve is described by the uniformity coefficient (C_u) defined as the 40% retained size of the sediment (d_{40}) divided by the 90% retained size (d_{90}). The lower the C_u value, the more uniform is the grading of the sediments. The general relation between grain-size distribution curves and the common range of aquifer and aquitard sediments is shown in Figure 8.3.

No reliable and accurate method for determining aquifer hydraulic conductivity from grain-size distribution curves is known. However, the approximate relation between aquifer hydraulic conductivity and grain-size distribution curves, developed by Patchick

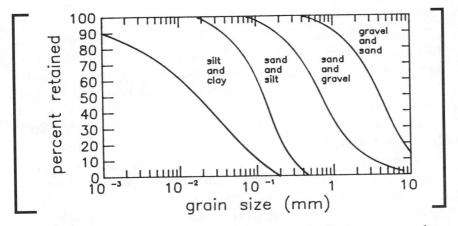

FIGURE 8.3　General relation between grain-size distribution curves and selected deposits.

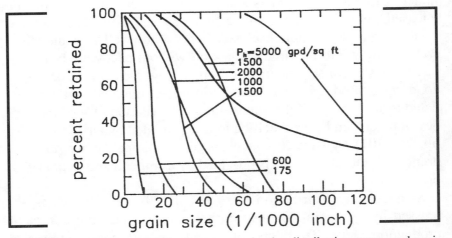

FIGURE 8.4　General relation between grain-size distribution curves and aquifer hydraulic conductivity.

(1967) shown in Figure 8.4, is useful in estimating the order of magnitude of aquifer hydraulic conductivity and the relative hydraulic conductivities of sediments. The hydraulic conductivity of fairly uniform ($C_u < 2$) sands (0.08 to 0.002 inch, 2.0 to 0.06 mm in grain size) in a loose state is estimated with the following equation (Hazen, 1893, 1911; see Shepherd, 1989, pp. 633-638):

$$P_h = cd_{90}^2 \tag{8.2}$$

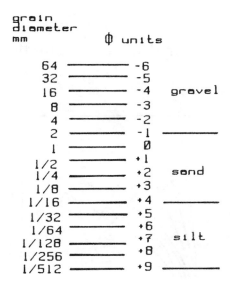

FIGURE 8.5   Phi scale of sediment size.

where P_h is the sediment horizontal hydraulic conductivity, c is a coefficient ranging from 90 to 150 and averaging 100, and d_{90} is the effective grain size.

Masch and Denny (1966, pp. 665-677) developed the following method for estimating the hydraulic conductivity of sand based on grain-size distribution curves. The method involves the median grain size (d_{50}) and a measure of the dispersion about the median grain size (inclusive standard deviation). The phi (ϕ) scale is used to classify grain size, where $\phi = -\log_2 d$ with d being the grain size diameter in mm. The ϕ scale of sediment size is illustrated in Figure 8.5. The expression for the inclusive standard deviation (σ) is (Masch and Denny, 1966, p. 667):

$$\sigma = (d_{84} - d_{16})/4 + (d_{95} - d_{5})/6.6 \qquad (8.3)$$

where d_{84}, d_{16}, d_{95}, and d_5 are expressed in ϕ units.

The curves in Figure 8.6, relating d_{50} (ϕ units) and σ to horizontal hydraulic conductivity were developed by Masch and Denny (1966, p. 673).

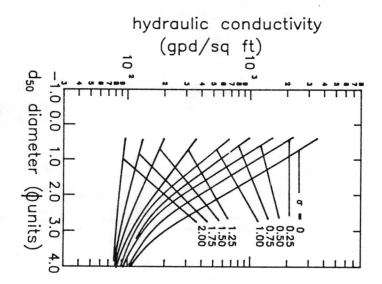

FIGURE 8.6   Semilogarithmic graphs of P_h vs d_{50} grain size diameter in ϕ units for selected values of σ.

There are two types of production well completion: natural-pack and artificial-pack. In a natural-pack well, the screen slot size is selected so that most of the finer formation materials near the borehole are brought into the screen and pumped from the well during development. In an artificial-pack well, the thin zone immediately surrounding the well screen is made more permeable by removing some formation material and replacing it with specially graded material.

Artificial-pack production wells in unconsolidated deposits are usually justified when the aquifer is homogeneous, has a C_u less than 3.0, and has a d_{90} less than 0.01 inch (0.254 mm) (Ahren, 1957). In addition, an artificial pack is sometimes needed to stabilize well-graded aquifers having a large percentage of fine materials in order to avoid excessive settlement of materials above the screen or to permit the use of larger screen slot openings. Artificial packs are often required to complete a production well in poorly cemented sandstone. Artificial-pack thickness usually ranges from 6 to 9 inches (152 to 229 mm) (Driscoll, 1986, pp. 443-447).

One of the important factors in well design is the screen slot size. The selection of the optimum screen slot size is based on the grain-size distribution curve and the hydrogeologic data base. With $d_{90} >$ 6 (a heterogeneous aquifer) and in the case where the sediments above the aquifer are fairly firm and will not easily cave, d_{30} on the grain-size distribution curve (sieve size that retains 30% of the sediments) is the optimum screen slot size in a natural-pack well. With $d_{90} > 6$ and in the case where the sediments above the aquifer are soft and will easily cave, d_{50} on the grain-size distribution curve is the optimum screen slot size (Walton, 1962, pp. 28-29). With d_{90} <3 (a homogeneous aquifer) and in the case where the sediments above the aquifer are fairly firm and will not easily cave, d_{40} on the grain-size distribution curve is the optimum screen slot size. With $d_{90} < 3$ and in the case where the sediments above the aquifer are soft and will easily cave, the d_{60} on the grain-size distribution curve is the optimum screen slot size (Walton, 1962, pp. 28-29).

Careful selection of an artificial pack grain size is important to prevent the clogging of the pack with fine sediments from the aquifer. The optimum artificial-pack C_u is approximately 2 (Driscoll, 1986, p. 439). The optimum ratio of the d_{50} sizes of the pack and aquifer is about 5 (Smith, 1954). The screen slot size is designed so that at least 90% of the size fractions of the artificial pack are retained.

A production well sometimes encounters several layers of sand and gravel having different grain sizes and gradations. If the d_{50} of the coarsest aquifer sediments is less than 4 times the d_{50} size of the finest aquifer sediments, the slot size and pack if necessary, is based on the grain-size distribution curve of the finest aquifer sediments. Otherwise, the screen slot size and pack is taylored to individual layer grain-size distribution curves (Ahrens, 1957).

Screens are usually installed in the lower part of the aquifer. Under water table conditions, optimum screen length is generally 0.3 to 0.5 times the aquifer thickness. Under artesian conditions, optimum screen length often is 75% or more of the aquifer thickness. Factors that affect the choice of screen length include: screen actual open area, aquifer hydraulic characteristics, screen cost, available drawdown, desired production well yield, and desired service life of the production well. Screen length selection is often a compromise between cost and hydrogeology considerations.

Screen length is based partly on the actual open area of the screen and an optimum screen entrance velocity which should not exceed 0.1 ft/sec (0.03 m/sec) (Walton, 1962, pp. 28-29 and Driscoll, 1986, p. 449). In slotted pipe, the actual open area ranges from 1% for 0.030-inch slots to about 12% for 0.250-inch slots. Punched and slotted screens have actual open areas ranging from 4 to 18%. Louvered screen actual open areas range from about 3% for 0.20-inch slots to about 33% for 0.200-inch slots. Cage-type, wire-wound screen actual open areas range from about 2% for 0.006-inch slots to 62% for 0.150-inch slots (Campbell and Lehr, 1963).

Representative actual open areas of continuous slot screens with selected radii are listed in Appendix I. Representative actual open areas of other types of screens are listed by Driscoll (1986, pp. 948-950). When a screen is installed in an aquifer, sediment settles around it and partially blocks slot openings. The amount of blocking depends largely on the shape and type of slot and the grain-size distribution curve for the aquifer and any artificial pack. On the average about one-half of the actual open area is blocked by sediments. Thus, the effective open area is about 50% of the actual open area.

Optimal screen entrance velocities between 1 and 6 fpm (0.005 to 0.03 mpm) insure a long production well service life by avoiding migration of fine sediments toward the screen and subsequent clogging of the well face and screen openings. A table of optimal screen entrance velocities for selected values of aquifer hydraulic conductivity is presented in Appendix I (Walton, 1962, pp. 28-29).

The following equation is used to determine the optimum screen length for a natural-pack production well (Walton, 1962, pp. 28-29):

$$L = Q/(AV) \qquad (8.4)$$

where L is the optimal screen length, Q is the production well discharge rate, A is the effective open area per unit length of screen, and V is the optimum screen entrance velocity. The average of the hydraulic conductivities of the aquifer and artificial pack is used to determine the optimum entrance velocity in the case of an artificial-pack production well.

The diameter of the production well casing is usually two nominal sizes larger than the bowl size of the pump to prevent the pump shaft from binding, to obtain the desired screen open area, to assure that the uphole velocity is 5 ft/sec (1.5 m/sec) or less, and to allow measurement of water levels in the well (Driscoll, 1986, pp. 414-417). The casing may be reduced below the maximum anticipated pump setting. Optimum casing diameters for selected discharge rates with highly efficient pumps (Driscoll, 1986, p. 415) are listed in Appendix I. Required nominal pump bowl diameters for pumps with other efficiencies and selected discharge rates (Campbell and Lehr, 1973, p. 274) are listed in Appendix I.

The spacing of production wells is often dictated by practical considerations such as property boundaries and existing distribution of pipe networks. Economically speaking, the farther apart production wells are spaced, the less their mutual interference but the greater the cost of connecting pipeline and electric equipment. In general, production wells are spaced at least 250 feet (76 m) apart. It is advisable to space production wells parallel to and as far away from barrier boundaries as possible and parallel to and as close as possible to a recharge boundary.

Hantush (1961b, pp. 350-364) developed the following equations for calculating the most economical spacing between interfering production wells arranged in selected patterns with nonleaky and leaky artesian conditions:

for nonleaky artesian aquifer and two production wells with different discharge rates

$$r_o = bQ_1Q_2t/(a\pi T) \tag{8.5}$$

for nonleaky artesian aquifer and three production wells on a line with different discharge rates

$$r_o = b(Q_1Q_2 + Q_1Q_3 + Q_2Q_3)t/(2a\pi T) \tag{8.6}$$

for nonleaky artesian aquifer and three production wells having the same discharge rate and forming an equilateral triangle

$$r_o = b3^{0.5}Q^2t/(a\pi T) \tag{8.7}$$

for nonleaky artesian aquifer and line of three equally spaced production wells having the same discharge rate

$$r_o = 3bQ^2t/(2a\pi T) \tag{8.8}$$

for leaky artesian aquifer and two production wells with different discharge rates

$$K_1(r_o/B) = a\pi TB/(bQ_1Q_2t) \tag{8.9}$$

for leaky artesian aquifer and three production wells forming an equilateral triangle with different discharge rates

$$K_1(r_o/B) = a\pi TB3^{0.5}/[(Q_1Q_2 + Q_1Q_3 + Q_2Q_3)bt] \tag{8.10}$$

for leaky artesian aquifer and four production wells forming a square with different discharge rates

$$[(Q_1Q_2 + Q_1Q_4 + Q_2Q_3 + Q_3Q_4)K_1(r_o/B) +$$

$$2^{0.5}(Q_1Q_3 + Q_2Q_4)K_1(2^{0.5}r_o/B)] = 2(2^{0.5})a\pi TB/(bt) \tag{8.11}$$

with

$$B = (Tm'/P')^{0.5} \tag{8.12}$$

where r_o is the optimum spacing between any two production wells, a is the capitalized cost in unit currency per unit time per unit length of connecting pipeline, for maintenance, depreciation, original cost of pipeline, etc; b is the cost in unit currency to raise a unit volume of water a unit height consisting of power and equipment charges; Q_n is the discharge rate of the n th production well; T is the aquifer transmissivity; t is the constant pumping period duration; $K_1(x)$ is

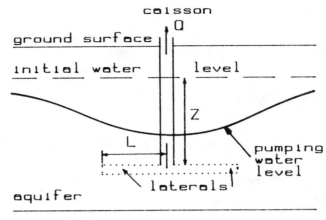

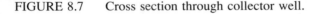

FIGURE 8.7   Cross section through collector well.

the first-order modified Bessel function of the second kind; P' is the aquitard vertical hydraulic conductivity; and m' is the aquitard thickness.

Equations 8.5 to 8.12 assume that the production wells fully penetrate a uniformly porous aquifer which is homogeneous, isotropic, infinite in areal extent, and of the same thickness throughout.

COLLECTOR WELLS

The yield of a collector well is often simulated as a fully penetrating vertical production well having the same specific capacity as the collector well. The effective radius of the vertical production well, assuming that the collector well has a radial lateral pattern covering its entire circumference and equal-length laterals, is about 75 percent of the average lateral length (Mikels and Klaer, 1956, pp. 232-242).

The drawdown in a collector well with four or more symmetrically located laterals (Figure 8.7) is governed by the following equation (Hantush, 1964, pp. 399-407):

$$s = Q\{W(a) + [(N-1)/L][L_1W(b) - r_cW(d)] + 2N + [m/2L)\ln(c)$$

$$+[4m(N-1)/(\pi L)]\sum_{m=1}^{M}\{(1/n)[\pi/2 - W(f,0)]\}$$

$$\cos(n\pi Z/m)\cos[n\pi(z+r_L)]/m\}/(4\pi P_h mN) \qquad (8.13)$$

with $t > 2.5m^2/(P_hm/S_y)$ and $> 5(r_c^2 + L^2)/(P_hm/S_y)$, $L > 0.5$ m, and $r_L < m/(2\pi)$

$$a = L^2/(4P_hmt/S_y) \qquad (8.14)$$

$$b = L_1^2/(4P_hmt/S_y) \qquad (8.15)$$

$$c = [m/(\pi r_L)]^2/2[1 - \cos\pi(2Z + r_L)/m] \qquad (8.16)$$

$$d = r_c^2/(4P_hmt/S_y) \qquad (8.17)$$

$$f = n\pi r_c/m \qquad (8.18)$$

$$L_1 = L + r_c \qquad (8.19)$$

where s is the drawdown in the caisson at time t; t is the time after pumping started in the caisson; P_h is the aquifer horizontal hydraulic conductivity; m is the aquifer thickness; N is the number of laterals; S_y is the aquifer specific yield; L is the lateral length; r_c is the caisson radius; r_L is the lateral radius; Q is the constant caisson discharge rate; Z is the vertical distance between laterals and the water table or the vertical distance between laterals and the aquifer top under artesian conditions; W(a), W(b), and W(d) are the well function W(u) (Equations 2.2 and 2.3); W(f,0) is another well function defined by Hantush (1964, p. 314); and M is an integer large enough so that $M > m/(2r_c)$.

Values of the well function W(f,0) presented by Hantush (1964, p. 319), over the practical range of f, are listed in Appendix G. Equation 8.13 assumes that the water table aquifer is uniformly porous, homogeneous, isotropic, infinite in areal extent, and the same thickness throughout. The radial laterals lie in the same horizontal plane, are equal in length, and are so oriented that each will drain an equal section of the aquifer. The constant discharge of all laterals is the same. The maximum drawdown of the water table is small relative to the original aquifer saturated thickness. The radius of the caisson is small compared with the lateral length.

The drawdown in a collector well under artesian conditions can be calculated with Equation 8.13 by substituting S for S_y, where S is the aquifer artesian storativity. The drawdown in a collector well near a stream can be estimated with Equation 8.13 and the image well theory. The recharging image well simulating the stream is assumed to be a fully penetrating vertical well.

CONSTRUCTION DEWATERING

Decisions regarding the spacing and depth of well points and/or deep wells and on pumps and piping for construction dewatering are based on estimates of the total dewatering system discharge rate and individual well-point and/or deep well discharge rates. These estimates are difficult to make because the saturated thickness of materials is seriously reduced during dewatering, complex anisotropic conditions prevail in near-surface materials, well points and deep wells usually only partially penetrate aquifers, and well points and deep wells are very closely spaced.

A well point is usually a 1.5 to 3.5-inch (38.1 to 88.9 mm) diameter well screen, 18 to 40 inches (0.46 to 1.02 m) long. The well point is attached to a riser pipe which may be 1.5 to 3 inches (38.1 to 76.2 mm) in diameter. Well points are typically spaced 3 to 12 ft (0.9 to 3.7 m) apart (Powers, 1981, pp. 290-300). Most well-point systems are designed on the basis of 15 ft (4.6 m) of suction lift and individual well-point discharges ranging from 10 to 35 gpm (40 to 140 L/min). Suction lift limitations are overcome by using separate dewatering systems installed at increasing depths in successive stages. Depth steps of 10 to 12 ft (3 to 3.7 m) per stage are common.

The discharge rates of individual well points in a well-point array vary because of mutual interference of well points. The lowest discharge rates occur at the center of the excavation. Discharge valves installed on each well point are adjusted so that the water level remains above the well-point screen thereby preventing air from entering the well point and causing the pump to break suction.

Deep-well dewatering systems are often required to lower artesian pressure thereby preventing blowthroughs (sand boils) from occurring in excavation floors. Design and construction techniques for water supply production wells apply to deep wells used for dewatering.

The arrays of well points used in construction dewatering are commonly simulated with a single large diameter well having an effective radius r_s (Figure 8.8). The simulation has the greatest validity with a circular well point system. If well points are closely spaced in a rectangular array and the ratio of the rectangle length and width is less than 1.5, the effective radius is calculated with the following equation (Powers, 1981, pp. 106-107):

$$r_s = (ab/\pi)^{0.5} \tag{8.20}$$

where r_s is the effective radius of a single well simulating a rectangular well-point array, a is the length of the rectangular array of well points, and b is the width of the rectangular array of well points.

Under water table conditions, the total discharge rate to maintain water levels at a desired position at construction sites (Figure 8.9) is commonly estimated with the following Dupuit equation (see Powers, 1981, p. 103):

$$Q = \pi P_h(h^2 - h_s^2)/[\ln(r_o/r_s)] \tag{8.21}$$

With (see Hantush, 1964, p. 363)

$$r_o = 1.5(Tt_c/S_y)^{0.5} \tag{8.22}$$

$$T = [(h + h_s)/2]P_h \text{ (water table aquifer)}$$

when the excavation is near a line source of water

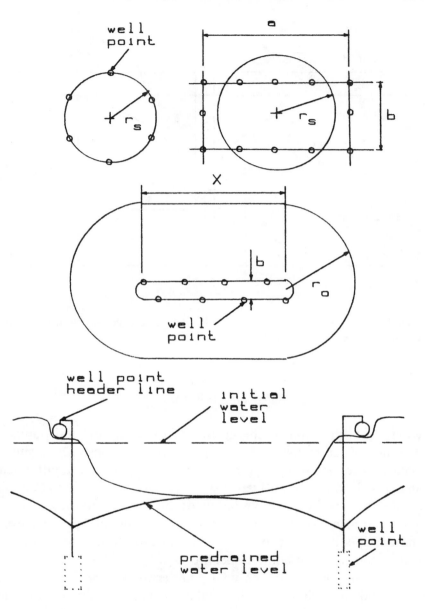

FIGURE 8.8   Simulation models for construction dewatering.

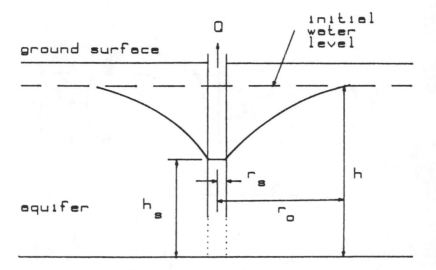

FIGURE 8.9   Cross section of a well simulating a construction dewatering well-point array in water table aquifer.

$$r_o = 2a_s \qquad (8.23)$$
$$\text{(Powers, 1981, p. 109)}$$

where Q is the total discharge rate of the array of well points to maintain water levels at a desired position at the construction site, P_h is the aquifer horizontal hydraulic conductivity, h is the aquifer head at r_o just beyond the influence of the excavation, h_s is the aquifer head at r_s, r_s is the effective radius of the well-point array, T is the aquifer transmissivity, t_c is the time after construction dewatering started, a_s is the distance from the excavation to the line source of water, and S_y is the aquifer specific yield.

For a long narrow trench well-point array, where the ratio a/b > 1.5 as shown in Figure 8.8, the total discharge rate to maintain water levels at a desired position at the trench excavation is commonly estimated with the following Dupuit-Forchheimer equation (Powers, 1981, p. 108):

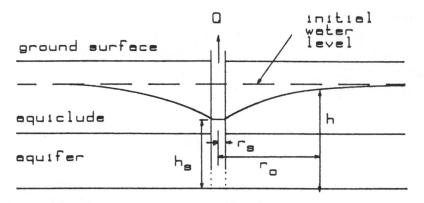

FIGURE 8.10   Cross section of construction dewatering well in artesian aquifer.

$$Q = 2[xP_h(h^2 - h_s^2)/(2r_o)]$$

$$+ \pi P_h(h^2 - h_s^2)/\ln[r_o/(b/2)] \tag{8.24}$$

where x is the length of the trench excavation and b is the width of the trench excavation.

Under artesian conditions, the preliminary total discharge rate to maintain artesian pressure at a desired level at the construction site (Figure 8.10) is estimated with the following Thiem equation (see Powers, 1981, p. 99):

$$Q = 2\pi P_h m(h - h_s)/\ln(r_o/r_s) \tag{8.25}$$

where Q is the total discharge rate to maintain artesian pressure at a desired level at the construction site, P_h is the aquifer horizontal hydraulic conductivity, m is the aquifer thickness, h is the aquifer head at r_o just beyond the influence of the excavation, h_s is the aquifer head at r_s, r_o is defined in Equation 8.22, and r_s is defined in equation 8.20.

Equations 8.21 to 8.23 assume there are no boundaries within the stabilized cone of depression and the volume of water initially in aquifer storage between the initial water level and the stabilized cone of depression has been drained prior to the construction period. The aquifer storage drainage rate which must be considered in addition to the discharge rates in Equations 8.21 to 8.23 is usually estimated with the following equation:

$$Q_m = 2\pi S_y \int_{r_s}^{r_o} (sr)dr / t_P$$

(8.26)

where Q_s is the aquifer storage drainage rate to deplete aquifer storage between the initial water level and the stabilized cone of depression, S_y is the aquifer specific yield, s is the drawdown at the distance r from the construction site, r_s is defined in Equation 8.20, r_o is defined in Equation 8.23, and t_p is the time prior to the construction period (usually several days or more) during which aquifer storage is depleted.

The integral in Equation 8.26 may be solved numerically with the Trapezoidal Rule, Simpson's Rule, or the Romberg method; data on the cone of depression profile; and the BASIC computer programs listed by Miller (1981, pp. 225-254) and Poole, et al. (1981, pp. 86 and 87). The numerical integration determines the area beneath the cone of depression curve.

The profile of the stabilized cone of depression (sr relationship) is defined with the following equations (see Powers, 1981, pp. 99-103):

for artesian aquifer

$$h_r = h - Q\ln(r_o/r)/(2\pi P_h m)$$

(8.27)

for water table aquifer

$$h_r = [h^2 - Q\ln(r_o/r)/(\pi P_h)]^{0.5}$$

(8.28)

where h_r is the aquifer head at a distance r from the large diameter well simulating an array of well points or deep wells, h is the aquifer head at r_o just beyond the influence of the excavation, Q is the total discharge rate of an array of well points or deep wells, r_s is defined by Equation 8.20, P_h is the aquifer horizontal hydraulic conductivity, and m is the aquifer thickness.

Frequently, the construction site is near a stream whose stage can change abruptly thereby causing a change in the total discharge rate of a dewatering system. The change in aquifer head at any distance from a straight-line fully penetrating stream due to an abrupt change in the stream stage is calculated with the following equation (Ferris, et al., 1962, pp. 126-128):

$$s = s_s D(u) \qquad (8.29)$$

with

$$u^2 = x^2 S/(4Tt) \qquad (8.30)$$

where s is the change in aquifer head at time t, s_s is the change in stream stage at time t, $D(u)$ is a drain function, x is the distance from the stream to the observation point, S is the aquifer storativity, T is the aquifer transmissivity, and t is the time after the abrupt change in stream stage.

Values of $D(u)$, for the practical range of u and u^2, presented by Ferris, et al. (1962, p. 127) are listed in Appendix G. Calculated values of s are added algebraically to the value of h in Equations 8.27 and 8.28 to estimate the total discharge of the dewatering system with a stream stage change.

Equations 8.29 and 8.30 assume that the stream occurs along an infinite straight line and fully penetrates a nonleaky artesian aquifer, the aquifer is semi-infinite in areal extent (bounded on one side by the stream), the head in the stream is abruptly changed from zero to s_s at time $t = 0$, and the direction of groundwater flow is perpendicular to the direction of the stream.

Drawdowns at critical well points and/or deep wells (at and near array center and corners) due to these discharges are calculated with a computer program and compared with required drawdowns to maintain water levels at the desired positions at the construction site. If the comparison is favorable, the dewatering system with the

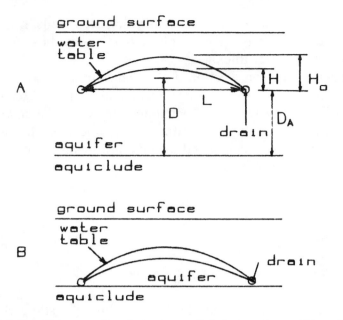

FIGURE 8.11   Cross section through parallel drains above (A) and on (B) aquifer base.

calculated discharge rates is declared valid. Otherwise the process is repeated until a favorable comparison is obtained.

A dewatering system may be simulated numerically as a constant head-variable discharge mine drift as described by Walton (1989b, pp. 32-36). FORTRAN mine drift programs are listed in OSM (1981) and Walton (1989b, pp. 112-117).

DRAINS

Underground drains are either placed above the aquifer base or rest on the aquifer base (see Figure 8.11). Flow is mostly horizontal if the drain rests on the aquifer base or if the aquifer base is a small distance below the drain and vertical components of flow are negligible. If the aquifer base is at an appreciable distance below the drain, vertical components of flow are appreciable and the effective aquifer transmissivity is less than it would be with only horizontal flow. The impacts of these vertical components of flows are

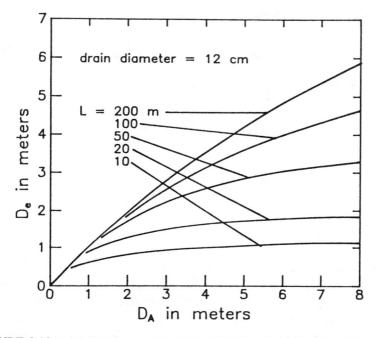

FIGURE 8.12   Arithmetic graphs of D_e vs D_A for selected values of L.

simulated by using an effective depth from the drain to the aquifer base (D_e) which is less than the actual depth from the drain to the aquifer base (D_A) in parallel drain equations. An arithmetic family of graphs of D_e vs. D_A, over the practical range of drain spacing and for a selected drain diameter based on a graph presented by Bouwer (1978, p. 296) and tables presented by Hooghoudt (1940, pp. 515-707), is shown in Figure 8.12. The effect of the drain diameter on D_e is small and can be neglected for practical purposes.

Drainage by parallel drains is governed by the following equations developed by R.D. Glover of the U.S. Bureau of Reclamation (see Luthin, 1966, pp. 164-169):

$$H = (192 / \pi^3) \sum_{n=1,3,5}^{\infty} (-1)^{(n-1)/2}[(n^2 - 8 / \pi^2) / n^5]$$

$$\exp(-\pi^2 n^2 at / L^2) \qquad (8.31)$$

for parallel drains resting on the aquifer base

$$q = 4P_hH_o^2/L \tag{8.32}$$

for parallel drains above the aquifer base

$$q = 2\pi P_hDH/L \tag{8.33}$$

with

$$D = D_e + H_o/2 \tag{8.34}$$

$$a = P_hD/S_y \tag{8.35}$$

where H is the water table height above the drains midway between drains at time t, t is the time after drainage starts, L is the drain spacing, P_h is the aquifer horizontal hydraulic conductivity, D is the average flow depth midway between drains, S_y is the aquifer specific yield, H_o is the water table height above the drains midway between the drains at the start of the drainage period, and q is the drainage rate per unit of drain length.

Solution of Equation 8.31 for the drain spacing involves the method of successive approximations. The drain depth, water table height above the drains midway between the drains at the start and end of the drainage period, and the length of the drainage period are estimated from the data base. An initial estimate of the drain spacing is made and Figure 8.12 is used to determine the effective depth D_e. This effective depth and known H_o are substituted in Equation 8.34 to calculate a. This value of a is substituted in Equation 8.31 to calculate a new value of the drain spacing. If the calculated drain spacing differs widely from the initial estimated drain spacing, a new drain spacing is assumed and a new value of D_e is obtained from Figure 8.12. The process is repeated until a favorable balance between assumed and calculated drain spacing is achieved. The semilogarithmic graph of H/H_o vs. $P_hDt/(S_yL^2)$ for drains which are above the aquifer base (Figure 8.13) and the semilogarithmic graph of H/H_o vs. $P_hHt/(S_yL^2)$ (see Luthin, 1966, pp. 165-166) for drains which are on the aquifer base (Figure 8.14) show the relationship

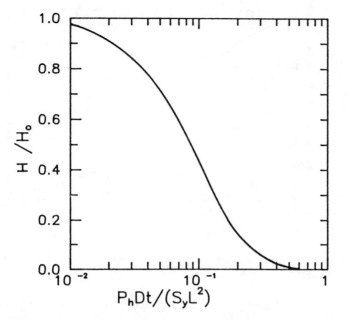

FIGURE 8.13   Semilogarithmic graph of H/H_o vs $P_h D_t/(S_y L^2)$ for parallel drains above aquifer base.

between the dimensionless parameters H/H_o and $P_h Dt/(S_y L^2)$ or $P_h H_o t/(S_y L^2)$.

The following empirical equations may be used to calculate D_e (Moody, 1966, pp. 1-6):

For $0 < D_A/L \leq 0.3$

$$D_A/D_e = 1 + D_A/L[8/\pi \ln(D_A/r) - a] \tag{8.36}$$

For $D_A/L > 0.3$

$$L/D_e = 8[\ln(L/r) - 1.15]/\pi \tag{8.37}$$

with

$$a = 3.55 - 1.6D_A/L + 2(D_A/L)^2 \tag{8.38}$$

where r is the drain radius.

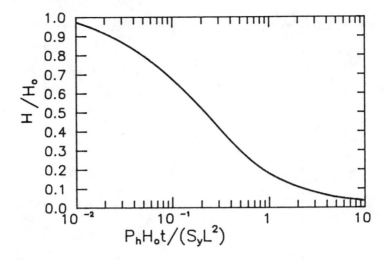

FIGURE 8.14   Semilogarithmic graph of H/H_o vs $P_hH_ot/(S_yL^2)$ for parallel drains on aquifer base.

Equations 8.29 and 8.30 may be used to estimate the drawdown distribution in the vicinity of a fully penetrating drain in a water table aquifer provided drawdown is small in relation to the initial aquifer thickness. The drain discharge rate is calculated with the following equation (Ferris, et al., 1962, p. 127):

$$q = [2s_s/(\pi t)^{0.5}](ST)^{0.5} \qquad (8.39)$$

where q is the drain discharge rate per unit length of drain due to groundwater contribution from both sides of the drain, s_s is the drain depth below the initial water level (abrupt change in head), t is the time after the drain was excavated, S is the aquifer storativity (specific yield in the case of a water table aquifer), and T is the aquifer transmissivity.

Aquifer boundaries near drains may be simulated with finite lines of closely spaced image wells in accordance with the image well theory and the method of successive approximations. Partially penetrating drains may be simulated with finite lines of partially penetrating production wells and the method of successive approximations.

MINES

Discharge from a mine box cut, pit, or drift and associated drawdowns may be estimated with well-array and successive approximation techniques (Davis and Walton, 1982, pp. 841-848) or methods developed by McWorter (1981, pp. 91-92). In the successive approximation technique, appropriate radial flow equations including those simulating fully and partially penetrating wells (mine drifts often partially penetrate an aquifer) in nonleaky artesian, leaky artesian, and water table aquifers are utilized. Mines are simulated with rectangular arrays of closely spaced wells which produce required drawdowns within specified mine areas. Well arrays are expanded consistent with the rate of mine excavation.

The discharge of the first well in an array is estimated based on the required drawdown within the first short segment of the mine. Additional wells are added to the array simulating mine expansion and the discharges of existing and new wells are mutually adjusted to compensate for well interference while maintaining required drawdowns in the mine. This process is repeated until expansion of mine areas is completed, thereafter, the discharge of the well array is adjusted to compensate for continued drawdown and discharge decline with time.

Well discharge is initially estimated for selected time increments. Drawdowns due to estimated discharges are calculated and compared with required drawdowns. If calculated and required drawdowns are equal, estimated discharge is declared valid. Otherwise, discharge is re-estimated and the process is repeated until calculated and required drawdowns are equal.

Equations governing the discharge to the mine box cut or pit in the uniformly porous nonleaky artesian aquifer illustrated in Figure 8.15, developed by (McWorter, 1981, pp. 91-92) are as follows:

When $t \leq L/R$

$$Q = 2R(S_y T m^2/12 + STH_o^2/4 + STH_o m/4)^{0.5} t^{0.5} \qquad (8.36)$$

when $t >= L/R$

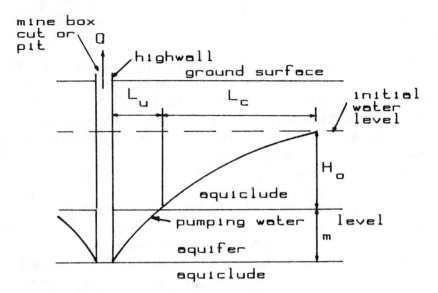

FIGURE 8.15   Cross section through mine box cut or pit.

$$Q = 2R(S_yTm^2/12 + STH_o^2/4 + STH_om/4)^{0.5}$$

$$[t^{0.5} - (t - L/R)^{0.5}] \qquad\qquad (8.37)$$

where Q is the discharge to one side of the mine box cut or pit at time t, t is the time measured from the initiation of mine box cut or pit excavation, R is the average rate of mine box cut or pit elongation, L is the maximum length of the mine box cut or pit, L/R is the time period during which the mine box cut or pit is advancing, S_y is the aquifer specific yield, T is the aquifer transmissivity, m is the aquifer thickness, S is the aquifer artesian storativity, and H_o is the height of the initial water level above the aquifer top.

Equations 8.36 and 8.37 are based on the following assumptions: the mine box cut or pit fully penetrates the aquifer; the length of the mine box cut or pit increases linearly with time; the mine box cut or pit has no storage capacity (mine cavity storage is ignored); the aquifer, underlain and overlain by aquicludes, is homogeneous, isotropic, infinite in areal extent, and constant in thickness throughout; flow is one-dimensional and in a direction normal to the lon-

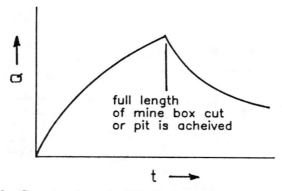

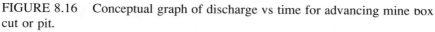

FIGURE 8.16   Conceptual graph of discharge vs time for advancing mine ʙox cut or pit.

gitudinal axis of the mine box cut or pit; inflow to the ends of the mine box cut or pit is negligible; and the portion of the aquifer which becomes unconfined increases with time. Discharge to the mine box cut or pit decreases sharply after the maximum length has been attained as shown in Figure 8.16. Equations 8.36 and 8.37 apply to a water table aquifer if H_o is set to zero.

The distance from the mine box cut or pit face to the outer edge of the cone of depression is governed by the following equation (McWorter, 1981, pp. 91-92):

$$D = D_u + D_c \tag{8.38}$$

with

$$D_u = Tm(t - L/R)^{0.5}/[2(S_yTm^2/12$$

$$+ STH_o^2/4 + STH_om/4)^{0.5}] \tag{8.39}$$

$$D_c = TH_o(t - L/R)^{0.5}/[2(S_yTm^2/12$$

$$+ STH_o^2/4 + STH_om/4)^{0.5}] \tag{8.40}$$

where D is the distance from the mine box cut or pit to the outer edge of the cone of depression at time t, t is the time measured from the

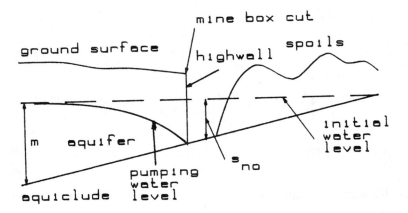

FIGURE 8.17   Cross section through advancing surface mine box cut on crop line.

initiation of mine box cut or pit excavation, T is the aquifer transmissivity, m is the aquifer thickness, R is the average rate of mine box cut or pit elongation, L is the maximum length of the mine box cut or pit, S_y is the aquifer specific yield, S is the aquifer storativity, and H_o is the distance between the initial water level and the aquifer top.

The rate of flow to an advancing surface mine box cut on a crop line as illustrated in Figure 8.17, ignoring the quantity of water released from storage within the mine cavity is calculated with the following equations (McWhorter, 1981, pp. 92-93):

for the first box cut when $t \leq L/R$

$$Q_1 = 2R(S_y T s_{1o}^2/12)^{0.5} t^{0.5} \tag{8.41}$$

for residual flow during the excavation of the second box cut in response to the increment of drawdown caused by the first cut when $t \geq L/R$

$$Q_1 = 2R(S_y T s_{1o}^2/12)^{0.5}[t^{0.5} - (t - L/R)^{0.5}] \tag{8.42}$$

for additional flow during the excavation of the second box cut when $L/R \leq t \leq 2L/R$

$$Q_2 = 2R(S_y Tm^2/12)^{0.5}(t - L/R)^{0.5} \qquad (8.43)$$

for residual flow during excavation of the third box cut that remains due to the increment s_o^2 when $t \geq 2L/R$

$$Q_2 = 2R(S_y Ts_{2o}^2/12)^{0.5}[(t - L/R)^{0.5}$$

$$- (t - 2L/R)^{0.5}] \qquad (8.44)$$

for flow to the final box cut n after it is completed when $t \geq nL/R$

$$Q_n = 2R(S_y Ts_{2o}^2/12)^{0.5}[t^{0.5} - (t - nL/R)^{0.5}] \qquad (8.45)$$

with $s_{no} = W\tan\phi$

where Q_1 is the discharge through the mine box cut highwall at time t due to the first cut, $Q_2(t)$ is the discharge through the mine box cut highwall at time t due to the second cut, Q_n is the discharge through the mine box cut highwall at the end of the final cut n, t is the time measured from the initiation of the mine box cut excavation, s_{no} is the drawdown at the mine box cut number n, T is the aquifer transmissivity, m is the aquifer thickness, R is the average rate of mine box cut elongation, L is the maximum length of the mine box cut, S_y is the aquifer specific yield, W is the width of the mine box cut, ϕ is the aquifer dip angle, and n is the mine box cut number.

The total discharge during construction of the first cut is calculated with Equation 8.41. During excavation of the second cut, the total discharge is calculated as the sum of Equations 8.42 and 8.43. The first cut is assumed to be made near the crop line and extends the full length along the strike of the bed being mined. The highwall is assumed to be advancing downdip with time. The drawdown s_{no} increases incrementally as the highwall advances downdip.

9

Modeling

Groundwater modeling (Domenico, 1972; Mercer and Faust, 1981; Wang and Anderson, 1982; Kinzelbach, 1986; and Bear and Verruijt, 1987) is concerned with the simulation of the dynamic behavior of aquifer systems. While recognizing the parts or sub-systems of an aquifer system and their functions, the ultimate concern of modeling is with the operation of the aquifer system as a whole in context with its surrounding environment. Models integrate fragmented knowledge of the aquifer system's component parts and represent the entire system with facility features such as production wells and waste disposal source areas.

Aquifer systems can not be completely comprehended or defined in their entireties. They are imagined to be simpler than they really are by considering only aspects which pertain to the problem at hand. Often, aquifer system and facility features are too complex to be modeled in actual detail. In this case, the aquifer system and facility features must be simplified before they can be modeled. This may require considerable ingenuity on the part of the modeler and consume a large amount of time.

Although a model may be less complex than the real system it represents, oversimplification is not justified. The degree of approximation is determined by the assumptions incorporated in a model after considering the model purpose, the status of available model theory, and the data base. An important aspect of modeling is the proper acknowledgement of the approximate nature of modeling through the clear description of model assumptions and limitations. Adequate documentation and appreciation of assumptions greatly assists the modeler and model user in keeping model result expectations within a realistic perspective.

Aquifer system facility feature data are commonly placed in primary data files after being collected in the field. These data files are frequently processed by data manipulation programs in various ways such as sorting, interpolation, aggregation, and statistical analysis before being placed in secondary data files for use in models. Output from data manipulation programs may be in the form of tables, contour maps, and plots. The U.S. Geological Survey operates and maintains a national water data storage and retrieval system (WATSTORE). An extensive groundwater data file is included in the system.

Different models require difference amounts and types of data (Moore, 1979). Associated with each level of model sophistication is a data base requirement. Generally, as the model becomes more sophisticated in order to more closely conform to reality, the associated data requirements increase rapidly. A point exists at which the benefits of a more realistic model are outweighed by the difficulty and expense of collecting the data necessary to adequately define such a model.

There is always a gap between data needs and data availability which increases with the complexity of the aquifer system and facility features. Filling data gaps often requires estimation, interpolation, or extrapolation based on available information and considerable scientific judgement of a subjective or intuitive nature. Model sophistication far beyond data availability is not warranted and is often misleading. The reliability and accuracy of model results can not exceed those of the data base. Adequate acknowledgement and documentation of data base limitations is an important aspect of modeling.

A geographic information system (GIS) commonly used by governmental agencies and private institutions to automate, manage, query, display, manipulate, and analyze geographic data is described by Broten, et al. (1987). A computerized GIS has been used as a central groundwater data base and linked with aquifer system modeling efforts (see Harris, et al., 1989, pp. 371-384). The linking process is awkward because groundwater model input data structures are usually complex and not standardized. Rumbaugh (1989, pp. 683-697) describes a data base approach similar to GIS which is taylored to the needs of an aquifer system modeler.

MODEL DEVELOPMENT

There are several phases in model development including: (1) collection and processing of the data base; (2) definition of the problem to be addressed; (3) design of a conceptual model; (4) formulation of a preliminary model based on conceptual model refinements; (5) programming the model; (6) specification of the structure of the model including its geometric features, dimensions, internal hydraulic characteristics, boundaries, and initial conditions; (7) validation of the model (Adrion, et al. , 1982 and Andriole, 1986); (8) testing the sensitivity of the preliminary model; (9) calibration of the preliminary model and finalization of the model; (10) design and execution of simulation exercises directed toward the solution of the problem; (11) analysis of simulation data; and (12) conclusion documentation. Before undertaking model development, the modeler and the user should agree upon the definition of the problem to be modeled and the scope of the project. An acceptable level of model confidence should also be defined. Model validation, calibration, and precision questions should be addressed and resolved at the beginning of the project. The scale of models should be in tune with the scale of decisions affected by model results (see Jousma, et al., 1989).

Model development is a sequential process often progressing from simple or relatively crude analytical conceptual models to increasingly refined, sophisticated, and realistic numerical representations of aquifer systems and facility features. In the beginning, a word description of a model and its purpose can prove most helpful in designing a conceptual model. An appreciation of the scope of a aquifer system is gained through conceptual model development. In addition, attention is focused on basic groundwater flow and contaminant migration theory. In refining conceptual models, an assessment of the aquifer system hydraulic characteristic and boundary data base sheds light on data base balance and factors limiting refinement. Uncertainties concerning parts of the aquifer system and the interrelations of parts should be documented and kept in mind when reviewing model results. Model refinements should not overkill the problem or accuracy warranted by the data base.

A model is said to be validated when there is a favorable comparison between model results and exact analytical model equation calculations. Unfortunately, exact analytical model equations are available only for fairly simple aquifer system and facility situations. In many cases, complex models go beyond available analytical model equation assumptions and complete validation is not possible. Validation of numerical models can be expensive and time consuming because high precision numerical model results require small and many time increments and small grid spacings.

Field validation refers to a model prediction made several years into the future which is later verified in the field. Convincing groundwater flow model field validation is rare partly because assumed future discharge scenarios are never completely realized. No contaminant migration model has been convincingly field validated to date (Anderson, 1986, pp.396-413). Rigorous field validation is practically impossible because of the expense and trouble of collecting sufficient data over a long period of time after the completion of a facility.

Model calibration involves a considerable degree of subjective judgement and is essentially a trial and error procedure. A conceptual model, based on the available data base, is used to estimate head, concentration, temperature, or subsidence during the simulation period in the area of interest. These estimates are compared with sets of field data on head, concentration, temperature, or subsidence. Differences between estimates and field data are noted and the conceptual model is modified in the appropriate direction to decrease the differences and thereby improve the fit between estimates and field data. This process is repeated until the differences are considered negligible and the model is said to be calibrated.

The ability of a calibrated model to duplicate past aquifer system behavior does not necessarily mean that the model can be used to predict future cause and effect relations. The calibrated model may have to be further modified because of anticipated factors such as future dewatering, changing aquifer system hydraulic characteristics, or shifting aquifer system boundaries. The longer the simulation period, the greater may be the need for modifications. Although some prediction models can not be calibrated, reference to case histories of similar field conditions can lend credibility to predictions.

Tests to determine the sensitivity of model results to changes in individual aquifer system hydraulic characteristics and boundaries and combinations thereof are often made during calibration. Sensitivity, geostatistical, and uncertainty methods for groundwater flow and mass transport modeling are presented by Buxton (1989). Sensitivity tests are influenced by such factors as model scale, relative boundary location, and calibration time period. Usually several sets of conceptual model modifications will fit field data because calibration is not a unique process. The designation of the best set of modifications is a matter of personal judgement concerning the likelihood or plausibility of final model ingredients in light of initial conceptual model decisions.

Some discretion is commonly used in selecting data against which to calibrate a model. For example, suppose the aquifer is stratified and there are head and concentration differences with depth. A two-dimensional model assumes that the head and concentration are uniform throughout the aquifer thickness. Thus, if three-dimensional field data are to be calibrated against calculations made with a two-dimensional model then head and concentration field data for wells terminating at different depths must be averaged before they can be used in the calibration process.

Field aquifer system data bases are commonly incomplete and uncertain. Major data deficiencies usually involve vertical hydraulic conductivity and well discharge rates in groundwater flow simulations and contaminant source strengths and the velocity field in contaminant migration simulations. Calibration should give adequate attention to initial conditions including long term transient antecedent trends which are often ill-defined.

Steady or transient state calibration are possible (Marsily, 1986, pp. 400-401). If steady state head data are available, steady state calibration should precede transient state calibration because it is not necessary to consider aquifer storativity during steady state simulation periods and one unknown drops out of model equations. Natural recharge, evapotranspiration, and discharge to surface bodies of water share importance with aquifer system hydraulic characteristics in the steady state calibration process.

A variety of groundwater models have been developed including: analog models- viscous fluid horizontal and vertical; electrical models-conductive liquid, conductive solid, resistance-capacitance networks, resistance networks; stretched membrane model; Moire pattern; thermal; blotting paper; resistance network-digital computer combination; electronic analog-digital computer combination; analytical calculator or digital computer; and numerical digital computer. Prickett (1975, pp. 1-143) and Todd (1980, pp. 384-408) discuss these model features and their applications in detail. This book is concerned only with analytical and numerical digital computer models.

Any given model is accurate for only a limited range of values for each model parameter. Most model authors have trouble enumerating all the caveats and bugs of which the user should be aware of when using the model. It is common to have instances of unrealistic results from extreme but legitimate data, convergence failure in iterative models due to oscillations with amplitudes greater than desired tolerances, and significant losses in water budgets due to accumulated round-off errors.

Analytical and numerical models simulating space- and time-dependent fluid flow, contaminant migration, heat transport, and deformation processes are structured to solve a partial differential equation, or a system of several partial differential equations, together with initial and boundary conditions. These basic equations (mathematical models) governing groundwater flow and mass transport are derived and defined in several textbooks including Bear (1979). Mathematical models may be deterministic, statistical, or a combination thereof. This book focuses on commonly used deterministic models.

Pinder (1988, pp. 119-134) presents an overview of the development of numerical groundwater modeling. Boutwell,et al. (1986) presents guidelines for developing numerical models of remedial measures at uncontrolled hazardous waste sites. A comprehensive survey of the development and application of numerical models is presented by van der Heijde, et al. (1985).

There are four types of numerical models: prediction models which simulate the behavior of an aquifer system and its response to stress, resource management models which integrate hydrologic prediction with explicit management decision procedures, identification models which determine input parameters for both of the above, and data manipulation codes which process and manage input data for the above models. Predictive models are the most numerous because they are the primary tools for testing hypotheses. Data manipulation codes received little attention until the recent development and application of preprocessors and postprocessors which simplify model data base input and facilitate production of model output graphic displays.

According to van der Heijde, et al. (1985), by 1983 about 399 numerical models developed in 19 countries had been surveyed. There were 203 flow models, 84 mass transport models, 22 heat transport models, 12 deformation models, 16 multi-purpose models, 33 management models, and 29 identification models. Water quantity flow models were operational and widely applied. Fractured rock flow models were rare and coupled groundwater -surface water flow models were scarce. There were some field applications of distributed water quality models and lumped water quality models were scarce. Some progress had been made in treating heterogeneous as well as multiphase and multicomponent transport phenomena, both linear and nonlinear. Developed management models were rarely utilized on a routine basis.

The transferability and flexibility of computer codes are often limited because of poor documentation thus restricting utilization of models. According to van der Heijde, et al. (1985, p. 42), proper documentation includes the following elements: a brief description of the model, providing information identifying the model, the author (or the person who provides model support), the organization where the model was developed, the date of completion of the first documentation, the version number or updates (if any), the programming language, the organization where the model may be obtained, and an abstract, or concise description, of the model; engineering documentation, including a description of the type of problems solved by the model, the basic theory and the method of solution, and its limitations and underlying assumptions; program documentation describing capabilities and limitations, different options available to users, and lists of input and output variables; system documentation containing information of the structure of the program, data structures of external files and the core storage required, a list of variables and subroutines, references to code listing, and a description of required computer hardware and communication equipment; and sample runs including both input and output files.

Aspects of quality assurance in groundwater modeling are presented by van der Heijde (1987, pp. 19-25). The International Ground Water Modeling Center has developed model review, verification, validation, and selection procedures to assist in quality assurance. Major criteria commonly used in selecting a model are that: the model is suited for the intended use, the model is thoroughly tested and validated for the intended use, and the model code and documentation are complete and user-friendly. The extensive groundwater model information retrieval system maintained and operated by the International Ground Water Modeling Center can assist modelers in the selection of a model. Guidance for the selection and use of models for evaluating the effectiveness of remedial actions at uncontrolled hazardous waste sites is presented by Boutwell, et al (1986).

Analytical microcomputer flow and mass transport programs are usually written in BASIC; numerical programs are usually written in FORTRAN; and graphics programs are usually written in BASIC, C, or Pascal. FORTRAN (Etter, 1987; Fuori, et al., 1986; Weinman and Kurshan, 1985) and BASIC (Microsoft QuickBASIC version 4.5 or later BASIC language reference, 1987) are the most utilized languages in the groundwater industry. BASIC programs are compiled with a variety of microcomputer compilers including QuickBASIC version 4.5 or later (trademark of Microsoft Corp.), Microsoft BASIC Compiler version 6.0 or later, and TURBO BA-

SIC version 1.1 or later (trademark of Borland International, Inc.) which allow numerical processing in the 8087/80287 math coprocessor environment. These compilers support structured programming and the calling of C and Assembly language routines.

Graphics programs may be speeded up by translating QuickBASIC into QuickC with the program B-TRAN (trademark of Software Translations, Inc., Newburyport, MA.). Available BASIC compilers and function and subprogram libraries (QuickPak Professional, QuickPak Scientific, and GraphPak-trademarks of Crescent Software, 11 Grandview Ave. Stamford, CT 06905) are reviewed in the PC Magazine, Vol. 8, No. 18, October 31, 1989, pp. 187-227.

FORTRAN microcomputer programs are compiled with a variety of compilers including the Microsoft FORTRAN compiler version 5.0 or later, Ryan-McFarland FORTRAN compiler version 2.4 or later (trademark of PC Brand), Lahey FORTRAN F77L-EM/32, F77L-EM/16, and F77L version 4.0 or later (trademark of Lahey Computer Systems, Inc.), and IBM Professional FORTRAN version 1.22 or later (trademark of International Business Machines Corp.). These compilers implement FORTRAN 77 (standard ANSI X309-1978), support Assembly and Pascal or C language routine calls, support 8087, 80287, 80387 math coprocessors, facilitate porting of mini- and mainframe computers, and offer IBM, VS, DEC VAX (trademark of Digital Equipment Corp.) extensions.

Some compilers break the DOS 640 K by addressing extended and/or virtual memory (see F77L-EM/16) and mainframe-sized programs with huge data arrays may be ported to a 386 PC running MS-DOS (see F77L-EM/32). A FORTRAN to C translator is distributed by PROMULA Development Corp., 3620 North High St., Suite 301, Columbus, Ohio 43214. A FORTRAN screen manager for creation of window, help, menu, and data entry screens is available from The West Chester Group, P.O. Box 1304, West Chester, Pennsylvania 19380. FORTRAN graphics and I/O libraries are distributed by IMPULSE Engineering, P.O. Box 190206, San Francisco, California 94119. Science/Engineering graphics FORTRAN-callable routines are available from Microcompatibles, 301 Prelude Drive, Dept. CL, Silver Spring, Maryland 20901.

Editors included with microcomputer compiler software are commonly used to write analytical and numerical groundwater programs. In addition, groundwater programs are developed with microcomputer commercial word processors such as WordPerfect (trademark of WordPerfect Corp.) which imports and exports ASCII files and places line numbers along the left edge of the text. WordPerfect supports automatic word wrap, search and replace, block commands, automatic rewrite, indexing, table of contents creation, outlining and paragraph numbering, mathematical functions, spelling checker, parallel column creation, thesaurus, sorting text or numbers, line drawing, and word search.

ANALYTICAL MODELS

Analytical models are most useful in the analysis of aquifer test data, uncomplicated aquifer system evaluation, and design and verification of numerical models (Walton, 1979; Clark, 1987; and Walton, 1989). The use of analytical models is usually justified when the groundwater flow field is fairly uniform and aquifer and facility feature geometries are fairly regular (Boutwell, et al., 1986, pp. 14-37). An inventory of available analytical models is presented by Walton (1979) and Boutwell, et al. (1986. pp. 79-183). A compilation of mass transport analytical models with computer source codes is presented by Wexler (1989).

Battelle Project Management Division, Performance Assessment Dept., Office of Nuclear Waste Isolation, 505 King Ave., Columbus, Ohio 43201 is custodian for a package of analytical models (VERTPAK-1) assembled to assist in the verification of numerical codes used to simulate fluid flow, rock deformation, and solute transport in fractured and unfractured porous media. Analytical modeling BASIC programs are listed by Walton (1989a).

The solution of analytical model equations commonly involves the method of complex variables (Strack, 1989, pp. 262-310) and the Laplace transform technique which often requires the application of integral transformation (Fourier, Hankel, etc.). A typical solution is in the form of an infinite series of algebraic terms, a double infinite series, or an infinite series of definite integrals. Only in simple aquifer systems can the solution be expressed in the form of a single analytical formula (Bear, 1979). Solution of analytical model equations also involves the more flexible Stehfest numerical method of inverting Laplace transform solutions Dougherty (1989, pp. 564-569) and Moench and Ogata (1984, pp. 146-170).

In analytical radial flow model development, any aquifer system boundaries and discontinuities, production and/or injection wells, and any boundary or discontinuity image wells are drawn to scale on a map. An area of interest covering the locations of these wells is selected for drawdown or recovery calculation. A grid is superposed over the area of interest, grid nodes are located at the intersections of grid lines, and grid lines are indexed. The origin of the grid (1,1) is usually placed at the upper left corner of the grid. Production, injection, image, or observation wells are located at grid nodes. The x and y coordinates of grid nodes are determined and the distances between production, injection, image, and observation well nodes are calculated using the Pythagorean equation.

Simulation period discharge and/or recharge rate schedules for production and/or injection wells and image wells are prepared. The time duration of the simulation period, aquifer system hydraulic characteristics, aquifer system geometric dimensions, and discharge and/or recharge rates are defined as the general data base. Appropriate analytical equations are chosen from those described in Chapters 2 through 4 of this book as approximately matching the general data base. These equations, the general data base, and the grid map constitute the analytical model. Equation well function variables are calculated based on the general data base, production well and/or injection well and image well discharge or recharge rates, and the distances between wells. Associated well function values are usually calculated using polynomial or other approximations. Individual production, injection, or image well drawdown and/or recovery impacts at nodes are calculated using appropriate analytical model equations and combined for the simulation period. Finally, nodal drawdown or recovery maps and observation well time-drawdown or time-recovery tables based on calculation results are displayed.

In addition to the image well theory, the principle of superposition and equivalent section, incremental, and successive approximation

techniques are utilized with analytical models to simulate complicated field conditions (Walton, 1984c, pp. 298-312). For example,
variable production well discharge rates are simulated by placing
several production and/or injection wells on top of one another at a
node with different discharge and/or recharge rates. Such features as
surface water bodies, mines, drains, waste disposal facilities, recharge basins, and constant flow boundaries are simulated by ar-
rangement of closely spaced wells on lines and in rectangles, circles,
and irregular areas.

Heterogeneous and anisotropic aquifer systems are simulated by
using average aquifer system hydraulic characteristics instead of
complex hydraulic characteristic variations in analytical models. For
example, an aquifer consisting of several horizontal layers, each
with different thicknesses and hydraulic conductivities, can be simulated approximately with an equivalent single-layer aquifer model.
The equivalent horizontal hydraulic conductivity of the single-layer
aquifer model is calculated as the sum of the products of individual
layer hydraulic conductivities and thicknesses divided by the total
aquifer thickness.

Analytical models require straight-line boundary and/or discontinuity demarcations and uniform width, length, and thickness. To
meet this this requirement, the variability of the areal extent of the
aquifer is converted to an equivalent uniform area. In addition,
boundaries and/or discontinuities are idealized to fit comparatively
elementary geometric forms such as aquifer wedges and infinite and
semi-infinite rectilinear strips.

Heterogeneous conditions are often simulated by varying aquifer
system hydraulic characteristics incrementally with time. Keeping
track of the the areal extend of the cone of depression with time, one
set of hydraulic characteristics is used to determine the drawdown at
the end of the first time increment and another set of hydraulic
characteristics is used to determine the change in drawdown between the first and second time increments, etc.

Flow to mines and drains with fixed drawdown are simulated with
the method of successive approximations. Drawdown due to first
trial discharges of the closely spaced wells simulating mine drainage
is calculated and compared with the required fixed drawdown. If the

comparison is favorable, the well discharges are declared valid; otherwise, well discharges are revised and a second trial drawdown is calculated and compared with the required fixed drawdown. This procedure is repeated until the comparison is favorable.

In analytical contaminant migration model development, contaminant point sources and other related features within the area of concern are drawn to scale on a map. A grid is superposed over the map, grid lines are indexed, and nodes are located at grid line intersections. The origin of the grid (1,1) is usually at the upper left corner of the grid. Monitoring wells are located at grid nodes not occupied by point sources which may also be located inside grid blocks. The x and y coordinates of grid nodes are determined and the distances between point sources and monitoring well nodes are calculated using the Pythagorean equation.

Simulation period mass injection rate schedules for point sources are prepared. The time duration of the simulation period, aquifer system hydraulic characteristics, aquifer system geometric dimensions, the uniform flow seepage velocity, and mass injection rates are defined as the general data base. Appropriate analytical equations are chosen from those described in Chapter 7 of this book as approximately matching the general data base. These equations, the general data base, and the grid map constitute the analytical model. Equation well function variables are calculated, if need be, based on the general data base, mass injection rates, and the distances between point sources and monitoring wells. Associated well function values are usually calculated using polynomial or other approximations. Individual point source impacts expressed as contaminant concentrations at nodes are calculated using appropriate analytical equations and combined for the simulation period. Finally, nodal concentration maps and monitoring well time-concentration tables based on calculation results are displayed.

Advection and dispersion from rectangular, circular, line, or irregular source areas are simulated with arrays of closely spaced point sources. The source area and associated contaminant mass are divided into equal parts and assigned to individual point sources. The cumulative impacts of individual point sources are then determined with the principle of superposition. Complicated continuous mass injection with variable rates is simulated by dividing the irregular continuous mass injection into several closely spaced slug injections. The cumulative impacts of individual slug injections are then determined with the principle of superposition.

In the late 1970's and early 1980's, Texas Instruments TI-59 and Hewlett-Packard HP-41C calculators (Garrison, 1982) and Radio Shack (trademark of Tandy Corporation) TRS-80 PC-1-4 pocket computers (Nowak, 1984) were commonly used to solve analytical model equations (Walton, 1984, pp. 298-312). Several HP-41C programs are available from the International Ground Water Modeling Center. Most available analytical model programs are written for use with microcomputers.

Radial flow analytical models usually describe drawdown or recovery distribution in uniformly porous aquifers and aquitards which are homogeneous, finite in areal extent, and of the same thickness throughout or in double-porosity fractured rocks. Isotrophy conditions are assumed to prevail as are isothermal conditions and no constant groundwater density and viscosity. Production and injection wells usually have finite diameters without wellbore storage. Both fully and partially penetrating wells are simulated with constant or variable production well discharge rates. Several analytical models simulating linear flow towards and from fully penetrating streams have been developed. Analytical models simulating vertical land subsidence due to artesian pressure decline are available.

Contaminant migration analytical models simulate one-, two-, and three- dimensional solute advection and dispersion from a slug or continuous source in a uniformly porous or fractured rock aquifer which is homogeneous, isotropic, infinite in areal extent, and of the same thickness throughout. Steady state flow with uniform regional flow and isothermal conditions are assumed to prevail. The density and viscosity of the contaminant are assumed to be the same as those of the native groundwater. Sorption and radioactive decay are simulated with analytical models. Fresh water injection in a nonleaky artesian aquifer with differences in density between the native groundwater and the injected water, upconing of salt water below a production well, and salt water intrusion also are simulated with analytical models.

Convection with or without dispersion from a heated-water injection well in uniformly porous nonleaky artesian aquifer systems which are homogeneous, isotropic, infinite in areal extent, and of the same thickness throughout are simulated with analytical models. Heat convection in the aquifer and heat conduction in the aquiclude are assumed to dominate heat flow conditions. Steady state uniform flow is assumed to prevail. The density and viscosity of the injected heated water are assumed to be the same as the those of the native groundwater.

NUMERICAL MODELS

Numerical flow models facilitate the simulation of irregular shaped aquifer system boundaries, the spatial variability of aquifer system characteristics, the nonuniformity of initial conditions, and the nonanalytic form of sources and sinks. A detailed classification of numerical flow models is presented by Mercer and Faust (1981). In numerical flow models, continuous (defined at every point in time and space) partial differential equations are approximated with a set of discrete equations in time and space. The aquifer area of interest is subdivided with a plan view grid into a large number of small discrete elements (usually rectangular in finite-difference methods and triangular in finite-element methods). The simulation time period is discretized into small time increments. An algebraic equation or set of equations is written for each element and time increment. This system of equations is solved using matrix methods and a digital computer.

All numerical models have a limited size and complexity solution domain. Consequently, more than one model may be required to simulate aquifer systems and facilities. Separate models for groundwater flow and contaminant migration are common. Models must communicate with one another via the transfer of data between models by hand or the use of external data management programs to indirectly link the models (Boutwell, et al., 1986, pp. 239-243).

Numerical models commonly use finite-difference or finite-element methods and the method of characteristics (Huyakorn and

Pinder, 1983). Collocation and boundary element methods are used less frequently. Strack (1989, pp. 589-605) describes the boundary integral equation method and its application. Some modelers prefer finite-element models because of their flexibility in representing irregular boundaries and in introducing irregularly spaced grids and their ease in representing tensorial concepts.

In the finite-difference flow model method (Kinzelbach, 1986, pp. 19-67), a set of algebraic water balance equations is generated based on the application of continuity and Darcy's law to the horizontal flow into and out of and the time rate of change of storage within aquifer or aquifer layer elements. Heads at finite-difference grid nodes or block centers represent average conditions over the entire aquifer or aquifer layer elements. Sources and sinks are simulated as vertical flows into or out of selected aquifer or aquifer layer elements. Boundaries are simulated by placing the active grid edge along barrier boundaries, prescribing constant heads along equipotential boundaries, and adding vertical sources or sinks along prescribed flux boundaries.

In developing a finite-difference flow model, maps are prepared with a common scale showing aquifer system hydraulic characteristic variation, aquifer system geometric dimensions including boundaries, and the location of any facilities such as production wells, injection wells, mines, and drains. A finite-difference grid is superposed over and fitted as closely as possible to the maps. Grid lines are indexed and nodes are located at grid line intersections or in the center of blocks between grid lines. Some simplification and idealization of actual conditions is required in fitting the grid to the maps. The simulation time period is divided into several time increments and a step discharge/recharge time-rate schedule for each facility is superposed over and fitted as closely as possible to the time increment schedule. The finite-difference model data base is then prepared based on the grid and discharge/recharge time-rate schedules and information concerning nodal aquifer system hydraulic characteristics and dimensions.

In the finite-element method (Kinzelbach, 1987, pp. 91-133), an integral representation of the mathematical model is derived usually with the method of weighted residuals or the variation method.

Heads at every point within the aquifer or aquifer layer element are approximated in terms of integral interpolation functions which can be approximated by a low-order linear, quadratic, or cubic polynomial. Time discretization is usually the same as in finite-difference methods. The integral flow relationship is expressed for each aquifer or aquifer layer element as a function of the coordinates of all node points of the aquifer or aquifer layer element. Values of the integrals are calculated for each aquifer or aquifer layer element and combined including boundary conditions as a system of first-order linear differential equations in time. Finite-difference techniques (Remson, et al. ,1965) are then employed to produce a set of algebraic equations.

Direct and iterative methods are used to numerically solve algebraic matrix equations (Kinzelbach, 1986; Huyakorn and Pinder, 1983; and Wang and Anderson, 1982). A sequence of operations is performed only once in direct methods, whereas, the process of successive approximations starting with an initial estimate of the matrix solution is employed in iterative methods. Commonly used iterative methods in finite-difference flow models are the successive overrelaxation procedure, alternating direction implicit procedure, iterative alternating direction implicit procedure, and the strongly implicit procedure. Among the widely used algorithms in finite-element models are the Gauss elimination and Choleski decomposition procedures (Huyakorn and Pinder, 1983, p. 26).

The method of characteristics (Reddell and Sunada, 1970) is widely used in mass transport models to avoid numerical dispersion or numerical oscillation with conventional finite-difference and finite-element contaminant migration simulation approaches (Kinzelbach, 1986, pp. 258-297). An equivalent system of ordinary differential equations are obtained by rewriting the transport equation using fluid particles as the point of reference. In two-dimensional transport with advection and dispersion, three equations for x-velocity, y-velocity, and concentration are generated knowing the average pore velocity distribution from flow model results. A set of moving particles (points) is traced within the stationary coordinates of a finite-difference grid. Particles placed in each grid aquifer or aquifer layer element are allowed to move a distance along streamlines proportional to the simulation time increment, the velocity at the point, and the dispersion coefficient.

Different particles having different concentrations enter and leave blocks to simulate advective transport. The remaining parts of the transport equation are then solved with finite-difference approximation methods (Kinzelbach, 1986, pp. 258-274) wherein dispersive fluxes are estimated with aquifer or aquifer layer element mass balance calculations, dispersion coefficients, grid spacings, aquifer or aquifer layer thicknesses, effective porosity, the time increment, and the concentrations of adjacent aquifer or aquifer layer elements. Sources and sinks are simulated by adding and destroying particles. Linear adsorption is simulated by dividing the average pore velocities, aquifer or aquifer layer dispersivities, and injected masses by the retardation factor. Radioactive decay is simulated by assigning every particle a mass which decreases in time. A review of the limitations and perspective on mass transport modeling is presented by Konikow (1988, pp. 643-662).

Another widely used technique for simulating mass transport is the random walk method (Kinzelbach, 1986, pp. 298-315) which tracks many individual particle paths (random walks). A large number of particles, each assigned the same fixed contaminant mass, are moved in the numerical flow model field. Movement has two superposed components: one component represents advective movement and is proportional to the simulation time increment and the average pore velocity at a point, and the second component represents dispersion and is a random movement with statistical properties that correspond to the properties of dispersion. Sources and sinks are simulated by adding and destroying particles. Linear adsorption is simulated by dividing the average pore velocities, aquifer or aquifer layer dispersivities, and injected masses by the retardation factor. Radioactive decay is simulated by assigning every particle a mass which decreases in time.

In developing a random walk mass transport model, maps are prepared with a common scale showing aquifer system hydraulic characteristic variation, aquifer system geometric dimensions including boundaries, head distribution during the simulation period, and locations of sources and sinks. A finite-difference grid is superposed over and fitted as closely as possible to the maps, grid lines are indexed, and nodes are located at grid line intersections. Some simplification and idealization of actual conditions is required in fitting the grid to the maps. Nodes may represent sources, sinks, or monitoring wells. Sources and sinks may be located within grid blocks. The random walk mass transport model data base is then prepared based on the grid and information concerning nodal aquifer system hydraulic characteristics, geometric dimensions, and head distribution.

MAINFRAME OR MINICOMPUTER SOFTWARE

FORTRAN numerical model mainframe or minicomputer groundwater flow and mass transport programs which are well documented, well tested, available at moderate cost, and in the public domain include MODFLOW, MOC, MOCNRC, SUTRA, MOCDENSE, USGS-2D-FLOW, HST3D, PLASM, and RANDOM WALK. These programs are distributed by the International Ground Water Modeling Center. The FORTRAN codes are implemented on a DEC VAX 11/780. Copies of the software are provided on a magnetic tape in user-specified formats. Most of these programs are supported by sets of pre- and post-processor programs which prepare input files, display both input and output results, and create data files for graphics software.

The input to these flow and mass transport programs without pre-processors is imported from formatted data files created by the modeler with an editor. Output from these flow and mass transport programs without post-processors is usually in simple tabular form and may require conversion by the modeler. For these reasons, model users must be thoroughly familiar with data types, data structures, data files, and file input/output considerations (Wagener, 1980, pp. 249-351).

One of two conventions for defining the configuration of aquifer system prisms with respect to the location of finite-difference grid nodes are used by most of these programs: the block-centered node system in which nodes are located at the center of blocks between grid lines and the point-centered node system in which nodes are located at grid line intersections. Points along a row are parallel to the x-axis, points along a column are parallel to the y-axis, and points along the vertical are parallel to the z-axis. The origin of the grid system (1,1,1) is the upper-left corner of the top most layer. The column index may be either i or j, the row index may be either i or j, and the layer index is k.

MODFLOW was developed by McDonald and Harbaugh (1984). It is a finite-difference block centered program which simulates quasi-three-dimensional constant density flow in multiple aquifer units (aquifer and adjacent aquitard) and/or aquifer layers under nonleaky artesian, leaky artesian, water table, or combined artesian and water table conditions. The aquifers or aquifer layers may be heterogeneous and anisotropic and have irregular boundaries. The following features are simulated: production well discharge or injection, drain discharge, areal recharge, evapotranspiration, and induced streambed infiltration. Finite-difference equations can be solved using either the strongly implicit procedure or the slice-successive overrelaxation procedure. The program calculates water level changes in time and space.

MOC was developed by Konikow and Bredehoeft (1978). It is a two-dimensional, finite-difference program that simulates contaminant migration in flowing groundwater caused by the processes of advective transport, hydrodynamic dispersion, and dilution from fluid sources. The program assumes that the solute is nonreactive and gradients of fluid density, viscosity, and temperature do not affect the velocity distribution. The aquifer or aquifer layer may be heterogeneous and/or anisotropic. A flow model USGS-2D-FLOW is coupled with a solute transport model which is based on the method of characteristics. The model uses a particle tracking procedure to simulate advection and a two-step explicit procedure to solve a finite-difference equation describing dispersion, sources and sinks, and divergence of velocity. The program calculates changes in contaminant concentration in time and space.

MOCNRC (Tracy, 1982) is a modified version of MOC to include radioactive decay and adsorption (linear, Langmuir and Freundich isotherm). MOCDENSE was developed by Sanford and Konikow (1985). It is a modified version of MOC that simulates solute transport of either one or two contaminants where there is two-dimensional, variable density flow. The code is commonly used to simulate salt water intrusion.

SUTRA was developed by Voss (1984). It is a two-dimensional hybrid finite-element and integrated-finite-difference program . SUTRA flow simulation is employed for areal and cross-sectional modeling of saturated aquifer systems, and for cross-sectional modeling of the unsaturated zone flow. Natural or man-induced chemical species transport including processes of solute sorption, production, and decay are simulated. In addition, variable density leachate movement, salt-water intrusion in aquifers with either dispersed or relatively sharp transition zones between fresh and salt waters, thermal regimes in aquifers, subsurface heat conduction, aquifer thermal energy storage systems, geothermal reservoirs, thermal pollution of aquifers, and natural hydrogeologic convective systems are simulated. Leakage through aquitards or multiple aquifer unit or aquifer layer conditions (quasi-three-dimensional flow) and simultaneous energy and solute transport are not simulated.

In SUTRA, hydraulic conductivities may be anisotropic and may vary both in direction and magnitude throughout the aquifer system as may most other aquifer and fluid properties. Non-equilibrium (kinetic) sorption models as well as the Langmuir and Freundlich equilibrium sorption models are simulated. Flow-direction-dependent longitudinal dispersion also is simulated. Boundary conditions, sources, and sinks may be time dependent. Options are available to print fluid velocities in the system, to print fluid mass and solute mass or energy budgets for the system, and to make temporal observations at points in the system.

SUTRA-PLOT was developed by Souza (1987). It is a graphical display program for SUTRA and features plots of the finite-element mesh; velocity vector plots; and contour plots of pressure and solute concentration, temperature or saturation. The program has a finite-element interpolator for gridding data prior to contouring. SUTRA-PLOT is written in FORTRAN 77 and runs on a PRIME 750 computer system. It requires version 9.0 or higher of the DISSPLA graphics library (computer Associates International, San Diego, CA.) to be present on the computer system used.

USGS-2D-FLOW was developed by Trescott, et al. (1976). It

simulates two-dimensional constant density flow in an artesian aquifer, a water table aquifer, or a combined artesian and water table aquifer. The aquifer may be heterogeneous and anisotropic and have irregular boundaries. The source term in the flow equation may include well discharge, constant areal recharge, leakage from aquitards in which the effects of storage are considered, induced streambed infiltration, and evapotranspiration. The finite-difference model features three numerical techniques: the strongly implicit procedure, the iterative alternating direction implicit procedure, and the line successive overrelaxation procedure. A USGS 3-D FLOW model modified to simulate variable water density and multiaquifer wells was developed by Kontis and Mandle (1988).

HST3D was developed by Kipp (1987). It couples three-dimensional variable density flow, solute transport, and heat transport equations to simulate subsurface waste injection, landfill leaching, saltwater intrusion, freshwater recharge and recovery, radioactive waste disposal, geothermal systems, or subsurface energy storage. The solute transport equation that is solved assumes a single solute species with possible linear adsorption and decay. The space discretization of the governing equation is based on node centered finite differences. Either a backward differencing or a centered-in-time time discretization may be used. Optional equation solution methods are: the alternating diagonal direct-equation solver method or the two-line successive-overrelaxation method. :

PLASM was developed by Prickett and Lonnquist (1971). It simulates one-, two-, and quasi-, three-dimensional nonsteady·state constant density flow in heterogeneous and anisotrophic aquifers under water table, nonleaky artesian, and leaky artesian conditions with irregular boundaries. Features simulated include: time varying production well discharge or injection, leakage through aquitards, induced streambed infiltration, areal recharge, evapotranspiration, storativity conversion, water table conditions, and multi-unit aquifer systems. A finite-difference modified alternating direction implicit method is used to generate head distribution in time and space.

RANDOM WALK was developed by Prickett, et al. (1981). It couples a two-dimensional finite-difference constant density flow model PLASM with a random walk constant density solute transport model. The program allows specification of chemical constituent concentrations of any segment of the model including: injection of contaminants by wells, vertically averaged saltwater fronts, leachate from landfills, leakage from overlying source beds, surface water sources such as contaminated streams and lakes, and well and surface water sinks. The solute transport portion of the code is based on a particle-in-a-cell technique for advection, and a random-walk technique for dispersion.

Other commonly used mainframe or minicomputer numerical model programs are: MOCMOD84, DEWATER, FTWORK, GEOTHER, COMPAC, FE3DGW, CFEST, SWIP, FEMWATER, FEMWASTE, SWSOR, SWIFT, SWIFT II, SEFTRAN, SATURN, AQUIFEM-1, ARMOS, FEDAR, and BIOPLUME II.

MOCMOD84 (Goode, et al., 1986) was developed for the Nuclear Regulatory Commission. It is an extended and updated version of MOC simulating retardation due to sorption and radioactive decay.

DEWATER was developed by the Analytical and Computational Research, Inc., Los Angeles. It is an integrated finite-difference program simulating pumping during surface and subsurface mining and construction operations. The modeled artesian or water table aquifer system may be anisotropic and heterogeneous.

FTWORK, (Sims, et al., 1989, pp. 821-841) developed by GeoTrans, Inc., is a three-dimensional, finite-difference model simulating constant density flow and solute transport processes in fully saturated porous media. The program is highly compatible with MODFLOW making transfer of input data between the two codes relatively easy. An efficient matrix solution technique is used based on Slice Successive Over-Relaxation. Transport calculations are obtained by incorporating dispersion cross-product terms. The Gauss-Newton technique for nonlinear least-squares parameter estimation is incorporated into the solution of the flow equation for automatic history matching. The complexities of more comprehensive codes such as HST3D (variable density, heat transport, deformation, etc.) are not considered. Solute processes considered include: dispersion (cross product terms), a linear equilibrium adsorption isotherm, and first order decay.

GEOTHER, developed by Faust and Mercer (1981, 1983), is a finite-difference model simulating the three-dimensional, single- and two-phase heat transport in groundwater. It simulates anisotropic and heterogeneous multi-layed porous medium under artesian conditions with two-phase mixtures (steam and water). Convection, conduction, condensation, and dispersion are coupled with flow. GEOTHER/VT4 is an improved version of GEOTHER developed at the Pacific Northwest Laboratory and is suitable for high-level radioactive waste repository applications.

COMPAC was developed by the U.S. Geological Survey and simulates aquifer system compaction due to changes in artesian pressure. One-dimensional vertical elastic and nonelastic deformation are coupled with flow in the finite-difference program.

FE3DGW, developed by Gupta, et al. (1984), is a three-dimensional, finite-element flow program. It simulates anisotropic and heterogeneous multi-layered porous media with compressible fluid and media.

CFEST, developed by Gupta, et al. (1981), is a three-dimensional, finite-element coupled fluid, energy, and solute transport program prepared for the U.S. Department of Energy. The Office of Nuclear Waste Isolation at Battelle Memorial Institute, Columbus, Ohio is custodian of the program. The program is supported by a set of pre- and post-processor programs to prepare input files for complex layered system, couple regional and local models, summarize fluxes at the exit boundaries, and analyze streamlines, pathlines and travel times. CFEST was used by Plomb (1989, pp. 955-978) to simulate drawdown caused by infiltration into a 32-foot diameter tunnel.

SWIP (Intercomp, 1976) is a three-dimensional, finite-difference model for simulating the transport and dispersal of heat and dissolved material. The model takes into account the effects of differences in groundwater density and viscosity on groundwater flow and has been used in modeling heat storage.

FEMWATER, developed by Yey and Ward (1980), is a finite-element model of flow through saturated-unsaturated porous media. FEMWASTE, developed by Yey and Ward (1981), is a finite-element model of waste transport through porous media. These two programs were developed by the Oak Ridge National Laboratory.

SWSOR was developed by the U.S. Geological Survey and simulates the head distribution and position of the fresh/salt-water interface. It is a two-dimensional finite-difference program.

SWIFT (Reeves and Cranwell, 1981; Finley and Reeves, 1982), developed by Sandia National Labortories, is a three-dimensional, finite-difference program simulating fluid flow, heat transport, and transport of trace or dominant (variable density) species. SWIFT II (Reeves, et al., 1986a,b,c) is a finite-difference model of flow and mass transport in a fractured rock (double porosity). SEFTRAN (Huyakorn, et al. 1984) is a finite-element flow and mass transport model and SATURN (Huyakorn, et al., 1984) is a model which simulates variable saturated flow conditions. AQUIFEM-1, developed by Townley and Wilson (1979) at Massachusetts Institute of Technology, is a two-dimensional flow model. ARMOS (Parker and Kaluarachchi, 1989, pp. 271-281), developed at the Virginia Polytechic Institute and State University, is a finite-element program . It simulates simultaneous flow of water and light hydrocarbon in an areal flow region of an unconfined aquifer. FEDAR (see Brody, 1989), developed at the University of California, is a finite-element program which simulates two-dimensional transport in a vertical plane of an immiscible lighter than water nonaqueous phase (NAPL) contaminant floating on the water table. BIOPLUME II, developed by Rifai, et al. (1988), is a two-dimensional model for simulating contaminant transport affected by oxygen limited biodegradation.

Kuiper (1986, pp. 705-714) describes methods for the solution of the inverse problem in two-dimensional flow mainframe or minicomputer models. Saltwater intrusion mainframe or minicomputer numerical models are documented by Segol, et al. (1975), Mercer, et al. (1980), and Sapik (1988). A mainframe finite-element model for simulating three-dimensional flow and solute transport in oil shale (dual-porosity) and associated hydrogeologic units is described by Glover (1987).

Nuclear waste disposal numerical mainframe models are described by Grove, et al. (1980), Onishi, et al. (1981), SAI (1981), and Thomas et al. (1982). A mainframe numerical model for simulating radionuclide transport through fractured media is described by Longsine, et al. (1987). A mainframe numerical model for simulating land subsidence due to pumpage from fully and partially penetrating wells is documented by Safai and Pinder (1977). WATEQF, BALANCE, and PHREEQE mainframe numerical programs for simulating geochemical reactions in regional flow systems are reviewed by Plummer (1984).

Limited attention has been given to the collection of the field data required by complicated and expensive variable density three-dimensional mainframe or minicomputer numerical models. Examples of field verification of the predictions made with these models are rare. In some cases, the use of averaged quasi-three dimensional or two-dimensional models instead of three-dimensional models may be justified based on available limited databases and the lack of experience with three-dimensional models.

MICROCOMPUTER SOFTWARE

Microcomputer analytical and numerical programs are marketed by the National Water Well Association and distributed by the International Ground Water Modeling Center. In addition, a variety of public domain and propriety microcomputer groundwater flow and mass transport programs are available from a number of vendors. A partial list of microcomputer program vendors is provided in Appendix J.

The Environmental Software Report (Donley Technology, Box 335, Department E9, Garrisonville, VA 22463) contains information on specialized microcomputer software packages, on line systems, and ways to use generic data base and spreadsheet software for environmental management. This newsletter (eight issues per year) brings a cross-section of computer application reviews and analysis of related issues. Applications include groundwater contamination.

A hands-on short course titled "IBM PC Applications In Ground Water Pollution And Hydrology", sponsored by The Association Of Ground Water Scientists And Engineers of the National Water Well Association, is designed for the professional with little or no previous microcomputer experience. Course topics include: PC/MS-DOS commands, BASIC and FORTRAN compilers, word processors, graphics packages, computer aided drafting, and creating PC electronic slides. Popular IBM PC hardware and software that has application in groundwater resource evaluation is comprehensively reviewed during the short course.

IBM PC BASIC (trademarks of International Business Machines Corporation) microcomputer analytical flow and mass transport programs which are well documented, well tested, and available from the International Ground Water Modeling Center at moderate cost include: AGU-10, SOLUTE, THWELLS, PUMPTEST, TGUESS, OPTP/PTEST, TIMELAG, THCVFIT, TSSLEAK, PLUME2D, PLUME, and CATTI. These programs are written for the IBM PC/XT/AT or compatibles and require the IBM Color Graphics Adapter or compatible graphic capability.

AGU-10 is a program package based on a series of semi-analytical mass transport models documented by Javandel (1984). SOLUTE (Beljin, 1985) is a program package of ten analytical models for solute transport in groundwater, THWELLS is a flow model for simulating uniformly porous homogeneous nonleaky artesian aquifers with boundaries and multiple wells. Results are displayed in tabular form, as time-drawdown curves, and as contour plots. PUMPTEST calculates aquifer transmissivity and storativity from time-drawdown, distance-drawdown, or recovery aquifer test data. TGUESS estimates aquifer transmissivity from production well specific capacity data. OPTP/PTEST determines optimum production well discharge using data from a step-drawdown test.

TIMELAG estimates aquifer hydraulic conductivity from time-lag test data using the Hvorslev slug method. THCVFIT is an interactive program for determining aquifer transmissivity and storativity from aquifer test data. There is a screen graphic rendition of the nonleaky artesian well function and the field drawdown data. The program has an option to read from and write to an external file. TSSLEAK fits the leaky artesian equation to aquifer test data and calculates values of aquifer transmissivity and storativity and aquitard vertical hydraulic conductivity by using the least-squares procedure.

PLUME2D calculates two-dimensional contaminant concentration distribution in a uniformly porous homogeneous nonleaky artesian aquifer with one-dimensional uniform flow field. The program simulates advection and dispersion from fully penetrating sources with optional sorption and/or radioactive decay. PLUME calculates three-dimensional contaminant concentration distribution in a uniformly porous homogeneous aquifer with continuous solute injection in a one-dimensional uniform flow field. CATTI is a program for tracer test interpretation. It calculates breakthrough curves with continuous or slug contaminant sources and is capable of automatic parameter identification.

A group of analytical/numerical models for groundwater flow and mass transport (Princeton Analyticals), developed by Princeton University, is available through the National Water Well Association. A Gaussian strip analytical model for evaluating the movement and impact of hazardous waste leachate migration and one- and two-dimensional groundwater transport models MYGRT 1 and 2 have been developed by M.J. Ungs at McLaren Engineering, Alameda, California. Analytical aquifer and slug test programs (AQTESOLV) and an interactive graphical program for creative design of a finite-difference flow model dataset (MODEL-CAD) have been developed by the Geraghty and Miller Modeling Group.

A variety of programs have been developed by TechMac for the Macintosh (trademark of Apple Computing Company) family of microcomputers. Models range from analytical to three-dimensional, finite-difference codes simulating flow, mass transport, and saltwater intrusion. A gridded network contour program allows "pasting" into the multitude of paint/draw programs. "Pull-down menus', "radio buttons", and "multiple windows" are supported. MacMODFLOW, MacSUTRA, MODGRAF, MOCGRAF, and PC SUTRA are distributed by the Scientific Software Group.

Turbo Pascal analytical programs which may be used to simulate the effects of production well discharge on water levels under nonleaky artesian, leaky artesian, and water table conditions; analyze aquifer test data; and generate screen and plotter graphics are presented by Clark (1988).

FORTRAN microcomputer numerical flow and mass transport programs which are well documented, well tested, available at moderate cost, and in the public domain include: MODFLOW, MOC, and SUTRA. These programs are IBM PC (trademark of

International Business Machines Corporation) or compatible versions of mainframe numerical programs developed by the U.S. Geological Survey. They are available from the International Ground Water Modeling Center and the Scientific Software Group. The input structure of these programs is designed to permit the data base to be gathered, as it is needed, from many different stored files. The user must provide a connection between a unit number and the names of files by use of job control statements.

Other popular FORTRAN numerical microcomputer programs include: PLASM, RANDOM WALK, and TRAFRAP-WT. PLASM and RANDOM WALK are microcomputer versions of mainframe programs developed by the Illinois State Water Survey. TRAFRAP-WT is a two-dimensional finite-element program for simulating flow and solute transport in single or dual-porosity media with or without discrete fractures. It can be used for areal or cross-sectional analysis of artesian or water table aquifer systems.

A package of IBM PC BASIC (trademarks of International Business Machines Corporation) finite-difference and finite-element programs documented by Bear and Verruijt (1987) is distributed by the International Ground Water Modeling Center under the name BAS23. These programs simulate steady and unsteady two-dimensional flow in nonhomogeneous aquifers, flow through dams, transport of contaminants by advection and dispersion, and saltwater intrusion. Steady-state and unsteady-state flow under nonleaky artesian, leaky artesian, and water table conditions is simulated by Aral (1990a, 1990b) in two finite-element IBM PC compatible programs.

Interactive textural and/or graphic pre-processors (see Van der Hyde and Srinivasan, 1983) to create, edit, display, and store on diskettes appropriate data base input files for popular models including MODFLOW, MOC, and SUTRA are commercially available. Pre-processor packages may have spread sheet capabilities, support a range of printers and plotters, and graphically display input data on the screen with or without a zoom utility (see PREPRO3FLOW developed by GeoTrans, Inc. and MODELCAD developed by Geraghty and Miller Modeling Group). In MODELCAD, aquifer characteristics, boundary conditions, and the model grid are defined graphically on the screen. A digitized base map may be displayed over the model grid. The graphic design may be translated into data base input files compatible with MODFLOW, PLASM, MOC, and RANDOM WALK.

The output structures of popular models including MODFLOW, MOC, and SUTRA are designed for printers and/or diskette storage. Post-processors are available to display output on the screen and provide graphics capability for programs.

Pre-processor and post-processor distributors include the International Ground Water Modeling Center; Scientific Software Group; Earthware; Geraghty and Miller Modeling Group; GeoTrans, Inc.; and Hall Groundwater Consultants, Inc.

Widely used proprietary microcomputer programs include: three-dimensional PLASM, three-dimensional RANDOM WALK, INTERSAT, INTERTRANS, SEFTRAN-PC, SWANFLOW, AQUA, PTC, FRACFLOW, HST3D and ARMOS.

Three-dimensional PLASM, developed and distributed by T.A. Prickett & Associates, is a three-dimensional, finite-difference node centered flow model; three-dimensional RANDOM WALK, developed and distributed by T.A. Prickett & Associates, is a three-dimensional random walk solute transport model; INTERSAT, developed by M.L. Voorhees and distributed by the Scientific Software Group, is a three dimensional finite-difference model similar to three-dimensional PLASM and integrated with a pre-processor and a post-processor; INTERTRANS, developed by M.L. Voorhees and distributed by Scientific Software Group, is a three-dimensional random walk solute transport model similar to three-dimensional RANDOM WALK and integrated with a pre-processor and a post-processor; SEFTRAN-PC, developed and distributed by Geotrans, Inc., is a two-dimensional finite-element flow and mass transport model; SWANFLOW, developed and distributed by GeoTrans, Inc., is a two-dimensional finite-difference immiscible flow model for water and a nonaqueous phase liquid (NAPL) in and below the unsaturated zone ; AQUA, developed and distributed by VATNAKSIL Consulting Engineers, solves two-dimensional groundwater flow and mass transport equations using the Galerkin finite-element method; PTC, developed and distributed by Princeton

University, is a combination of finite element-Galerkin techniques and a finite difference scheme for simulating three-dimensional mass transport; FRACFLOW, developed and distributed by GeoTrans, Inc., is a finite-element model simulating two-dimensional (areal) flow in fractured and karst aquifers based on a double porosity approach; ARMOS is a finite-element model simulating areal migration of separate phase hydrocarbon in unconfined aquifers distributed by Environmental Systems & Technologies, Inc.; PREARM is an interactive program to create input data files for ARMOS; and POSTARM is an interactive graphical post-processor for ARMOS.

Microcomputer software marketed by the National Water Well Association includes the following programs: GWAP, SUTRA, PC SUTRA PLOT, FASTEP, WELLCOST, HJ-MATCH, PAPADOP, PTDPS I, PTSPS II, STEP-MATCH, WHIP, FINITE, FLOWPATH, PORFLOW-2D, PORFLOW 3-D VADOSE, CAPTURE, FRACQUAL, GRIDZO, KON, and LOGGER.

GWAP, developed by Groundwater Graphics, is an aquifer test analysis package containing the standard graphical curve matching technique; SUTRA, developed by Hunter/HydroSoft Inc., is a preprocessor which allows quick and easy cursor driven data file input for the program SUTRA; PC SUTRA PLOT, developed by Hunter/ HydroSoft Inc, is a menu-driven graphics post-processing package for use with the program SUTRA; FASTEP, developed by Ulrich & Associates, analyzes step-drawdown test data ; WELLCOST, developed by Ulrich & Associates, analyzes the costs of a production well; HJ-MATCH, developed by In-Situ Inc., matches type curves to leaky artesian aquifer test data; PAPADOP, developed by In-Situ Inc., analyzes aquifer test data taking into account directional transmissivities; PTDPS I and PTDPS II, developed by Irrisco, analyzes confined and unconfined aquifer test data; STEP-MATCH, developed by Insitu Inc., automates the processing and analysis of data from slug tests; WHIP, developed by Hydro Geo Chem Inc., analyzes complex aquifer tests with time variable discharge rates and multiple production wells; FINITE, developed by Koch and Associates, simulates inflow to mines and dewatering schemes with an analytical finite length sink model; FLOWPATH, developed by

Waterloo Hydrogeologic, analyzes capture zones for wellhead protection; PORFLOW-2D and PORFLOW-3D, developed by Analytic & Computational Research Inc., simulate fluid flow, heat, salinity, and chemical species transport in porous or fractured media; VADOSE, developed by Insitu Inc., simulates unsaturated flow in a porous media; CAPTURE, developed by Data Services, plots flow paths and areal extent of capture zones; FRACQUAL, developed by Koch & Associates, simulates mass transport in a single fracture; GRIDZO, developed by RockWare Inc., grids and contours surface data on a customized basis; KON, developed by Data Services, plots contours of water elevation and chemical concentration; and LOGGER, developed by RockWare Inc., plots well logs.

Many microcomputer software libraries include at least the following programs: MODFLOW, MOC, SUTRA, GWPATH, GWAP, THWELLS, AGU-10, Microsoft QuickBASIC version 4.5 or later compiler, and Microsoft FORTRAN 77 version 4.0 or later compiler. Hardware recommendations or requirements of these programs are: IBM AT (trademark of International Business Machines Corporation) or compatible 286-based PC w/40 meg hard drive/5 1/4" 360 Kb diskette drive/640 Kb RAM/math co-processor/1 parallel port/2 serial ports/EGA color monitor and graphic card/IBM Graphics Printer (trademark of International Business Machines Corporation)/Hewlett-Packard HP-7475A plotter.

GRAPHICS

Most computer programs have statements or post-processors that convert numerical calculation results into tabular screen and/or printer displays. In addition, calculation results are usually stored in a disk file for later use. The tabular numeric output is sufficient for many analyses, however, graphics catch the eye, make comparisons, and convey information far more quickly and effectively than can a series of numbers. For this reason, graphics programs are frequently used in the groundwater industry.

BASIC language graphics program theory is presented by Hearn and Baker (1983) and Goldstein (1984, pp. 9-70). Screen graphics microcomputer programs use special graphics mode commands to turn on or off or change the color of video display "pixels" (individual dots of light) in point and line patterns representing pictures. Images in any rectangular area of the screen or entire screen images can be saved on a diskette and recalled with appropriate statements. The contents of the screen can be printed on paper (hard copy) at the keyboard by pressing keys dedicated to achieve a screen dump or by using a graphics screen dump subprogram (Goldstein, 1984, pp. 243-247). Usually hard copies are made with a IBM Graphics Printer (trademark of International Business Machines, Inc.) or compatible dot-matrix printer which are "raster" devices and draw by connecting thousands of dots in a matrix (Sandler, 1984, pp. 79-88).

A plotter is commonly used to produce higher quality graphs than those produced by dot-matrix printers. Graphic commands which move the pen in an up or down position to desired locations anywhere on a page are sent to a plotter via a communications file opened and closed with input/output (I/O) statements and functions. Most plotters are "vector devices" which draw by connecting two designated points with their own built-in ROM graphics languages (see Sandler, 1984, pp. 59-77). The ROM contains a set of graphics command subroutines called for with I/O statements in high-level languages including BASIC, FORTRAN, PASCAL, and C. The Hewlett-Packard Graphics Language (HP-GL) is a two-letter-mnemonic graphics language (Hewlett-Packard Interfacing and Programming Manual) understood by Hewlett-Packard plotters which is commonly used in the groundwater industry. A BASIC program for drawing a semilogarithmic time-drawdown graph with a Hewlett-Packard 7470A plotter and HP-GL is listed by Walton (1984, Third Edition, pp. 486-489).

The trend graph displays a single data series over a period of time or space interval (see BASIC program listed by Walton, 1989a, pp. 111-132). In trend graph microcomputer programs (Hearn and Baker, 1983, pp. 54-69; Korites, 1982, LIN/LIN, LOG/LIN, LOG/ LOG, AND LINFIT) axes with either arithmetic or logarithmic scales are drawn together with scale interval "ticks" and labels. Data points are identified and lines or curves are drawn between or through points. Axes and the graph are provided with appropriate annotations. A BASIC program which draws a screen graphics semi-logarithmic time-drawdown graph is listed by Walton (1989a, pp. 113-132). A linear XY graph BASIC program is listed by Hearn and Baker (1983, p. 62). Linear, semi-log, and log-log XY graph BASIC programs are listed by Korites (1982, pp. Lin/Lin 6-14, Log/Lin 3-12, and Log/Log 2-8).

It is often desirable to find the best-fit trend line through aquifer test time-drawdown semilog data points so that values of aquifer transmissivity and storativity may be determined with predicted values of drawdown along the line. The regression least-squares method employs differential calculus to arrive at the slope and the y-intercept of the best-fit line using the following equations (Davis, 1986, p. 180):

$$c_1 = [\sum_{i=1}^{n} X_i Y_i - (\sum_{i=1}^{n} X_i \sum_{i=1}^{n} Y_i) / n] / \tag{9.1}$$

$$[\sum_{i=1}^{n} X^2_i - (\sum_{i=1}^{n} X_i)^2 / n]$$

$$\tag{9.2}$$

$$c_0 = \sum_{i=1}^{n} Y_i / n - c_i (\sum_{i=1}^{n} X_i) / n$$

where c_1 is the slope of the best-fit line, c_o is the zero drawdown-intercept of the best-fit line, Y_i is the specified value of drawdown, X_i is the specified value of time, and n is the number of specified X_i and Y_i values.

Values of drawdown for specified times along the best-fit line may be calculated knowing c_1 and c_0 with the following equation of a line (Davis, 1986, p. 179):

$$Y_t = c_o + c_1 X_t \tag{9.3}$$

where Y_t is the calculated value of drawdown along the best-fit line and X_t is the specified value of time along the best-fit line.

Data for any two points on the best-fit line may be substituted in the following equation to calculate aquifer transmissivity (Cooper and Jacob, 1946, pp. 526-534):

$$T = 2.30 Q \log(t_2/t_1)/[4\pi(s_2 - s_1)] \qquad (9.4)$$

where T is the aquifer transmissivity, Q is the constant production well discharge rate, s_1 is the drawdown at selected point 1, s_2 is the drawdown at selected point 2, t_1 is the time at selected point 1, and t_2 is the time at selected point 2.

The calculated value of aquifer transmissivity and data for any point on the best-fit line may be substituted in the following equation to calculate aquifer storativity (Cooper and Jacob, 1946, pp. 526-534):

$$S = 4Tt/[r^2 e^{\ln(1/u)}] \qquad (9.5)$$

with

$$\ln(1/u) = 4\pi sT/Q + 0.5772 \qquad (9.6)$$

where s is the drawdown at the selected point, T is the calculated aquifer transmissivity, Q is the constant production well discharge rate, S is the aquifer storativity, t is the time at the selected point, r is the distance between the production well and the observation well, and e = 2.71828183.

Lines frequently connect data points in arithmetic trend graphs. Continuous curves may be drawn through data points by using Lagrange interpolation or a spline program listed by Fowler (1984, pp. 301-305). Curvilinear regression analysis (Davis, 1986, pp. 186-189) is used to find the best-fit curve through data points. In this analysis, polynomial equations are fitted to data points by least-square methods. A BASIC program for curve fitting is listed by Poole, et al. (1981, pp. 151-153).

Contour maps represent paths of lines or curves connecting points having common Z values on a two-dimensional XY surface (Jones, et al, 1986). Contouring requires interpolating contour positions by gridding or triangulation between control points. Occasional saddle

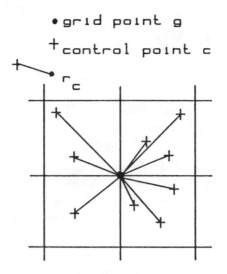

FIGURE 9.1   Grid for sorting control data points on a surface.

points, situations of low data surrounded by high data, and missing
control point values are handled in programs. Contours are drawn by
fitting lines or curves to contour positions (Davis, 1986, pp. 353-
376). Tracing the same contour twice and taking a shortcut across a
different contour is avoided in programs. Contour lines are labeled,
axes are drawn together with scale interval ticks and labels, and the
map is provided with appropriate annotations.

Computer programs that generate contour maps and three-dimensional plots require XYZ data in a regular spaced (gridded) format.
This format may be created by interpolating between grid points and
irregular and scattered control data points with triangulation or
kriging techniques (see Davis, 1986, pp. 353-405).

In the triangulation technique, interpolation is commonly based
on inverse distance squared weighted average methods which incorporate the principle that the closer the control data points are to each
other, the closer the measured Z values (see BASIC program listed
by Walton, 1989a, pp. 111-132). Distances between grid points and
control data points (Figure 9.1) are calculated with the following
Pythagorean equation (Davis, 1986, p. 367):

$$r_c = [(x_c - x_g)^2 + (y_c - y_g)^2]^{1/2} \qquad (9.7)$$

where r_c is the distance between a grid point g and control data point c, x_c is the control data point x-coordinate, x_g is the grid point x-coordinate, y_c is the control data point y-coordinate, and y_g is the grid point y-coordinate.

Distances are sorted in ascending order and the 6 control data points closest to the grid point of interest are identified. In the sorting algorithm, each distance is compared to all remaining unsorted distances. If a particular pair of distances is found to be out of order, the two distances are interchanged. This process is repeated $n - 1$ times where n is the number of distances to be sorted. BASIC sorting subroutines are listed by Miller (1981, pp. 143-160) and Cuellar (1984, pp. 68-77) and FORTRAN sorting subroutines are listed by Press, et al (1986, pp. 226-237).

The grid point Z value is calculated using the Z values for the 6 sorted control points and the following weighting function equation (Davis, 1986, pp. 367-371):

$$Z_g = \sum_{c=1}^{6} (F_c / F_a)(Z_c / \pi_c) / [\sum_{c=1}^{6} (F_c / F_a)(1 / \pi_c)] \qquad (9.8)$$

with

$$F_c = (1 - r_c/r_6)^2/(r_c/r_6) \qquad (9.9)$$

$$F_a = \sum_{c=1}^{6} F_c \qquad (9.10)$$

where Z_g is the calculated Z value at grid point g, Z_c is the Z value at control data point c, r_6 is the distance between grid point g and the farthest control data point, F_c is the inverse distance-squared weighting function for control point c (weightings vary according to the distances between the grid node of interest and the control data points).

The procedure is repeated for each grid point. A FORTRAN subroutine for calculating Z data in a regular spaced (gridded) format by interpolating between grid points and the 6 closest scattered control data points with triangulation is listed by Davis (1973, pp. 317-318).

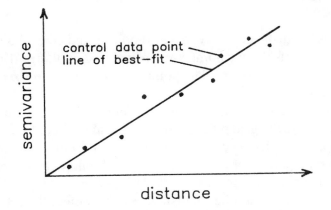

FIGURE 9.2   Linear semi-variogram.

Kriging procedures involving a semi-variogram are outlined by Davis (1986, pp. 353-405), Marsily (1986, pp. 284-337), Virdee and Kottegoda (1984, pp. 367-387), and Olea (1975). Kriging is a method for optimizing the estimation of a magnitude, which is distributed in space and is measured at a network of points. A semi-variogram (Figure 9.2) is an arithmetic plot of semivariance along the y-axis versus distance along the x-axis. Semivariance is a measure of the degree of spatial dependence of Z values between control data points and is defined for a pair of data control points by the following equation (Davis, 1986, p. 240):

$$\gamma_r = (Z_i - Z_{i+r})^2/2 \qquad (9.11)$$

where γ_r is semivariance, Z_i is a known Z value at control data point i and Z_{i+r} is a known Z value at another control data point at a distance r from the control data point i.

A semi-variogram is commonly developed by calculating the distances between many pairs of control data points using the Pythagorean equation. The semivariances of these pairs of control data points are calculated with known values of Z and Equation 9.11. Semivariances are plotted against corresponding distances and the best-fit line through the semi-variogram is found with the regression least-squares method. The slope and r-intercept of the best-fit line are calculated with the following equations (Davis, 1986, p. 180):

$$c_1 = [\sum_{i=1}^{n} r_i S_i - (\sum_{i=1}^{n} r_i \sum_{i=1}^{n} S_i)/n]/[\sum_{i=1}^{n} r_i^2$$

$$-(\sum_{i=1}^{n} r_i)^2/n]$$

(9.12)

$$c_0 = \sum_{i=1}^{n} S_i/n - c_1 \sum_{i=1}^{n} r_i/n$$

(9.13)

where c_1 is the slope of the best-fit line, c_0 is the semivariance-intercept of the best-fit line, r_i is the distance between a pair of points i, S_i is the semivariance of the pair of control data points i, and n is the number of pairs of control data points.

A BASIC subroutine for fitting a line to a set of control data points is listed by Poole, et al (1981, pp. 145-146).

Punctual kriging (Davis, 1986, pp. 383-393) is the simplest and most commonly used form of kriging employed in generating Z data in a gridded format. In punctual kriging, it is assumed that the variable being mapped is statistically stationary, or free of drift. Commonly, the linear model with points below a sill value (flat region) is chosen to represent the semi-variogram and the semi-variogram is assumed to go through the origin (nugget effect is zero).

Distances between control data points and grid nodes are calculated using the Pythagorean equation and these distances are sorted in ascending order. Semivariances for the three closest control data points to each grid node are calculated using the slope and semivariance-intercept of the best-fit line through the semi-variogram and the following equation (Davis, 1986, p. 179):

$$S_t = c_0 + c_1 r_t$$

(9.14)

where S_t is calculated semivariance, c_0 is the semivariance intercept of the best-fit line, c_1 is the slope of the best-fit line, and r_t is the known distance.

The value of Z at each grid node is estimated as a weighted average of the known Z values for the three closest control data points to each grid node with the following equation (Davis, 1986, p. 386):

$$Z_g = W_1 Z_1 + W_2 Z_2 + W_3 Z_3 \qquad (9.15)$$

where Z_g is the calculated value of Z at the grid node; W_1, W_2, and W_3 are weights for the three closest control data points to the grid node; and Z_1, Z_2, and Z_3 are known Z values at the three closest control data points to the grid node.

Optimum values for the weights are found by solving the following set of simultaneous kriging equations (Davis, 1986, p. 385):

$$W_1 \gamma(r_{11}) + W_2 \gamma(r_{12}) + W_3 \gamma(r_{13}) + \lambda = \gamma(r_{1g}) \qquad (9.16)$$

$$W_1 \gamma(r_{12}) + W_2 \gamma(r_{22}) + W_3 \gamma(r_{23}) + \lambda = \gamma(r_{2g}) \qquad (9.17)$$

$$W_1 \gamma(r_{13}) + W_2 \gamma(r_{23}) + W_3 \gamma(r_{33}) + \lambda = \gamma(r_{3g}) \qquad (9.18)$$

$$W_1 + W_2 + W_3 + 0 = 1 \qquad (9.19)$$

where $\gamma(r_{ij})$ is the semivariance over a distance corresponding to the separation between control data points i and j, $\gamma(r_{ig})$ is the semivariance corresponding to the separation between control data point i and the grid node g, and λ is a Lagrange multiplier.

Solving simultaneous equations involves matrix calculations (Weinman and Kurshan, 1985, pp. 191-198; Miller, 1981, 53-114). A BASIC subroutine for solving simultaneous equations is listed by Poole, et al (1981, pp. 100-101).

Some sophisticated computer kriging programs (for example Skrivan and Karlinger, 1980) calculate a theoretical semi-variogram, validate the theoretical semi-variogram, and interpolate Z values at grid control data points. In the validation algorithm, individual control data point Z values are suppressed and then estimated using the remaining control data point Z values. The errors in estimating control data point Z values are averaged and a calculated and theoretical variance are determined and averaged. The semi-variogram equation is adjusted until the kriged average error is approximately zero, the mean error is made as small as practical, and the average ratio of theoretical to calculated variance (reduced mean squared error) is near unity.

Multi-variate geostatistical estimation "co-kriging" is commonly used in the groundwater industry. The complexities of co-kriging are described by Marsily (1986, pp. 284-337). A co-kriging computer program is described by Carr, et al. (1985, pp. 11-127). Clark, et al. (1989, pp. 473-493) developed a variation in co-kriging to improve estimation in areas where few variables have been measured.

Contouring algorithms in screen graphics programs often consist of the detection of contour lines intersecting the four sides of grid-squares and the drawing of contour lines through the grid-square blocks. Two grid-node values along a grid-square side are compared with the value of a contour. If the contour value lies between the grid-node values, the point location of the crossing is calculated by linear interpolation using the following equation (Davis, 1986, p. 147):

$$C_i = [(C_2 - C_1)/(S_2 - S_1)](S_i - S_1) + C_1 \qquad (9.20)$$

where S_i is the value of the contour, S_1 is the grid-node 1 surface value Z, S_2 is the grid-node 2 surface value Z, C_1 is the grid-node 1 X or Y coordinate, C_2 is the grid-node 2 X or Y coordinate, and C_i is the X or Y coordinate of the contour point location.

This procedure is repeated for all of the grid-square sides. The grid-square is then subdivided into smaller squares and intermediate point locations of the contour line within the grid-square block are interpolated with Equation 9.20. Selection of smaller squares is based on the desired density of intermediate point locations of the contour. The small square size relates to the pixel (dot) size on the screen. Interpolation through both X and Y directions is performed to ensure that contours parallel to both axes are considered. Screen pixels are turned on at point locations of the contour. With suffi-ciently fine small square sizes, dots representing point contour locations merge together to look like continuous curves. A BASIC screen graphics contour map program is listed by Walton (1989a, pp. 113-132). Contour subroutines, written in BASIC or QuickBasic (trademark of Microsoft, Inc.), are listed by Simons (1983, pp. 487-492), Kinzelbach (1986, pp. 68-75), and Bourke (1987, pp. 143-150).

Text, lines, and special symbols may be interactively superposed on contour maps with the BASIC LABELER program listed by Korites (1982, pp. Labeler 1-13). That program features a moving cursor (dot) for positioning and the creation of fonts. It is capable of drawing lines between two designated points and may be used to erase previously created fonts or contents of the contour map. The cursor is moved up, down, right, or left a selected number of spaces by choosing the appropriate option from a menu. The cursor algorithm erases the dot at its previous position and replots it in its new position. Text is input as a string variable. A loop interprets each character of the string and calls a font subroutine to plot it. The cursor is then advanced to the position of the next number or character. In font subroutines, lines are drawn between appropriate points to define characters and numbers.

It is often desirable to view XYZ surface data in three-dimensional fishnet plots thereby adding greatly to the realism of data. Fishnet plots may consist of a deformed rectangular grid or mesh with spacings between mesh lines corresponding to equal increments of X and Y. The mesh is shifted to the right and up on the screen to conform to values of Z thus spreading surface curves out and giving a three-dimensional surface appearance (Hearn and Baker, 1983, pp. 246-249). Hidden lines are eliminated by drawing the mesh from front to back and not drawing mesh segments that would overlap with previously drawn lines. Axes and labeling for the three coordinate directions and a skirt around the surface may be added. Fishnet plots can be displayed at various angles and rotations using the principles of three-dimensional graphics transformations (Hearn and Baker, 1983, pp. 252-268). A BASIC program for drawing three-dimensional surface fishnet plots is listed by Fowler (1984, pp. 283-299).

Programs that generate three-dimensional surface plots require

XYZ data in a regularly spaced (gridded) format. This format may be created by interpolating between grid points and irregular and scattered control data points with triangulation or kriging techniques (Davis, 1986, pp. 353-405).

Commercial microcomputer software graphics packages are available for producing graphs, contour maps, and three-dimensional plots from output data generated by groundwater flow and mass transport computer programs. For example, GRAPHER (trademark of Golden Software, Inc., P.O. Box 281, Golden, CO 80402) displays black and white or color linear-linear, log-linear, linear-log, and log-log XY graphs. It supports the IBM PC, XT, or AT (or compatibles) and a wide variety of printers and plotters. GRAPHER has five different types of curve fitting to control points: linear, logarithmic, exponential, power, and cubic spline. Grid lines may be solid or dashed and multiple scales may be displayed on a single graph. Axes may be positioned at any location and axis tic marks and labels are user-defined. Lines through data points may be solid or dashed. Other features include: user-defined data point symbols, data file (free ASCII) import or keyboard spreadsheet data entry, data point labels, superposition of graphs, zoom and panning, text selected from user-defined symbols sets, and text rotated to any angle and scaled to any size. GRAPHER supports CAD drawing exchange format (DXF) files.

SURFER (trademark of Golden Software, Inc.) creates contour maps and three-dimensional surfaces. It imports ASCII data files or creates a spreadsheet data base from the keyboard. Inverse distance, minimum curvature, or kriging techniques are available for interpolation between irregularly spaced data points. SURFER displays black and white or color contour maps with: in-line contour labels scaled to any size and user-defined symbol sets; bold, dashed, or normal contour lines; hachure marks; smoothed or unsmoothed contour lines; irregular contour intervals; user-defined scaling and sizing; location posted by user-defined symbol sets and labeled with numeric or text strings; axes with tic marks and labels on any side of the map; and multiple shaped boundaries. Black and white or color three-dimensional surface plots may be created with: stacked contours or fish net surfaces; perspective or orthographic projections; hidden line removal; rotation to any angle; smoothed or unsmoothed lines; irregular stacked contour intervals; user-defined scaling and sizing; locations posted by any symbol and labeled with numeric or text strings; three-dimensional axes with tic marks and labels; multiple shaped boundaries; and the top, bottom, or both of the surface. A surface plot from a user-defined function may be displayed. SURFER supports IBM PC, XT or AT (or compatibles), a wide variety of printers and plotters, and CAD drawing exchange format (DXF) files.

GRAPHER and SURFER have an "import" option to load (X,Y) or (X,Y,Z) data from ASCII (text) files generated under other programs or created with a commercial word processor with line numbering capability or an editor such as IBM's EDLIN editor on the PC-DOS (trademark of International Business Machines Corp.) diskette. Files with a .DAT filename extension must contain one (X,Y) or (X,Y,Z) entry per line; values on a single line must be separated by either a space, tab, or comma.

Many groundwater flow and mass transport computer programs contain options to create files with formats matching those required by GRAPHER and SURFER. The (X,Y,Z) data file has the following line by line format: $1,1,Z(1,1)$; $2,1,Z(2,1)$; ... up to $NX,1,Z(NX,1)$; then $1,2,Z(1,2)$; $2,2,Z(2,2)$; ... up to $NX,2,Z(NX,2)$. These entries are continued through $NX,NY,Z(NX,NY)$, where NX is the largest x-coordinate and NY is the largest y-coordinate. Commonly, data files generated by groundwater flow and mass transport programs with the following row to row format have to be converted to the (X,Y,Z) data file format before they can be used with graphics programs: $Z(1,1)$, $Z(2,1)$, ... up to $Z(NC,1)$, then $Z(1,2)$, $Z(2,2)$, ... up to $Z(NC,2)$. These entries are continued through $Z(NC,NR)$, where NC is the number of columns and NR is the number of rows.

Examples of graphics created with GRAPHER and SURFER are presented in Figures 9.3 to 9.5.

PLOT88 (trademark of Plotworks, Inc., Dept. W-9, 16440 Eagles Crest Road, Ramona, CA 92065) contains a library of graphic subroutines which may be called for by programs written in Microsoft, Ryan McFarland, Lahey F77L, or IBM Professional FORTRAN when constructing XY graphs, contour maps, and three-dimensional mesh drawings. PLOT 88 supports IBM PC, XT, or AT (or compatibles) and a wide range of printers and plotters. It contains fully integrated sets of subroutines for generating contour maps and three-dimensional mesh surfaces. Subroutines draw: a linear or logarithmic axis with title, tic marks, and tic labels; an arithmetic, log-log, or semi-log line between a set of XY points; a smooth curve between a set of XY points; characters and special centered symbols; two-dimensional surface contours based on a grid of control points; and a three-dimensional mesh surface based on a grid of control points. Other subroutine generate straight line segments; define graph enlargement or reduction, line width, width-to-height ratio of characters, format of axis tic, color index, shading pattern, and line type and symbol characteristics; selects a character set; determine logarithmic scale factors; perform two-dimensional interpolation to construct a grid of data points; smooths gridded data; determine contours levels; and marks and labels randomly spaced data points.

The program OMNIPLOT (trademark of MICROCOM-PATIBLES, 301 Prelude Dr., Silver Spring, MD. 20901) is a popular stand alone graphics library that drives a microcomputer's screen and pen plotters. OMNIPLOT (S) produces screen graphics to preview plots; a hard copy can be obtained via a dump of the screen to a dot-matrix printer. OMNIPLOT (P) produces a similar set of graphics commands for pen plotters with high resolution.

The following graphic formats are supported: tabular, bar, pie, contour, and three-dimensional wire frame. Choices include: pen speed, paper size, text size, plotting region, marker symbol type, line type, cubic spline interpolation to data, least squares fit to data, standard plots, semi-log plots, log-log plots, gridding, tic mark label format, axis labels, plot labels, number and numerical value of contours, mesh size defining contour data, contour labels, and choice of size and viewing direction for three-dimensional plot. Data are entered in response to menu prompts or imported from a diskette file prepared with a text editor such as IBM's EDLIN editor on the PC-DOS diskette or a BASIC program included with the package.

OMNIPLOT requires contour map and three-dimensional plot data to be defined on a regular rectangular grid. Z values on an NX

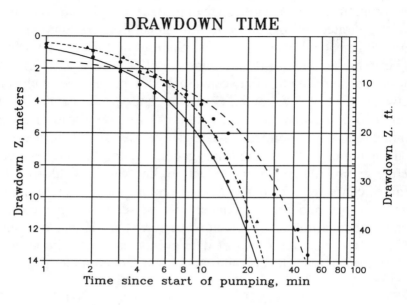

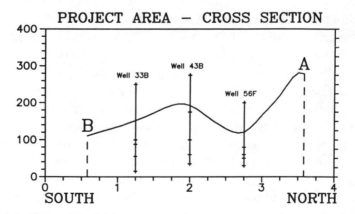

FIGURE 9.3   Example trend graph (A) and cross section (B) created with microcomputer graphics program and plotter.

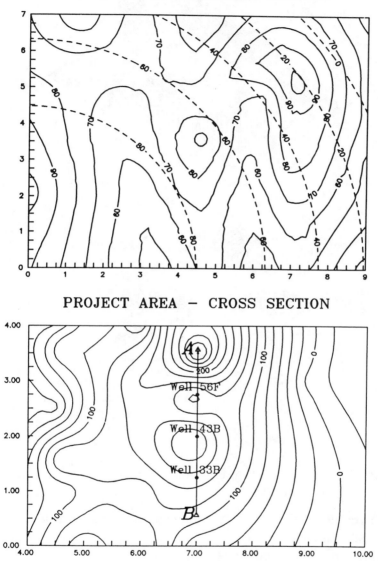

FIGURE 9.4   Example contour maps created with microcomputer graphics program and plotter.

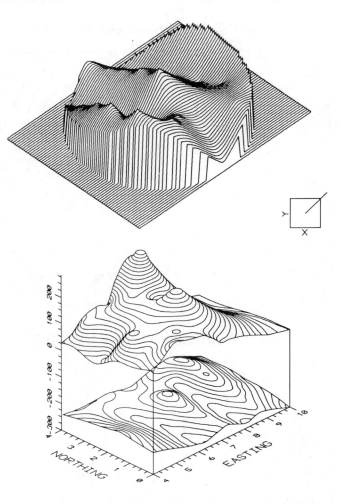

FIGURE 9.5   Example 3-D fishnet surface plots created with microcomputer graphics program and plotter.

by NY (column by row) mesh of data points are user-defined. A new line is started each time the entry for a new row begins. Individual Z values are separated by a space or comma. An interpolation program is provided for filling out a nonregular array of data points. Many groundwater flow and mass transport computer programs contain an option to create a file of Z values with a format matching that required by OMNIPLOT which may be integrated with other required pieces of formatted data (characters, integers, and real numbers) utilizing the DOS EDLIN Transfer (T) command.

OMNIPLOT supports the IBM PC (trademark of International Business Machines, Inc.) and fully IBM-compatable PC's. Screen dumps are to an IBM dot-matrix or compatible printer. Hewlett-Packard (HP-GL) or Houston Instrument (DMP) x-y pen plotters are supported. Codes can utilize 8087/80287 math coprocessors.

Lithology, well construction, and geologic cross-section micro-computer graphics programs developed by Hall Groundwater Consultants, Inc. are available from Earthware. A plotter is required for these programs which display borehole lithology with up to twenty symbols; well construction features such as multiple casings, screens or slotted pipe, packers, washdown valve, submersible pump, and static and pumping water level; geologic cross sections with up to 24 well logs and up to 20 formation patterns; monitor wells and piezometers with multiple screened zones, frac sand, and cemented zones; and grain-size distribution diagrams. These programs support the IBM PC or compatibles, CGA and EGA screens, and Houston Instruments DMP Series (trademark of Houston Instruments Division, Bausch & Lomb, Inc.) or Hewlett Packard Company Series plotters.

Reviews of 14 commercial microcomputer scientific graphics software products appear in the March 14, 1989 issue of *PC Magazine* (Volume 8, Number 5, pp. 259-286). Some graphics program information is also provided in the National Water Well Association Publications Catalog.

Commonly, groundwater technical illustrations and drawings are prepared with the aid of a computer (Voisinet, 1986). Commercial CAD (computer-aided drafting) microcomputer programs are available for creating clean and precise drawings, revising drawings without having to redraw them from scratch, having frequently used forms and shapes duplicated automatically, and making instant copies of drawings at selected scales. Generic CADD Level 3 (trademark of Generic Software, Inc.) is a popular CAD microcomputer program (Freeman, 1989) which supports the IBM PC (trademark of International Business Machines, Inc.) and compatables and a wide range of mouses, printers, plotters, and digitizers (graphics tables) such as SketchPro (trademark of Hewlett-Packard Company).

Generic CADD Level 3 contains the basic drawing and editing commands; has dimensioning capabilities and sophisticated geometric snapping features; creates text with seven fonts; supports zoom views, hatching, filling, automated double lines, and Bezier curves; scales drawings; prints Generic CADD drawing files with a PostScript printer; converts Generic CADD drawing files to encapsulated PostScript files to use with desktop publishing and word processing programs; converts Generic CADD drawing files to a pixel image format for use with paint or draw programs that use pixel-based images; and has a printing utility that prints crisp, high-resolution drawings with a dot matrix printer.

SketchPro features a four-button cursor and stylus for defining points to be digitized and selecting and performing software functions. It comes complete with overlay softkeys which provide several graphics functions when selected, an IBM PC adapter cable, and a Microsoft Mouse driver file. SketchPro can be configured to a wide variety of CAD programs including Generic CADD Level 3.

10

Finite-Difference Flow Model

The finite-difference flow model described herein is based on the finite-difference flow model (PLASM) developed by Prickett and Lonnquist (1971). Simulation techniques are similar to those incorporated in the finite-difference flow model (MODFLOW) developed by McDonald and Harbaugh (1984). In the model, a rectangular grid is superposed over a plan view single aquifer or aquifer layer map (Figure 10.1) thereby subdividing the aquifer or aquifer layer into discretized volumes with dimensions m, Δx, Δy where m is the aquifer or aquifer layer thickness, Δx is the grid spacing in the x-direction, and Δy is the grid spacing in the y-direction. The intersections of the grid lines are called nodes and are referenced with a column (i) and row (j) coordinate system colinear with the x and y directions, respectively. The grid origin (1,1) is at the upper left corner of the grid. i coordinates increase left to right and j coordinates increase top to bottom. The simulation time is subdivided into discretized increments with the dimension Δt (Figure 10.1). Grid spacings and time increments Δt are made small in comparison to the area underlain by the aquifer or aquifer layer and the simulation time, respectively.

The algebraic finite-difference form of the partial differential equation governing the nonsteady state two-dimensional constant density flow in a nonhomogeneous aquifer or aquifer layer underlain by an aquiclude simulated in the model can be derived from Darcy's law and the principle of the conservation of mass. The derivation is based on the instantaneous flow balance at the end of a discrete time

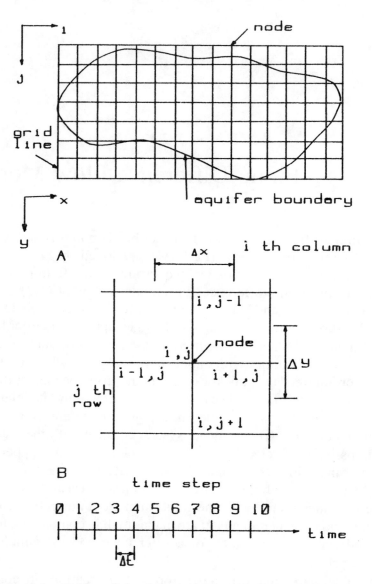

FIGURE 10.1   Uniform finite-difference grid (A) and time increments (B).

increment around discretized finite aquifer or aquifer layer volumes centered at nodes and extending the full aquifer or aquifer layer thickness (Prickett, 1975, pp. 91-97).

It is assumed that vertical components of flow within the aquifer or aquifer layer are negligible, that only components of flow in the plane of the aquifer or aquifer layer need be considered, and the grid coordinate axes are aligned with the principal axes of the transmissivity tensor. Flow balance components are: horizontal flows into and out of the finite aquifer or aquifer layer volume, flow due to changes in water storage within the finite aquifer or aquifer layer volume, and vertical flows due to sinks and/or sources through the top of the finite aquifer or aquifer layer volume.

With a square grid, the finite-difference flow balance nodal equation is (Prickett, 1975, p. 101) :

$$T2_{i-1,j}(h_{i-1,j} - h_{i,j}) + T2_{i,j}(h_{i+1,j} - h_{i,j})$$

$$+ T1_{i,j}(h_{i,j+1} - h_{i,j}) + T1_{i,j-1}(h_{i,j-1} - h_{i,j})$$

$$= (S_{i,j}\Delta x^2/\Delta t)(h_{i,j} - h\phi_{i,j}) + q_{s(i,j)} \tag{10.1}$$

with

$$q_{s(i,j)} = \sum_{n=1}^{N} q_{n(i,j)} \tag{10.2}$$

possible component sink (+) or source (−) flow rates included in $q_{s(i,j)}$ are:

$$+q_{1(i,j)}, -q_{2(i,j)}, +q_{3(i,j)}, +q_{4(i,j)}, +q_{5(i,j)}, +q_{6(i,j)},$$
$$-q_{7(i,j)}, +q_{8(i,j)}, \pm q_{9(i,j)}, \pm q_{10(i,j)}, -q_{11(i,j)}, \text{ and } -q_{12(i,j)}$$

and

$$q_{9(i,j)} = q_{a(i,j)} + q_{t(i,j)} \tag{10.3}$$

where $T1_{i,j}$ is the aquifer or aquifer layer transmissivity of the vector volume extending the full aquifer or aquifer layer thickness between nodes i,j and i,j + 1 with horizontal dimensions defined in Figure 10.2; $T2_{i,j}$ is the aquifer or aquifer layer transmissivity of the vector

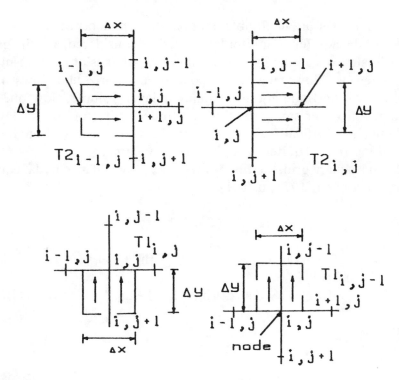

FIGURE 10.2   Vector volume definition for transmissivity.

volume extending the full aquifer or aquifer layer thickness between nodes i,j and i + 1,j with horizontal dimensions defined in Figure 10.2; $T1_{i,j-1}$ is the aquifer or aquifer layer transmissivity of the vector volume between nodes i,j and i,j − 1; $T2_{i-1,j}$ is the aquifer or aquifer layer transmissivity of the vector volume between nodes i,j and i − 1,j; $h_{i,j}$ is the calculated head at the end of a time increment Δt at node i,j; $h_{i-1,j}$ is the calculated head at the end of a time increment Δt at node i − 1,j; $h_{i+1,j}$ is the calculated head at the end of a time increment Δt at node i + 1,j; $h_{i,j+1}$ is the calculated head at the end of a time increment Δt at node i,j + 1; $h_{i,j-1}$ is the calculated head at the end of a time increment Δt at node i,j−1; $S_{i,j}$ is the aquifer or aquifer layer storativity of the vector volume centered at node i,j with horizontal dimensions defined in Figure 10.3; Δx is the grid spacing; Δt is the time increment elapsed since the last calculation of heads; $h\phi_{i,j}$ is the calculated head at node i,j at the end of the previous time increment;

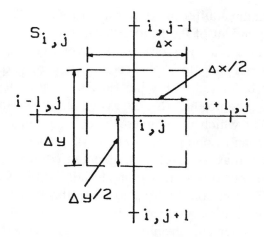

FIGURE 10.3   Vector volume definition for storativity.

$q_{s(i,j)}$ is the algebraic sum of source and sink flow rates at node i,j; N is the number of source and/or sink flow rates at node i,j; $q_{1(i,j)}$ is a production well sink flow rate (discharge rate); $q_{2(i,j)}$ is an injection well source flow rate (injection rate); $q_{3(i,j)}$ is a flowing well sink flow rate; $q_{4(i,j)}$ is a multi-aquifer production well sink flow rate; $q_{5(i,j)}$ is a drain sink flow rate; $q_{6(i,j)}$ is a mine sink flow rate; $q_{7(i,j)}$ is an areal recharge source flow rate (recharge rate); $q_{8(i,j)}$ is an evapotranspiration sink flow rate; $q_{9(i,j)}$ is a leakage through an overlying and/or underlying aquitard sink or source flow rate (Trescott, et al., 1976, pp. 4-7) which includes steady leakage due to the initial gradient across the aquitard $q_{a(i,j)}$ and transient leakage due to the release of water in storage within the aquitard $q_{t(i,j)}$; $q_{10(i,j)}$ is an inter-aquifer flow transfer between two adjoining aquifer layers source or sink flow rate (McDonald and Harbaugh, 1984, p. 13); $q_{11(i,j)}$ is a release of water due to delayed gravity drainage of aquifer or aquifer layer pores under water table conditions sink flow rate (Rushton and Redshaw, 1979, pp. 259-261); and $q_{12(i,j)}$ is an infiltration through a surface water body bed source flow rate.

Source or sink flow rates [$q_{3(i,j)}$, $q_{4(i,j)}$, $q_{5(i,j)}$, $q_{8(i,j)}$, $q_{9(i,j)}$, $q_{10(i,j)}$, $q_{11(i,j)}$, and $q_{12(i,j)}$] may be dependent on the head at the source or sink node but independent of all other heads in the aquifer or aquifer layer. They may be dependent on the head at the source or sink node and other heads in the aquifer or aquifer layer [$q6_{(i,j)}$]. They also may be entirely independent of the head at the source or sink node and all other heads in the aquifer or aquifer layer [$q_{1(i,j)}$, $q_{2(i,j)}$, and $q_{7(i,j)}$]. Some source and sink flow rates such as $q_{3(i,j)}$, $q_{5(i,j)}$, $q_{6(i,j)}$, $q_{11(i,j)}$, and $q_{12(i,j)}$ can not be mixed with other flow rates at nodes.

Commonly, the product ($S_{i,j} \Delta x^2$) is called the storativity simulation factor $S_{f(i,j)}$ which is assigned to each aquifer or aquifer layer node (Prickett and Lonnquist, 1971, p. 8). Transmissivity values in Equation 10.1 may be averages, geometric means, or harmonic means of the transmissivities of materials within vector volumes (Butler, 1957; Huntoon, 1974; and Kinzelbach, 1986, pp. 22-24). Using transmissivities between nodes allows the simulation of aquifer or aquifer layer anisotrophy provided grid coordinate axes coincide with the principal axes of the transmissivity tensor. If the grid coordinate axes do not coincide with the principal axes of the transmissivity tensor, four more water balance components representing the contributions of the off-diagonal elements are added to Equation 10.1 (Kinzelbach, 1986, pp. 62-63).

A large set of simultaneous algebraic equations are solved to determine unknown aquifer or aquifer layer heads $h_{i,j}$ because there is an equation of the same form as Equation 10.1 for every grid node. There are many procedures available for solving large sets of simultaneous equations including the alternating-direction implicit procedure, strongly implicit procedure, and line successive overrelaxation procedure (Trescott, et al., 1976, pp. 14-29; Kinzelbach, 1986, pp. 29-61). The modified iterative alternating-direction implicit (MIADI) procedure described herein was developed by Prickett and Lonnquist (1971, pp. 3-7) and is based on a combination of the iterative alternating-direction implicit (IADI) procedure of Peaceman and Rackford (1955), the Gauss-Seidel iterative procedure (Tyson and Weber, 1964), and a preliminary head predictor procedure.

In the MIADI procedure, Equation 10.1 is rearranged and written in the following coefficient forms (Prickett and Lonnquist, 1971, p. 5):

For calculations by columns

$$AA_j h_{i,j-1} + BB_j h_{i,j} + CC_j h_{i,j+1} = DD_j \tag{10.4}$$

with

$$AA_j = -T1_{ij-1} \tag{10.5}$$

$$BB_j = T2_{i-1j} + T2_{i,j} + T1_{i,j} + T1_{i,j-1} + S\Delta x^2/\Delta t \tag{10.6}$$

$$CC_j = -T1_{i,j} \tag{10.7}$$

$$DD_j = (S\Delta x^2/\Delta t)h\phi_{i,j} - q_{s(i,j)} + T2_{i-1j}h_{i-1,j}$$

$$+ T2_{i,j}h_{i+1,j} \tag{10.8}$$

For calculations by rows

$$AA_i h_{i-1,j} + BB_i h_{i,j} + CC_i h_{i+1,j} = DD_i \tag{10.9}$$

with

$$AA_i = -T2_{i-1,j} \tag{10.10}$$

$$BB_i = T2_{i-1,j} + T2_{i,j} + T1_{i,j}$$

$$+ T1_{i,j-1} + S\Delta x^2/\Delta t \tag{10.11}$$

$$CC_i = -T2_{i,j} \tag{10.12}$$

$$DD_i = (S\Delta x^2/\Delta t)h\phi_{i,j} - q_{s(i,j)}$$

$$+ T1_{i,j-1}h_{i,j-1} + T1_{i,j}h_{i,j+1} \tag{10.13}$$

The solution of sets of column and row equations with three head unknowns for each node (tridiagonal matrix) involves Gauss elimination using the Thomas algorithm (Kinzelbach, 1986, pp. 54-56). Equation coefficients are assembled with G and B arrays, defined as follows: (Prickett and Lonnquist, 1971, pp. 5-7):

For calculations by columns

$$G_j = (DD_j - AA_jG_{j-1})/(BB_j - AA_jB_{j-1}) \qquad (10.14)$$

$$B_j = CC_j/(BB_j - AA_jB_{j-1}) \qquad (10.15)$$

For calculations by rows

$$G_i = (DD_i - AA_iG_{i-1})/(BB_i - AA_iB_{i-1}) \qquad (10.16)$$

$$B_i = CC_i/(BB_i - AA_iB_{i-1}) \qquad (10.17)$$

In the processing of columns and rows, AA_n is set equal to zero at the first node of a column or row and CC_n is set equal to zero at the last node of the column or row where $n = i$ for row calculations and $n = j$ for column calculations.

In the course of assembling G and B arrays for the nodes of a column or row in order of increasing j or i, the head at the last node of a column or row is calculated. By reverse substitution, all other heads in the column or row in order of decreasing j or i are then calculated with the following equations (Prickett and Lonnquist, 1971, p. 7)

For column calculations

$$h_{i,j} = G_j - B_jh_{i,j+1} \qquad (10.18)$$

For row calculations

$$h_{i,j} = G_i - B_ih_{i+1,j} \qquad (10.19)$$

This process is repeated for all columns and rows and one iteration step is completed. Iterations continue alternating between column and row calculations until convergence is reached.

The preliminary head predictor increases the convergence rate by predicting future heads based on past head changes (Prickett and

Lonnquist, 1971, p. 11). It is assumed that the ratio of present to past differences in head will equal the ratio of the future to present differences in head. Convergence is declared when the sum or absolute values of changes in head for all nodes has not changed more than an error tolerance during an iteration. The error tolerance (see Prickett and Lonnquist, 1971, pp. 14-15) is commonly determined with the following equation:

$$e/a = d \qquad (10.20)$$

where e is the error tolerance, a is the number of active nodes, and d is the desired accuracy at individual nodes which is usually set at 0.01 ft (0.00308 m).

Frequently, the water balance summed for all nodes (OSM, 1981, p. II-28) is used as second measure of convergence. If water balance is used as a measure of convergence, a high precision of water balances at all nodes is required. However, precision tends to be low at nodes where large simulation factors or storativities are multiplied by very small head losses or differences (Prickett, 1987, oral communication). This can cause severe program problems. For this reason, some programs do not use the water balance to measure convergence but instead only display water balances when requested to do so by the user.

A FORTRAN MIADI procedure program is listed by Prickett and Lonnquist (1971, p. 23) and a BASIC IADI procedure program is listed by Kinzelbach (1986, pp. 59-61).

The computer model generates head values at grid nodes which are observation points. If the location of the point of interest does not coincide with a grid node and is within a grid block (Figure 10.4) then interpolation from the head values at the four corners of the grid block in which the point occurs is necessary. The interpolation is commonly performed with the following Lagrange interpolating function which generates the algebraic polynomial of lowest degree with known head values at the four block corners (Kinzelbach, 1986, p. 68):

$$h(x,y) = (x - x_{i+1})(y - y_{j+1})h_{ij}/[(x_{i+1} - x_i)(y_{j+1} - y_j)]$$

$$- (x - x_i)(y - y_{j+1})h_{i+1,j}/[(x_{i+1} - x_i)(y_{j+1} - y_j)]$$

$$- (x - x_{i+1})(y - y_j)h_{i,j+1}/[(x_{i+1} - x_i)(y_{j+1} - y_j)]$$

$$+ (x - x_i)(y - y_j)h_{i+1j+1}/[(x_{i+1} - x_i)(y_{j+1} - y_j)] \qquad (10.21)$$

where x,y are the coordinates of the interpolation point within the grid block; (x_i,y_j), (x_i,y_{j+1}), (x_{i+1},y_j), and (x_{i+1},y_{j+1}) are the coordinates of the four node corners of the grid block; and h_{ij}, $h_{i,j+1}$, $h_{i+1,j}$, $h_{i+1,j+1}$ are the head values at the four corners of the grid block at the end of the time increment Δt.

A FORTRAN subroutine for calculating head values inside grid blocks is listed by Press, et al (1986, pp. 95-101).

TIME INCREMENTS AND GRID SPACING

Time increments and grid spacings are interdependent and are usually jointly specified so that head changes at nodes adjacent to sink/source nodes are appreciable during the initial time increment. Commonly, the number of time increments is less than 51 and non-uniform time increments are specified. The proper choice of time increments is important for high precision head calculations. Small time increments are specified each time there is a change in sink/source or aquifer system boundary conditions such as production well discharge starting or stopping. Between changes, time increments may be programmed to gradually increase in a stepwise fashion as a geometric progression of ratio 1.2 (see FORTRAN program listed by Prickett and Lonnquist, 1971, pp. 9). Variable time increments may also be interactively specified from the keyboard at the beginning of each time increment (see FORTRAN program listed by Walton, 1989b, pp. 67-156).

Time increments are kept small in comparison to the length of the simulation time. To insure high precision head calculations, six time increments without significant changes in sink/source and aquifer system boundary conditions should precede the time of interest (Ghislain de Marsily, 1986, pp. 398-400 and Prickett and Lonnquist,

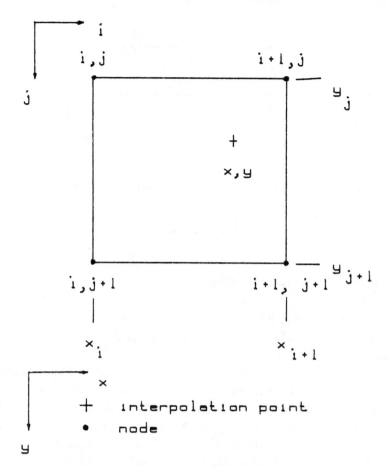

FIGURE 10.4   Grid for Lagrangian interpolation of z values inside grid blocks.

1971, pp. 13-14). In general, the smaller the time increment the greater the precision of head calculations.

It is often desirable to use variable grid spacings instead of uniform square grid spacings to reduce needed computer core storage and execution time. A comparison of finite-difference equations for uniform and variable grids (OSM, 1981, p. II-26) indicates that the algorithm for solving large sets of finite-difference equations based on a square grid can be utilized with variable grids provided $T2_{i,j}$ and $T2_{i-1,j}$ are multiplied by the arithmetic mean expression $2\Delta y_j/(\Delta x_{i+1} + \Delta x_i)$ and $T1_{i,j}$ and $T1_{i,j-1}$ are multiplied by $2\Delta x_i/(\Delta y_{j+1} + \Delta y_j)$ (Figure

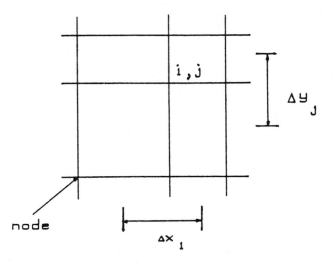

FIGURE 10.5   Variable finite-difference grid.

10.5). A FORTRAN program containing an algorithm for adjusting square grid transmissivities for a variable grid is listed by Prickett, et al. (1981, pp. 19-20).

The grid spacing is kept small in comparison to the simulation area to insure high precision head calculations. Smallest grid spacings are specified where the head gradient varies significantly such as in the vicinity of production or injection wells, rivers, mines, and drains. Grid spacings are gradually increased outward from these features. It is frequently desirable to specify variable grid spacing with large grid spacings along grid borders. Grid spacings are commonly uniform along individual columns and rows. The ratio of adjacent grid spacings is usually less than 5 except for high precision head calculations when it is 2 or less (Trescott, et al, 1976, p. 30). Most nodes are active with proper grid spacing.

In simulating contaminant migration, it is often desirable to perform a three-phased modeling study involving telescopic three grid spacings (Ward, et al., 1987). A large grid spacing is used in phase one to calculate regional head distribution. A smaller grid spacing is used in phase two with local head distribution in context with the regional head distribution. A smaller grid spacing is used in phase three with site head distribution in context with the local head distribution to obtain site contaminant migration patterns. When switching from one phase to another, boundary conditions for the local and site modeling efforts are obtained from the results of the regional modeling effort.

INITIAL WATER LEVEL CONDITIONS

Commonly, the simulation of drawdown or recovery caused by a specified stress on the aquifer or aquifer layer is desired and zero initial water levels are specified for all nodes so that no flow occurs in the aquifer or aquifer layer at the start of the simulation time. If heads are simulated, initial water level conditions constant in space are specified so that no initial flow occurs in the aquifer or aquifer layer at the start of the simulation time, or initial water level conditions varying in space are specified so that an appropriate flow occurs in the aquifer or aquifer layer at the start of the simulation time (Trescott, et al., 1976, p. 30).

If initial heads vary in space at the start of the simulation time, water levels change during the simulation time not only in response to the specified stress but also due to the prescribed initial flow conditions. Steady-state initial heads can be calculated by omitting man-made stresses such as production well discharge, setting all storage factors to zero, and simulating only the natural flow system with appropriate boundary conditions (Franke, et al., 1985 and Trescott, et al., 1976, p. 30). Steady-state conditions may also be simulated by specifying a large time increment (Prickett and Lonnquist, 1971).

BOUNDARIES

The edges of the finite-difference grid represent barrier boundaries located one half of the grid spacing beyond border nodes (Prickett and Lonnquist, 1971, p. 17). An effective infinite aquifer or aquifer layer may be simulated by making the number of columns

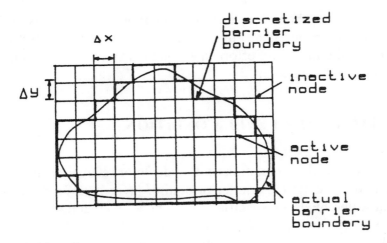

FIGURE 10.6   Discretized barrier boundary.

and rows and grid spacings sufficiently large so that the heads at interior nodes of interest are not affected by the finite-difference grid boundary (Prickett and Lonnquist, 1971, p. 13). Usually, aquifer or aquifer layer boundaries are discretized along grid lines as shown in Figure 10.6.

Barrier (no flow Neumann type) boundaries are simulated by assigning zero transmissivities to boundary nodes. Recharge (constant head Dirichlet type) boundaries are simulated by assigning large storage factors to boundary nodes (Franke, et al, 1985, pp. 1-26). Semipervious (leakage) boundaries may be simulated by specifying a linear combination of head and flux at a boundary.

Without transmissivity, storage factor, and simulation factor modifications along grid borders, aquifer or aquifer layer boundaries are located midway between adjacent nodes outside and inside boundaries. This location may be changed and irregular boundaries may be more precisely simulated with "vector volume" techniques described by Karplus (1958). The following equations are used to adjust transmissivities, storage factors, and simulation factors (Walton, 1989b, pp. 19-20):

$$T_{xa} = T_{xu}[(\Delta y/2 + y_b)/(\Delta x/2 + x_b)] \qquad (10.22)$$

$$T_{ya} = T_{yu}[(\Delta x/2 + x_b)/(\Delta y/2 + y_b)] \tag{10.23}$$

$$S_{fa} = S_{fu}[(\Delta x/2 + x_b)(\Delta y/2 + y_b)/(\Delta x \Delta y)] \tag{10.24}$$

$$R_a = R_u[(\Delta x/2 + x_b)(\Delta y/2 + y_b)/(\Delta x \Delta y)] \tag{10.25}$$

where T_{xa} is the aquifer or aquifer layer transmissivity in the x-direction adjusted for the irregular boundary location, T_{ya} is the aquifer or aquifer layer transmissivity in the y-direction adjusted for the irregular boundary location, T_{xu} is the aquifer or aquifer layer transmissivity without the boundary in the x-direction, T_{yu} is the aquifer or aquifer layer transmissivity without the boundary in the y-direction, S_{fa} is the aquifer or aquifer layer storativity simulation factor adjusted for the irregular boundary adjacent to the node, S_{fu} is the aquifer or aquifer layer storativity simulation factor without the boundary, R_a is the simulation factor adjusted for the irregular boundary adjacent to the node, R_u is the simulation factor without the boundary, x is the grid spacing in the x-direction, y is the grid spacing in the y-direction, x_b is the distance between the grid line and the boundary in the x-direction, y_b is the distance between the grid line and the boundary in the y-direction.

If there is no boundary in the y-direction $y_b = \Delta y/2$. If there is no boundary in the x-direction $x_b = \Delta x/2$. With a square grid $\Delta x = \Delta y$.

PRODUCTION WELL DISCHARGE AND HEAD

Production well discharge is simulated as a sink flow rate assigned to the production well node. Discharge rates may be programmed to automatically change at the beginning of time increments in a stepwise discretized fashion (Figure 10.7) to simulate variable pumping rates (see FORTRAN program listed by Prickett and Lonnquist, 1971, p. 29). Variable discharge rates may also be interactively specified from the keyboard at the beginning of time increments (see FORTRAN program listed by Walton, 1989b, pp. 67-156).

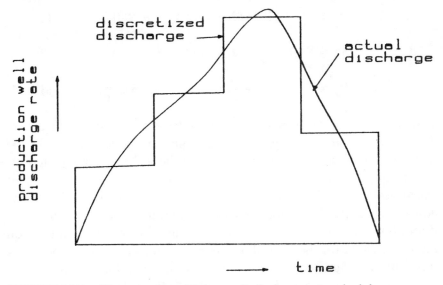

FIGURE 10.7   Discretized production well discharge rate schedule.

A production well is simulated by imposing a discharge rate on a grid block centered at a node. Grid block dimensions are usually much larger than the production well diameter. The head numerically calculated at the production well node is an average head for the grid block not the head in the production well (Beljin, 1987, pp. 340-351). The concept of equivalent well block radius with a square (uniform) or rectangular (variable) grid and anisotropic aquifer hydraulic conductivity (Peaceman, 1983, pp. 531-543) based on the Thiem equation may be used to calculate the head in the production well.

The radius at which the head in the aquifer is equal to the numerically calculated head for the production well grid block is estimated with the following equation (Peaceman, 1983):

$$r_e = 0.28[(T_{yy}/T_{xx})^{0.5}\Delta x^2 + (T_{xx}/T_{yy})^{0.5}\Delta y^2]^{0.5}/$$

$$[(T_{yy}/T_{xx})^{0.25} + (T_{xx}/T_{yy})^{0.25}] \qquad (10.26)$$

where r_e is the equivalent well block radius, T_{xx} is the aquifer or aquifer layer transmissivity in the x-direction at the production well node, T_{yy} is the aquifer or aquifer layer transmissivity in y-direction at the production well node, Δx is the production well block grid spacing in x-direction, and ΔY is the production well block grid spacing in the y-direction.

The head in the production well based on the head numerically calculated for the production well block is estimated with the following equation (Peaceman, 1983):

$$h_{w(i,j)} = h_{b(i,j)} - q_{1(i,j)}\ln(r_e/r_w)/[2\pi(T_{xx}T_{yy})^{0.5}] \qquad (10.27)$$

where $h_{w(i,j)}$ is the head in the production well at the end of the time increment Δt at node i,j; $h_{b(i,j)}$ is the average head in the well block at the end of the time increment Δt at node i,j; $q_{1(i,j)}$ is the constant production well discharge rate at node i,j; and r_w is the production well effective radius.

Commonly, the grid spacing at the production well node is square and usually the ratio of grid spacings in the x- and y-directions is less than 5 at the production well node. A method developed by Pedrosa and Aziz (1986, pp. 611-621) may be used to estimate the head in the production well with greater precision than can be attained with Equations 10.26 and 10.27.

The production well head may be adjusted for the effects of turbulent flow (well loss) with Equation 6.31.

PARTIALLY PENETRATING WELLS

Equation 10.1 assumes wells fully penetrate the aquifer or aquifer layer. Whether or not partial penetration impacts are appreciable and should be simulated is determined with Equation 2.12. Partially penetrating wells in aquifers are simulated with Equations 2.21 to 2.27. The drawdown due to the effects of partial penetration is algebraically summed with the production well head or drawdown calculated with Equation 10.27. A FORTRAN program simulating partially penetrating wells is listed by Walton (1989b, pp. 118-122).

WELLBORE STORAGE

Equation 10.1 assumes wellbore storage is negligible. Whether or not wellbore storage should be simulated is determined with Equation 2.10 or 2.11. Wellbore storage may be simulated (see FORTRAN program listed by Walton, 1989b, pp. 67-156) with an iterative procedure described by Huyakorn and Pinder (1983, p. 127-128). The approximate procedure assumes that discharge derived from the aquifer or aquifer layer varies linearly instead of exponentially during the simulation time and uses the principle of superposition. Total well discharge is the sum of discharge derived from the aquifer or aquifer layer and discharge derived from storage within the wellbore. Drawdown with wellbore storage impacts is the product of drawdown without wellbore storage impacts and the discharge derived from the aquifer or aquifer layer divided by total production well discharge.

The equation relating the total constant discharge from the production well, discharge derived from the aquifer or aquifer layer, and discharge derived from storage within the wellbore assuming constant well geometry is as follows (Huyakorn and Pinder, 1983, pp. 127-128):

$$Q = Q_a + \pi(r_w^2 - r_c^2)s_{ws}/t \qquad (10.28)$$

where Q is the constant production well discharge rate, Q_a is the average discharge derived from the aquifer or aquifer layer during the time increment Δt, r_w is the production well effective radius, r_c is the pump column-pipe radius, s_{ws} is the drawdown with wellbore storage impacts at the end of time increment Δt, and t is the time at the end of time increment Δt.

The iterative procedure starts with a first trial value of s_{ws} based on the numerically calculated drawdown without wellbore storage impacts. Equation 10.28 is used to calculate a first trial value of Q_a. A second trial value of s_{ws} is calculated by multiplying drawdown without wellbore storage impacts (s) by the ratio first trial value of Q_a divided by total discharge from the well as shown in the following equation (Walton, 1989b, p. 26):

$$S_{ws} = sQ_a/Q \qquad (10.29)$$

The first and second trial values of s_{ws} are compared and if the difference exceeds an error tolerance the iteration is repeated. If the difference is less than or equal to an error tolerance the estimated values of Q_a and s_{ws} are declared valid. The method (successive bisection) involves deciding whether estimated values of s_{ws} are too high or too low in relation to the known value of drawdown without well storage impacts.

FLOWING WELL

A flowing well is simulated as a sink flow rate assigned to the flowing well node. Upon removal of the cap from the flowing well in Figure 10.8, the head in the well declines to the top of the casing and remains constant. Discharge from the well is time variable and decreases with declining heads in the aquifer or aquifer layer. If the head in the aquifer or aquifer layer declines below the top of the casing, flow ceases. Flowing well simulation is based on the Thiem equation (Todd, 1980, p. 117) defining steady state flow from a production well as follows:

$$Q = 2\pi T(h_2 - h_1)/[\ln(r_2/r_1)] \qquad (10.30)$$

where Q is the steady state flow from a production well; T is the aquifer or aquifer layer transmissivity; r_1 and r_2 are the distances from flowing well to observation points 1 and 2, respectively; and h_1 and h_2 are the heads at observation points 1 and 2, respectively. Nonsteady state discharge from a flowing well is simulated numerically by applying the Thiem equation to successive time increments (see FORTRAN program listed by Walton, 1989b, pp. 102-103). h_1 is assumed to be the constant head $h_{f(i,j)}$ at the flowing well node i,j with r_1 equal to the effective flowing well radius (r_w). h_2 is assumed to be the head $h_{i,j}$ calculated at the flowing well node i,j at the end of the time increment Δt and r_2 is the equivalent flowing well block radius (r_e). With rectangular grid spacing and anisotropic

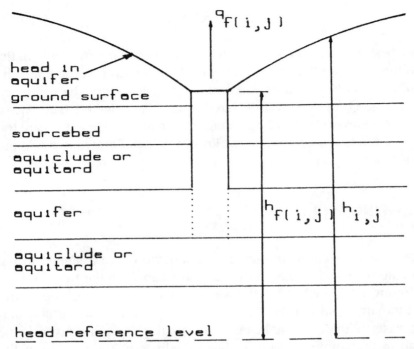

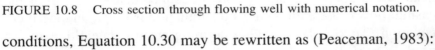

FIGURE 10.8   Cross section through flowing well with numerical notation.

conditions, Equation 10.30 may be rewritten as (Peaceman, 1983):

$$Q = 2\pi(T_{xx}T_{yy})^{0.5}(h_2 - h_1)/\ln(r_e/r_w) \qquad (10.31)$$

with r_e defined by Equation 10.26, where T_{xx} is the aquifer or aquifer layer transmissivity in the x-direction at the flowing well node and T_{yy} is the aquifer or aquifer layer transmissivity in the y-direction at the flowing well node.

Substitution of $h_{1(i,j)}$, r_w, $h_{2(i,j)}$, and r_e in Equation 10.31 results in the expression (Walton, 1989b, p. 31):

$$q_{3(i,j)} = [2\pi(T_{xx}T_{yy})^{0.5}/\ln(r_e/r_w)][h_{(i,j)} - h_{f(i,j)}] \qquad (10.32)$$

A flowing well simulation factor $R_{f(i,j)}$ is assigned to the flowing well node i,j through the DD terms in Equations 10.8 and 10.13 and is defined as:

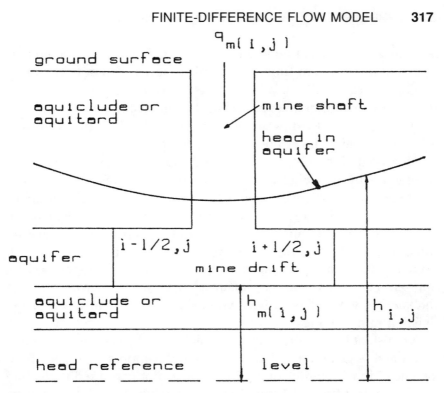

$q_{m(i,j)}$

ground surface

aquiclude or
aquitard

mine shaft

head in
aquifer

aquifer

$i-1/2,j$ $i+1/2,j$

mine drift

aquiclude or
aquitard

$h_{m(i,j)}$ $h_{i,j}$

head reference   level

FIGURE 10.9   Cross section through mine with numerical notation.

$$R_{f(ij)} = 2\pi(T_{xx}T_{yy})^{0.5}/\ln(r_e/r_w) \qquad (10.33)$$

The discharge rate $q_{3(i,j)}$ from the flowing well at node i,j at the end
of the time increment Δt is defined in terms of the simulation factor
by the following equation (Walton, 1989b, p. 31):

$$q_{3(ij)} = R_{f(ij)}h_{(ij)} - R_{f(ij)}h_{f(ij)} \qquad (10.34)$$

MINES

A mine shaft and drift illustrated in Figure 10.9 is simulated as a
constant head-variable discharge sink flow rate assigned to the mine
node. (OSM, 1981, pp. A21-A-22). The mine shaft in Figure 10.9
has sealed walls and is centered at the mine node. The drift cavity
extends the full aquifer or aquifer layer thickness to distances mid-

way between adjacent nodes. The grid at and near the mine node is assumed to be square (mine cavity plan view is square). The drift-aquifer or drift-aquifer layer surface is uniform and perpendicular to the aquifer or aquifer layer base. The drift intercepts all flow towards the mine node. At least two nodes separate individual mines and any aquifer or aquifer layer boundaries.

An unsealed shaft may be simulated with flowing well simulation techniques by replacing the top of casing elevation with the shaft floor elevation. Irregular mine drifts may be simulated by subdividing the total mine drift length into appropriate parts which are assigned to adjacent mine nodes. A FORTRAN program simulating mines is listed by Walton (1989b, pp. 112-117). Mine cavities which do not have square plan views and/or uniform thicknesses may be approximately simulated as square plan view mines with uniform thicknesses which have cavity volumes equivalent to those of the existing mine cavities.

A constant head is created at the mine node by setting the storage factor at a large number. The mine floor elevation $h_{m(i,j)}$ is specified. The head in the aquifer or aquifer layer is not permitted to decline below the drift floor. A form of Darcy's law useful in defining mine discharge rates is as follows (Ferris, et al, 1962, p. 73):

$$Q_d = TIL \tag{10.35}$$

with

$$I = \Delta h/D_f \tag{10.36}$$

where Q_d is the flow rate through an aquifer or aquifer layer cross section, T is the aquifer or aquifer layer cross section transmissivity, I is the hydraulic gradient, h is the head loss between adjacent nodes, L is the width of aquifer or aquifer layer cross section through which flow occurs, and D_f is the flow distance.

In numerical simulation, flow distances are equal to the grid spacing. However, at mine nodes the drift-aquifer or drift-aquifer layer face is midway between adjacent nodes and the effective flow distance is 1/2 of the grid spacing. Since numerical flow simulation rates at all nodes are based on head losses and distances between nodes (grid spacing), the transmissivity at mine nodes is divided by a mine node transmissivity adjustment factor (T_m) equal to 0.5 to compensate for the drift-aquifer or drift-aquifer layer face position and to satisfy Equation 10.35.

The total mine discharge rate from the four sides of the square cavity based on equation 10.35 is defined by the following equation (Walton, 1989b, p. 34):

$$q_{6(ij)} = T2_{i-1j}(h_{i-1j} - h_{m(ij)}) + T2_{ij}(h_{i+1j} - h_{m(ij)})$$

$$+ T1_{ij}[h_{ij+1} - h_{m(ij)}) + T1_{ij-1}(h_{ij-1} - h_{m(ij)})$$

$$(10.37)$$

where $q_{6(i,j)}$ is the mine discharge rate at node i,j at the end of the time increment t; $T1_{i,j}$ is the aquifer or aquifer layer transmissivity within the vector volume between nodes i,j and i,j+1; $T2_{i,j}$ is the aquifer or aquifer layer transmissivity within the vector volume between nodes i,j and i+1,j; and $h_{m(i,j)}$ is the mine floor elevation at node i,j.

Drainage of the mine cavity is not included in numerical calculations and is estimated with the following equation and added uniformly to $q_{6(i,j)}$ during the mine drift excavation time increment (Walton, 1989b, p. 34):

$$q_{c(i,j)} = m \, \Delta x \, \Delta y S_y / \Delta t_m \qquad (10.38)$$

where $q_{c(i,j)}$ is the mine cavity drainage rate at node i,j, m is the aquifer or aquifer layer thickness, Δx is the grid spacing in the x-direction at the mine node, Δy is the grid spacing in the y-direction at the mine node, S_y is the aquifer or aquifer layer specific yield at the mine node, and t_m is the mine drift excavation time period.

A partially penetrating mine drift similar to that illustrated in Figure 10.9 may be simulated by further adjustment of T_m based on the degree of mine drift penetration into the aquifer or aquifer layer and the ratio of the horizontal and vertical hydraulic conductivities (P_h/P_v) of the aquifer or aquifer layer (Prickett, oral communication, 1983). Flow into the mine drift decreases as the degree of penetration decreases and the P_h/P_v ratio increases because of variable vertical partial penetration head losses in addition to horizontal head losses with full penetration.

A separate finite-difference cross-sectional aquifer slice model (see Wang and Anderson, 1982, pp. 19-22) was used to calculate the values of T_m under selected partial penetration conditions in Appendix I. T_m is the ratio of the flow into a mine node versus the flow into a non-mine node with a constant head loss and selected mine penetrations and conductivity ratios. The depth of penetration into the aquifer or aquifer layer is substituted into Equation 10.38 instead of m to calculate mine drift cavity drainage under partial penetrating conditions.

Mine drift stepwise advancement may be simulated by creating constant heads sequentially in time at adjacent mine nodes. Small time increments are specified each time there is a change in the mine schedule. After mining is complete, constant heads are released and T_m is reset to 1.0. Mine drift cavities backfilled with spoil materials may be simulated. The hydraulic conductivity and storativity of the aquifer or aquifer layer at the mine nodes are set equal to those of the spoil materials during reclamation times.

DRAINS

A drain is simulated as a sink flow rate assigned to the drain node. Seepage into the drain illustrated in Figure 10.10 is simulated in a manner analogous to the simulation of induced infiltration of surface water (see FORTRAN program listed by Walton, 1989b, pp. 94-95). The head in the drain is assumed to be the elevation of the drain top. A drain layer represents drain tile openings with any chemical precipitation and the backfill around the tile. The effective horizontal width of the drain layer is assumed to be the drain tile diameter plus the drain layer thickness multiplied by 2. It is further assumed that the aquifer or aquifer layer head does not decline below the drain top and the saturated drain layer has negligible storativity. The total drain length is subdivided in several lengths which are assigned to appropriate nodes. Drain lengths may vary from node to node and differ from grid spacings.

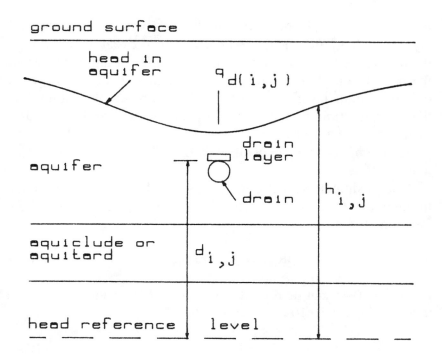

FIGURE 10.10 Cross section through drain with numerical notation.

According to Darcy's law, the drain discharge rate is directly proportional to the drain horizontal area (effective horizontal width of drain layer multiplied by drain length), the vertical hydraulic conductivity of the drain layer, and the head difference between that in the aquifer or aquifer layer and in the drain; it is inversely proportional to the drain layer thickness as indicated in the following equation (McDonald and Harbaugh, 1984, pp. 288-292):

$$q_{5(ij)} = (P_{d(ij)}/m_{d(ij)})\Delta h_{ij}A_{d(ij)} \tag{10.39}$$

with

$$A_{d(ij)} = W_{d(ij)}L_{d(ij)} \tag{10.40}$$

where $q_{5(i,j)}$ is the drain discharge at node i,j at the end of the time increment Δt, $P_{d(i,j)}$ is the vertical hydraulic conductivity of the drain layer at node i,j, $m_{d(i,j)}$ is the drain layer thickness at node i,j, $\Delta h_{i,j}$ is the difference between the head in the aquifer or aquifer layer and the head in the drain at the end of time increment t, $W_{d(i,j)}$ is the effective drain layer width (drain tile diameter plus drain layer thickness multiplied by 2) at node i,j, $L_{d(i,j)}$ is the length of the drain assigned to node i,j (the total drain length is subdivided into node centered lengths which may differ from grid spacings and vary from node to node).

A drain simulation factor $R_{d(i,j)}$ is assigned to all drain grid nodes through the DD terms in Equations 10.8 and 10.13 as follows (Walton, 1989b, p. 37):

$$R_{d(ij)} = (P_{d(ij)}/m_{d(ij)})A_{d(ij)} \qquad (10.41)$$

Equation 10.41 can be rewritten in terms of $R_{d(i,j)}$ and the head notation in Figure 10.10 as (Walton, 1989b, p. 37):

$$q_{5(ij)} = R_{d(ij)}h_{ij} - R_{d(ij)}d_{ij} \qquad (10.42)$$

where $q_{5(i,j)}$ is the drain discharge rate at node i,j at the end of time increment Δt; $R_{d(i,j)}$ is the drain simulation factor at node i,j; $h_{i,j}$ is the aquifer or aquifer layer head at node i,j at the end of the time increment Δt; and $d_{i,j}$ is the drain top elevation at node i,j.

LEAKAGE THROUGH AQUITARDS

Leakage through an aquitard is simulated as source/sink flow rates assigned to nodes where leakage occurs. Leakage through an aquitard (Figure 10.11) with negligible storativity is proportional to the aquitard vertical hydraulic conductivity, the area of the aquitard through which leakage takes place, and the vertical hydraulic gradient. It is inversely proportional to the aquitard thickness. Commonly, the leakage rate through the aquitard is assumed to reach a maximum value and becomes independent of the aquifer or aquifer layer head when the aquifer or aquifer layer head declines below the

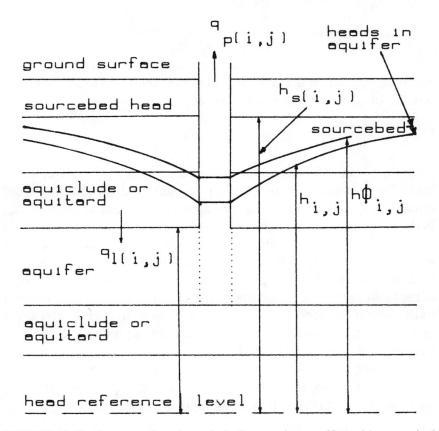

FIGURE 10.11 Cross section through leaky artesian aquifer with numerical notation.

aquitard base and maximum vertical hydraulic gradients are established (Trescott, et al., 1976, p. 11). With this assumption, leakage may be simulated with the following equation which is a modified form of Darcy's law (Prickett and Lonnquist, 1971, pp. 30-32):

$$q_{a(i,j)} = (P'_{i,j}/m'_{i,j})\Delta h_{i,j}A_{l(i,j)} \qquad (10.43)$$

with

$$A_{l(i,j)} = \Delta x_{i,j}\Delta y_{i,j} \qquad (10.44)$$

For $h_{i,j} > h_{a(i,j)}$

$$\Delta h_{i,j} = h_{s(i,j)} - h_{i,j} \tag{10.45}$$

For $h_{i,j} \leq h_{a(i,j)}$ when leakage rates are maximum and independent of aquifer or aquifer layer head changes

$$\Delta h_{i,j} = h_{s(i,j)} - h_{a(i,j)} \tag{10.46}$$

where $q_{a(i,j)}$ is the leakage rate through the aquitard at node i,j at the end of the time increment Δt; $P'_{i,j}$ is the aquitard vertical hydraulic conductivity at node i,j; $m'_{i,j}$ is the aquitard thickness at node i,j; $A_{li,j}$ is the aquitard area through which leakage occurs at node i,j; $\Delta x_{i,j}$ is the grid spacing in the x-direction at node i,j; $\Delta y_{i,j}$ is the grid spacing in the y-direction at node i,j; $\Delta h_{i,j}$ is the vertical head loss associated with the leakage at node i,j at the end of the time increment Δt; $h_{s(i,j)}$ is the head in the sourcebed above the aquitard at node i,j at the end of the time increment Δt; $h_{i,j}$ is the head in the aquifer or aquifer layer at node i,j at the end of the time increment Δt; and $h_{a(i,j)}$ is the elevation of the aquitard base at node i,j.

A leakage simulation factor $R_{1(i,j)}$ is assigned to each node where the aquitard exists through the DD terms in Equations 10.8 and 10.13 as follows (Prickett and Lonnquist, p. 30):

$$R_{1(i,j)} = (P'_{i,j}/m_{i,j}')A_{1(i,j)} \tag{10.47}$$

The leakage rate through the aquitard $q_{a(i,j)}$ in terms of the leakage factor is given by the following equation:

$$q_{a(i,j)} = R_{1(i,j)}\Delta h_{i,j} \tag{10.48}$$

The vector volume of the portion of the aquitard represented at each node is illustrated in Figure 10.12. The vector volume is centered at each node, extends the full depth of the aquitard m', and has horizontal dimensions of $\Delta x \Delta y$. A FORTRAN program for simulating leakage through aquitards is listed by Prickett and Lonnquist (1971, p. 32).

Equation 10.1 assumes that aquitard storativity is negligible.

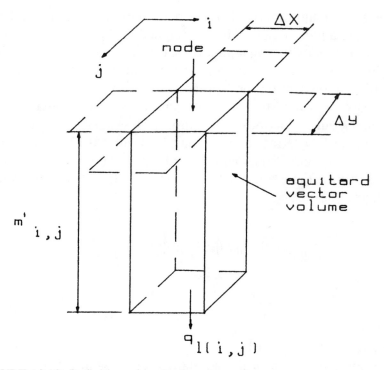

FIGURE 10.12 Definition of aquitard vector volume.

Sometimes it is necessary to simulate the release of water stored in aquitards (see FORTRAN program listed by Walton, 1989b, pp. 67-156). The time period during which aquitard storativity impacts are appreciable and should be simulated can be calculated with the following equation (Bredehoeft and Pinder, 1970, pp. 887-888; Ghislain de Marsily, 1986, p. 363):

$$t_a = S'm'/(5P') \qquad (10.49)$$

where t_a is the time after the last pumping rate change during which release of water stored in aquitard is appreciable and simulation of aquitard storativity is important, S' is the aquitard storativity, m' is the aquitard thickness, and P' is the aquitard vertical hydraulic conductivity.

In some cases, one-half of the aquitard storativity is assigned to the aquifer or aquifer layer beneath an underlying aquitard and one-half of the aquitard storativity is assigned to the aquifer or aquifer layer above an overlying aquitard. The aquitard may be subdivided into three layers, 2 aquitards and 1 aquifer, with the aquifer storativity representing the aquitard storativity.

Release of water stored in an aquitard may be simulated more accurately during periods when time is less then t_a by assigning through the DD terms in Equations 10.8 and 10.13 to each node where the aquitard exists an aquitard storativity simulation factor $R_{t(i,j)}$ as follows (Trescott, et al, 1976, pp. 4-7; Bredehoeft and Pinder, 1970, pp. 883-888):

$$R_{t(ij)} = P'_{ij}A_{ij} / \{[\pi P'_{ij} t_a m'_{ij} / (3m'^2_{ij} S'_{ij})]^{0.5}$$

$$m'_{ij}\}\{1 + 2\sum_{n=1}^{200} \exp(-n^2 / [P'_{ij} t_a m'_{ij} / (3m'^2_{ij} S'_{ij})]\}\} \tag{10.50}$$

where $R_{t(i,j)}$ is the aquitard storativity simulation factor at node i,j; $A_{i,j}$ is the horizontal area of the aquitard vector volume within which water is released from storage at node i,j; $S'_{i,j}$ is the aquitard storativity at node i,j; $m'_{i,j}$ is the aquitard thickness at node i,j: $P'_{i,j}$ is the aquitard hydraulic conductivity at node i,j; and t_a is the time after last pumping rate change.

It is assumed that transient leaky affects from previous pumping periods have dissipated (each pumping rate period exceeds t_a in the case of multi-pumping rate schedules). The aquitard storage release rate $q_{t(i,j)}$ at node i,j is governed by the following equation (Walton, 1989b, p. 41):

$$q_{t(ij)} = R_{t(ij)}h\phi_{ij} - R_{t(ij)}h_{ij} \tag{10.51}$$

where $h_{i,j}$ is the head in the aquifer or aquifer layer at the end of the time increment Δt and at node i,j; $h\phi_{i,j}$ is the head in the aquifer or aquifer layer at the end of the previous time increment at node i,j.

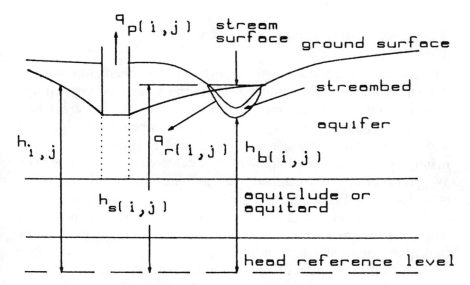

FIGURE 10.13 Cross section through streambed adjacent to production well with numerical notation.

INDUCED STREAMBED INFILTRATION

Induced streambed infiltration is simulated as source flow rates assigned to streambed nodes. Similar to leakage through aquitards, leakage through streambeds (induced streambed infiltration in Figure 10.13) or other surface water beds is usually simulated with Darcy's law as follows (Prickett and Lonnquist, 1971, pp. 33-36):

$$q_{12(i,j)} = (P'_{i,j}/m'_{i,j})\Delta h_{i,j}A_{r(i,j)} \tag{10.52}$$

with

$$A_{r(i,j)} = W_{s(i,j)}L_{s(i,j)} \tag{10.53}$$

For $h_{i,j} > h_{b(i,j)}$

$$\Delta h_{i,j} = h_{s(i,j)} - h_{i,j} \tag{10.54}$$

For $h_{i,j} \leq h_{b(i,j)}$ when the leakage rate through the streambed is assumed to be maximum and independent of aquifer head changes

$$\Delta h_{i,j} = h_{s(i,j)} - h_{b(i,j)} \qquad (10.55)$$

where $q_{12(i,j)}$ is the leakage rate through the streambed at node i,j at the end of the time increment Δt; $P'_{i,j}$ is the streambed vertical hydraulic conductivity assigned to node i,j; $m'_{i,j}$ is the streambed thickness assigned to node i,j; $A_{r(i,j)}$ is the streambed area assigned to node i,j (the total irregular streambed area is simulated approximately by a number of component node centered rectangular areas); $W_{s(i,j)}$ is the effective streambed width assigned to node i,j; $L_{s(i,j)}$ is the effective streambed length assigned to node i,j; $\Delta h_{i,j}$ is the aquifer or aquifer layer head loss at node i,j at the end of the time increment Δt; $h_{s(i,j)}$ is the stream surface elevation at node i,j; $h_{i,j}$ is the head in the aquifer or aquifer layer at node i,j at the end of time increment Δt; $h_{b(i,j)}$ is the elevation of the streambed base at node i,j.

Equation 10.52 assumes that streamflow is sufficient to balance induced infiltration, the streambed does not go dry as the result of induced infiltration, streambed storativity is negligible, the drawdown in the aquifer or aquifer layer is small in comparison to the aquifer or aquifer layer thickness; the impacts of streambed partial penetration are negligible; the surface water temperature and viscosity remain constant during the time increment; the materials beneath the stream are saturated; atmospheric pressure conditions exist beneath the streambed regardless of the position of aquifer or aquifer layer heads; and the streambed vertical hydraulic conductivity, thickness, width, and length remain constant during the time increment.

$W_{s(i,j)}$, $L_{s(i,j)}$, and $A_{r(i,j)}$ usually differ from grid spacings and vary from node to node.

An induced streambed infiltration simulation factor $R_{r(i,j)}$ is assigned to each node where the streambed exists through the DD terms in Equations 10.8 and 10.13 using the following equation (Prickett and Lonnquist, 1971, p. 33):

$$R_{r(i,j)} = (P'_{i,j}/m'_{i,j})A_{r(i,j)} \qquad (10.56)$$

The leakage flow rate through the streambed $q_{12(i,j)}$ expressed in terms of the simulation factor is as follows:

$$q_{12(ij)} = R_{r(ij)}\Delta h_{ij} \tag{10.57}$$

The streambed vector volume is centered at each node, extends the full depth of the streambed m', and has horizontal dimensions of $W_{s(i,j)}$ and $L_{s(i,j)}$ (Figure 10.14).

A FORTRAN program for simulating induced streambed infiltration is listed by Prickett and Lonnquist (1971, p. 36). The streambed vertical hydraulic conductivity, thickness, width and length, the stream surface elevation, and the streambed base elevation may be programmed to automatically change at the beginning of each time increment in stepwise fashion to simulate changes in the characteristics and geometric dimensions of the streambed due to stream stage changes. Variable streambed characteristics and geometric dimensions may also be interactively specified from the keyboard at the beginning of each time increment (FORTRAN program listed by Walton, 1989b, pp. 134-136).

In semi-arid to arid climates, the streambed may go dry as the result of induced infiltration. Rovey (1975) describes a method for simulating induced infiltration when the streambed goes dry.

When heads in the aquifer or aquifer layer are lowered below the stream, the streambed and underlying materials may become unsaturated. Equations 10.52 and 10.54 do not consider seepage of water through unsaturated materials. Under unsaturated conditions, calculated aquifer or aquifer layer heads are lower than actual and the calculated leakage rate through the streambed is less than actual. Induced infiltration rates calculated with Equations 10.52 and 10.54 may be incorrect because the influence of negative pressure heads in unsaturated materials are ignored (Peterson, 1989, pp. 899-927).

Under unsaturated conditions, aquifer or aquifer layer head changes continue to affect the leakage rate through the streambed even though heads are below the streambed base. Therefore, the leakage rate through the streambed is not at a maximum and independent of the aquifer or aquifer layer head when $h_{i,j} \leq h_{b(i,j)}$. Instead, the leakage rate through the streambed continues to increase until further reduction in the moisture content in the unsaturated zone beneath the stream creates a condition in which unsaturated materials are incapable of conveying water at the same rate as stream losses. Peterson (1989, pp. 899-927) describes a method for simulating unsaturated conditions beneath a stream. In many cases, inaccuracies in estimating the streambed vertical hydraulic conductivity and thickness may overshadow errors due to ignoring unsaturated conditions.

AREAL RECHARGE

Areal recharge is simulated as source flow rates assigned to nodes where recharge occurs. Areal recharge from precipitation is usually assumed to be uniformly distributed over areas centered at nodes and independent of the head in the aquifer or aquifer layer. With these assumptions, the areal recharge flow rate $q_{7(i,j)}$ is usually simulated with the following equation (McDonald and Harbaugh, 1984, p. 241):

$$q_{7(i,j)} = I_{i,j} \Delta x_{i,j} \Delta y_{i,j} \tag{10.58}$$

where $q_{7(i,j)}$ is the areal recharge flow rate from precipitation at node i,j at the end of the time increment t; $I_{i,j}$ is the areal recharge rate from precipitation per unit area at node i,j at the end of the time increment Δt; $\Delta x_{i,j}$ is the grid spacing in the x-direction at node i,j; and $\Delta y_{i,j}$ is the grid spacing in the y-direction at node i,j.

The areal recharge flow rate from precipitation may be programmed to automatically change at the beginning of each time increment in a stepwise fashion to simulate variable recharge from precipitation. Variable areal recharge flow rates from precipitation may also be specified from the keyboard at the beginning of each time increment (FORTRAN program listed by Walton, 1989b, pp. 67-156).

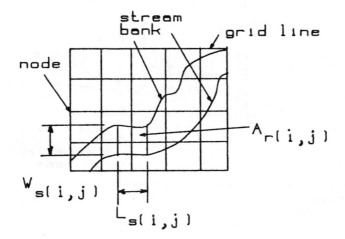

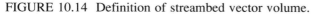

FIGURE 10.14 Definition of streambed vector volume.

EVAPOTRANSPIRATION

Evapotranspiration is simulated as sink flow rates assigned to nodes where evapotranspiration occurs. The flow rate of evapotranspiration from the water table $q_{e(i,j)}$ (Figure 10.15) is usually simulated as a linear function of the difference between the elevations of the ground surface and the water table (Prickett and Lonnquist, 1971, pp. 37-39). Other functions may also be used to simulate evapotranspiration (Trescott, et al., 1976, p. 8). It is assumed that evapotranspiration ceases when the water table declines below a critical depth. Evapotranspiration is usually simulated with the following equations (Prickett and Lonnquist, 1971, p. 37 and Trescott, et al. ,1976, pp. 7-8):

when $h_{i,j} = g_{i,j}$

$$q_{8(i,j)} = q_{x(i,j)} \tag{10.59}$$

when $h_{i,j} \leq c_{i,j}$

$$q_{8(i,j)} = 0 \tag{10.60}$$

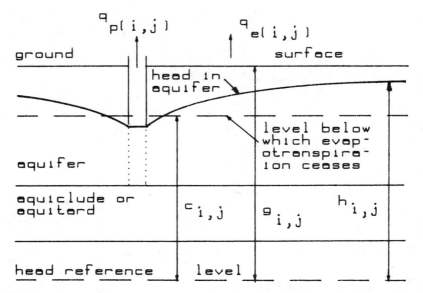

FIGURE 10.15 Cross section through aquifer with evapotranspiration losses and numerical notation.

when $h_{i,j} < g_{i,j}$ and $h_{i,j} > c_{i,j}$

$$q_{8(i,j)} = q_{x(i,j)} - R_{e(i,j)}(g_{i,j} - h_{i,j}) \qquad (10.61)$$

with

$$R_{e(i,j)} = q_{x(i,j)}/(g_{i,j} - c_{i,j}) \qquad (10.62)$$

where $h_{i,j}$ is the head in the aquifer or aquifer layer at node i,j at the end of the time increment Δt; $g_{i,j}$ is the ground surface elevation at node i,j; $q_{8(i,j)}$ is the evapotranspiration flow rate at node i,j at the end of the time increment Δt; $q_{x(i,j)}$ is the maximum evapotranspiration rate which occurs when the water table is at ground surface at node i,j; $c_{i,j}$ is the elevation below ground surface at which evapotranspiration ceases at node i,j; and $R_{e(i,j)}$ is the evapotranspiration simulation factor at node i,j which is assigned to each node where evapotranspiration occurs through the DD terms in Equations 10.8 and 10.13.

A FORTRAN program for simulating evapotranspiration is listed

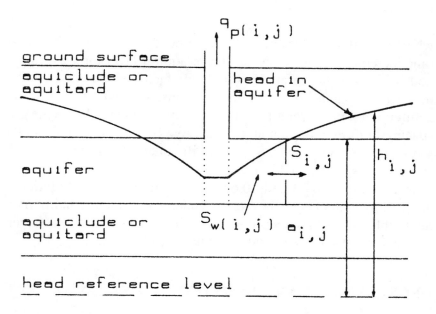

FIGURE 10.16 Cross section through aquifer undergoing storativity conversion with numerical notation.

by Prickett and Lonnquist (1971, p. 39).

STORATIVITY CONVERSION

Equation 10.1 assumes that the aquifer or aquifer layer storativity is constant during the time increment Δt. When water levels decline below the aquifer or aquifer layer top there is a storativity conversion from artesian to water table (Figure 10.16). This condition is simulated with the following equations (Prickett and Lonnquist, 1971, p. 40 and Trescott, et al., 1976, p. 11):

For $h_{i,j} \geq a_{i,j}$

$$S_{f(i,j)} = S_{i,j}\Delta x_{i,j}\Delta y_{i,j} \qquad (10.63)$$

For $h_{i,j} < a_{i,j}$

$$S_{f(ij)} = S_{w(ij)}\Delta x_{ij}\Delta y_{ij} \qquad (10.64)$$

where $h_{i,j}$ is the aquifer or aquifer layer head at node i,j at the end of time increment Δt; $a_{i,j}$ is the aquifer or aquifer layer top elevation at node i,j; $S_{f(i,j)}$ is the storativity simulation factor which is assigned to node i,j through the DD terms in Equations 10.8 and 10.13; $S_{i,j}$ is the aquifer or aquifer layer artesian storativity at node i,j; $\Delta x_{i,j}$ is the grid spacing in the x-direction at node i,j; $\Delta y_{i,j}$ is the grid spacing in the y-direction at node i,j; and $S_{w(i,j)}$ is aquifer or aquifer layer water table storativity (specific yield) at node i,j.

In Equations 10.63 and 10.64, it is assumed that the aquifer or aquifer layer thickness dewatered under water table conditions is negligible. A FORTRAN program for simulating storativity conversion is listed by Prickett and Lonnquist (1971, p. 42).

DECREASING TRANSMISSIVITY

Transmissivity is assumed to be constant during the time increment Δt in equation 10.1. However, under water table conditions (Figure 10.17) transmissivity is variable because gravity drainage of pores decreases the aquifer or aquifer layer saturated thickness as the head declines with time (Prickett and Lonnquist, 1971, pp. 43-45; McDonald and Harbaugh, 1984, pp. 136-137; and Kinzelbach, 1986, pp. 63-65). In a fully implicit scheme, transmissivity is defined at the start of each iteration as a function of the head from the present iteration meaning that Equation 10.1 would become nonlinear (Kinzelbach, 1986, pp. 63-65). To avoid nonlinearity, transmissivity is usually approximated as a function of heads from the preceding iteration with the following equations (Butler, 1957):

$$T2_{i,j} = P2_{(i,j)}[(h_{i,j} - b_{i,j})(h_{i+1,j} - b_{i+1,j})]^{0.5} \qquad (10.65)$$

$$T1_{i,j} = P1_{(i,j)}[(h_{i,j} - b_{i,j})(h_{i,j+1} - b_{i,j+1})]^{0.5} \qquad (10.66)$$

where $T2_{i,j}$ is the transmissivity of the aquifer or aquifer layer vector volume between i,j and i + 1,j; $P2_{i,j}$ is the horizontal hydraulic conductivity of the aquifer or aquifer layer vector volume i,j and i

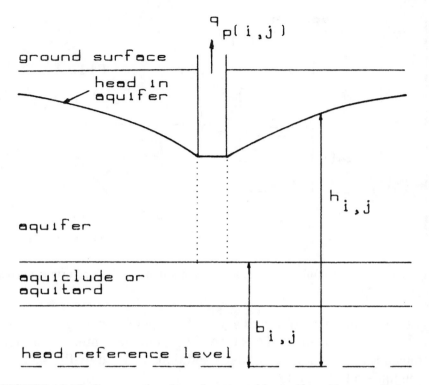

FIGURE 10.17 Cross section through water table aquifer with numerical notation.

+ 1,j with the same horizontal dimensions as those of $T2_{i,j}$ defined in Figure 10.2; $T1_{i,j}$ is the transmissivity of the aquifer or aquifer layer vector volume between i,j and i, j + 1 with the same horizontal dimensions as those of $T1_{i,j}$ defined in Figure 10.2; $P1_{i,j}$ is the horizontal hydraulic conductivity of the aquifer or aquifer layer vector volume between i,j and i,j + 1; $h_{i,j}$ is the head at node i,j at the end of the time increment Δt; and $b_{i,j}$ is the constant elevation of the aquifer base at node i,j.

In Equations 10.65 and 10.66 transmissivity is calculated as the geometric average saturated thickness within a wedge shaped aquifer or aquifer layer vector volume multiplied by the aquifer or aquifer layer horizontal hydraulic conductivity (Figure 10.18). Equations 10.65 and 10.66 assume that the impacts of delayed

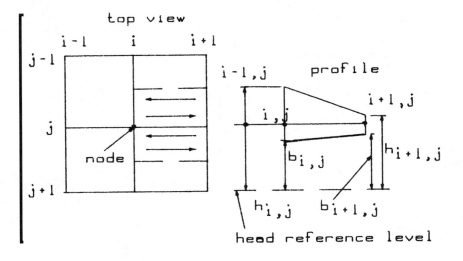

FIGURE 10.18 Definition of vector volume for transmissivity adjustment under water table conditions.

gravity drainage of aquifer or aquifer layer pores is negligible during the time increment. A FORTRAN program to simulate variable transmissivity under water table conditions is listed by Prickett and Lonnquist (1971, p. 45).

Transmissivity must always have a positive residual value to prevent erroneous changes to no flow conditions and to allow refilling of the aquifer or aquifer layer and vertical inflows. Commonly, heads are set equal to the aquifer or aquifer layer base elevation plus 0.01 ft (0.0308 m) in cases where the head has declined to or below the aquifer or aquifer layer base (Prickett and Lonnquist, 1971, p. 43). Computer programs simulating a multi-layer aquifer in which a production well causes drawdown below the base of the upper aquifer layer may experience difficulty with no flow conditions and the simulation may degenerate (McDonald and Harbaugh, 1984, p. 149).

Potter and Gburek (1987, pp. 722-732) developed equations for simulating zones of seepage contiguous and noncontiguous to streams, lakes, or other points of discharge with variable transmissivity conditions.

DELAYED GRAVITY DRAINAGE

Equation 10.1 assumes that delayed gravity yield under water table conditions is negligible during the time increment Δt. The time period during which the impacts of delayed gravity yield are appreciable and should be simulated is determined with Equation 2.13. Delayed gravity yield is simulated by assigning to each active node an effective storativity $S_{e(i,j)}$ and an effective recharge source flow rate $q_{11(i,j)}$ defined as (Rushton and Redshaw, 1979, pp. 259-262):

$$S_{e(i,j)} = S_{i,j} + S_{y(i,j)}[1 - \exp(-a_{i,j}t)] \tag{10.67}$$

$$q_{11(ij)} = a_{ij}S_{y(ij)} \sum_{i=1}^{n-1} \Delta s_m \exp[-a_{ij}(t_n - t_m)]A_{ij} \tag{10.68}$$

with

$$a_{i,j} = 3P_{v(i,j)}/(S_{y(i,j)}m_{i,j}) \tag{10.69}$$

(Neuman, 1979, pp. 899-908; Streltsova, 1972, pp. 1059-1066)

where $S_{e(i,j)}$ is the effective aquifer or aquifer layer storativity with delayed gravity yield during the current time increment at node i,j; $S_{i,j}$ is the aquifer or aquifer layer storativity at node i,j; $S_{y(i,j)}$ is the aquifer or aquifer layer specific yield at node i,j; $a_{i,j}$ is the reciprocal of the delayed index at node i,j; t is the time after pumping started; Δs_m is the drawdown during the m th time increment; t_m is the time at the end of the m th time increment; t_n is the time at the end of the n th time increment; n is the current time increment number; $q_{11(i,j)}$ is the effective recharge flow rate arising from delayed gravity yield due to previous drawdown (during m th time increment at a drawdown Δs_m occurred which produces a delayed gravity yield contribution at a later time t_n) at node i,j; $A_{i,j}$ is the horizontal area of the aquifer or aquifer layer vector volume in which delayed gravity yield occurs at node i,j; $P_{vi,j}$ is the aquifer or aquifer layer vertical hydraulic conductivity at node i,j; $h_{i,j}$ is the head in the aquifer or aquifer layer at end of the current time increment at node i,j; and $m_{i,j}$ is the aquifer or aquifer layer thickness at node i,j.

Small time increments (minutes or less) and grid spacings near the production well (usually tens of feet or several meters) and a small nodal head convergence error (usually 0.001 ft or 0.0003 m) are required for precise simulation of delayed gravity yield. Other simulation factors dependent on head variations cannot be assigned to delayed gravity yield nodes. A FORTRAN program simulating delayed gravity yield is listed by Walton (1989b, pp. 67-156).

MULTI-AQUIFER SYSTEMS

Many groundwater systems consist of several aquifers interconnected by aquitards (Figure 10.19) or several adjoining aquifer layers with interaquifer flow transfer. It is theoretically possible to numerically solve transient flow equations in three space dimensions including those for aquitards but a large number of nodes and a large calculation time are required. Commonly, multi-aquifer or aquifer layer systems are simulated by coupling two-dimensional flow equations for stacked aquifer-aquitard units or adjoining aquifer layers via source/sink flow rates $q_{9(i,j)}$ (see equations 10.43-10.51) or $q_{10(i,j)}$ (Bredehoeft and Pinder, 1970, pp. 883-888 and Prickett and Lonnquist, 1971, p. 46).

In this quasi three-dimensional simulation, horizontal flow in aquitards is assumed to be negligible and aquitard heads are not calculated. A FORTRAN program simulating a multiaquifer system is listed by Prickett and Lonnquist (1971, p. 48).

In multi aquifer-aquitard units or aquifer layers, units or layers are commonly numbered from land surface downward using the K notation colinear with the z-direction. The origin of the system is usually the upper-left corner of the topmost unit or layer. Grid spacing is usually the same in all aquifer-aquitard units or layers. The number of aquifer-aquitard units and/or aquifer layers is usually 10 or less. The flow exchange between stacked aquifer-aquitard units or adjoining aquifer layers is dependent on the relative heads in the aquifers or aquifer layers (Kinzelbach, 1986, p. 65).

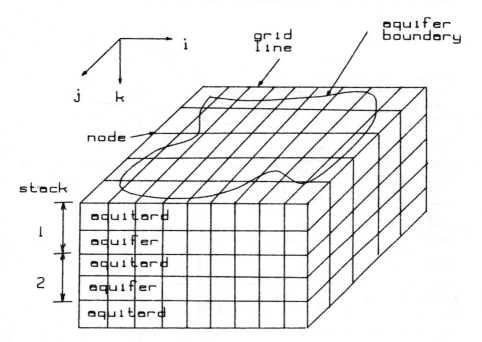

FIGURE 10.19 Discretized multi-aquifer system.

AQUIFER LAYERING

Interaquifer flow transfer between ajoining aquifer layers is simulated as source/sink flow rates assigned to layered aquifer nodes. The layered aquifer illustrated in Figure 10.20 is simulated by assigning to each node through the DD terms in Equations 10.8 and 10.13 an aquifer layering simulation factor $R_{y(i,j)}$ defined as follows (McDonald and Harbaugh, 1984, pp. 130, 134, 142):

$$R_{y(i,j)} = C_1 C_2/(C_1 + C_2) \qquad (10.70)$$

with

$$C_1 = P1_{v(i,j)} A_{y(i,j)}/(m1_{i,j}/2) \qquad (10.71)$$

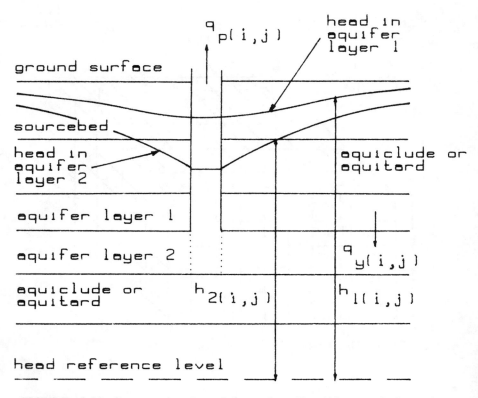

FIGURE 10.20 Cross section through layered aquifer with numerical notation.

$$C_2 = P2_{v(i,j)}A_{y(i,j)}/(m2_{i,j}/2) \qquad (10.72)$$

where $P1_{v(i,j)}$ is the aquifer layer 1 vertical hydraulic conductivity at node i,j; $P2_{v(i,j)}$ is the aquifer layer 2 vertical hydraulic conductivity at node i,j; $A_{y(i,j)}$ is the horizontal area of inter-aquifer flow transfer centered at node i,j; $m1_{i,j}$ is the aquifer layer 1 thickness at node i,j; and $m2_{i,j}$ is the aquifer layer 2 thickness at node i,j.

The rate of flow between aquifer layers $q_{10(i,j)}$ is as follows (McDonald and Harbaugh, 1984, p. 130):

$$q_{10(ij)} = R_{y(ij)}(h_{1(ij)} - h_{2(ij)}) \qquad (10.73)$$

where $h_{1(i,j)}$ is the head in aquifer layer 1 at the end of the time increment Δt and $h_{2(i,j)}$ is the head in aquifer layer 2 at the end of the time increment Δt.

It is usually assumed that water levels in an aquifer layer do not decline below the base of that layer. Storativity in an aquifer layer is artesian unless the head in the layer is less than the layer top elevation then it is water table. A FORTRAN program simulating aquifer layering is listed by Walton (1989b, pp. 67-156).

MULTI-AQUIFER WELL

A multi-aquifer production well is simulated as source/sink flow rates assigned to production well nodes with an approximate method developed by Bennett, et al (1982, pp. 334-341). The method allows calculation of the water level in the production well and individual aquifer or aquifer layer discharges to the production well. A nonpumping case is covered in which there may be flow through the production well between aquifers or aquifer layers. A multi-aquifer observation well is simulated as a nonpumping multi-aquifer production well. A multi-aquifer recharge well (negative discharge) also may be simulated.

A multi-aquifer production well in a multi-aquifer system is illustrated in Figure 10.21. The discharge rates into the production well from aquifers or aquifer layers open to the production well which appear as $q_{4(i,j)}$ in Equation 10.2 (Prickett and Lonnquist, 1971, pp. 46-48) are calculated with the following equation (Bennett, et al, 1982, pp. 334-341)

$$q_{4(i,j,b)} = 2\pi T_{i,j,b} h_{i,j,b} / \ln(r_s / r_w) - [2\pi T_{i,j,b} /$$

$$\ln(r_s / r_w)]\{ \sum_{b=m}^{n} [T_{i,j,b} h_{i,j,b} / \ln(r_s / r_w)] / \sum_{b=m}^{n} [T_{i,j,b} /$$

$$\ln(r_s / r_w)]\} + T_{i,j,b} q_{1(i,j)} / \ln(r_s / r_w) \sum_{b=m}^{n} [T_{i,j,b} / \ln(r_s / r_w)]\}$$

$$(10.74)$$

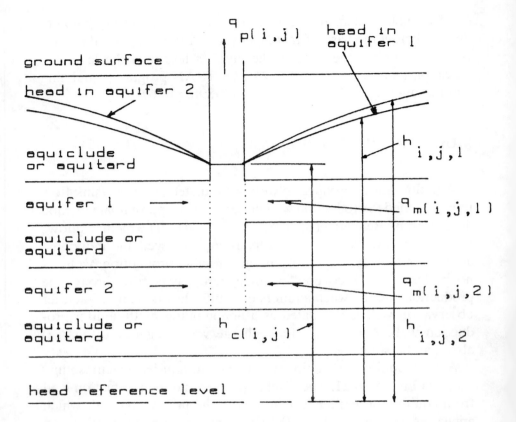

FIGURE 10.21 Cross section through multi-aquifer production well with numerical notation.

with r_e defined by Equation 10.26, where $T_{i,j,b}$ is the average aquifer or aquifer layer transmissivity at the production well node i,j in aquifer or aquifer layer b; $h_{i,j,b}$ is the head in aquifer or aquifer layer b at node i,j at end of the time increment t; r_e is the equivalent production well block radius defined in equation 10.26; r_w is the production well effective radius, $q_{1(i,j)}$ is the constant production well discharge rate at node i,j; m,n is the range of summation which includes each aquifer or aquifer layer open to the production well; $q_{4(i,j,b)}$ is the constant flow rate from aquifer or aquifer layer b into or out of the production well at node i,j.

Values of $h_{i,j,b}$ are generated by iterative techniques and substituted into the following equation to calculate the composite water level in the production well $h_{c(i,j)}$ (Bennett,et al, 1982, p. 337):

$$h_{c(i,j)} = \sum_{b=m}^{n} [T_{i,j,b} h_{i,j,b} / \ln(r_s / r_w)] / \sum_{b=m}^{n} [T_{i,j,b} /$$

$$\ln(r_s / r_w)]\} - q_{1(i,j)} / \{2\pi \sum_{b=m}^{n} [T_{i,j,b} / \ln(r_s / r_w)]\} \tag{10.75}$$

Then, values of $h_{c(i,j)}$ and $h_{i,j,b}$ are substituted in the following equation to estimate the rate of flow from each aquifer or aquifer layer $[q_{4(i,j,b)}]$ into or out of the production well (Bennett, et al, 1982, p. 336):

$$_{4(i,j,b)} = 2\pi T_{i,j,b}(h_{i,j,b} - hc_{i,j})/\ln(r_e/r_w) \qquad 10.76$$

FORTRAN program simulating a multiaquifer well is listed by alton (1989b, p. 67-156).

11

Random Walk Mass Transport Model

The random walk method (Spitzer, 1964) has been applied to the analysis of mass transport by Ahlstrom, et al. (1977), Schwartz and Crowe (1980), Smith and Schwartz (1980, pp. 303-313), Prickett, et al. (1981), Uffink (1985, pp. 103-114), Kinzelbach (1988, pp. 227-245), and Ackerer (1988, pp. 475-486). The random walk mass transport model described herein is based on the RANDOM WALK model developed by Prickett, et al. (1981). The model employs particle tracking to describe both advective and dispersive contaminant migration. Advection is simulated as particle movement along streamlines of the groundwater flow velocity field and dispersion is simulated as a random particle movement the statistical properties of which are related to the size of the dispersion coefficient. The groundwater flow velocity field is generated from an aquifer or aquifer layer head data base commonly obtained with a finite-difference flow model.

The random walk model is approximate because of the general roughness of calculated concentration distributions in time and space (Figure 11.1) due to statistical fluctuations (Kinzelbach, 1988, pp. 231-233). Smoothing procedures are often applied to model results. The model is more accurate in defining the position of the concentration front than it is in calculating the concentration value at a point. In the case of two-dimensional contaminant migration, the model assumes a vertically averaged concentration distribution in the aquifer or aquifer layer. Usually, uniform aquifer or aquifer layer dispersivities and negligible density-induced advection are as-

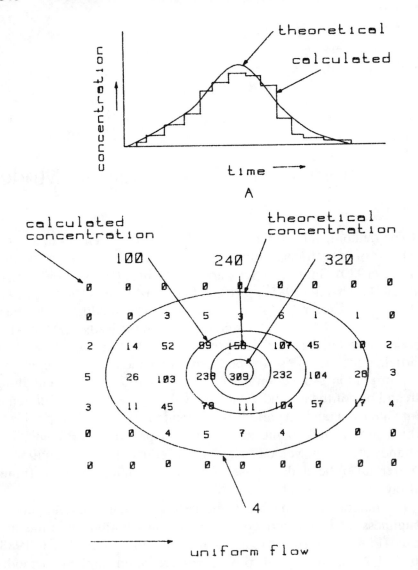

FIGURE 11.1 Conceptual relation between values of time-concentration (A) and distance-concentration (B) calculated with exact analytical equations and random walk method

sumed. It is further assumed that advection patterns are not dependent on the chemical composition or temperature of the solute thereby implying the presence of contaminants in trace concentrations. This assumption is valid for systems that are nearly isothermal and that contain relatively low contaminant concentration (Ahlstrom, et al., 1977, pp. II-6).

A FORTRAN random walk model program simulating two-dimensional contaminant migration is listed by Prickett, et, al. (1981, pp. 15-42) and Walton (1989b, pp. 157-246). BASIC random walk model programs simulating two-dimensional contaminant migration are listed by Walton (1984, pp. 459-465), Kinzelbach (1986, pp. 307-315), and Bear and Verruijt (1987, pp. 339-343).

In the random walk model, the contaminant mass reaching the aquifer or aquifer layer either as a slug or continuous source is divided into a large number of equal parts and assigned to a large number (usually less than 5000) of particles. These particles are assigned coordinates (initial positions in designated source areas) within a finite-difference grid. Any sinks present in the flow field are also assigned coordinates so that decontamination as well as contamination conditions are simulated. During successive small time increments, particle coordinates are advanced from initial (old) to secondary (new) positions towards any sinks in context with the groundwater flow head data base and associated velocity field. At the end of time increments, particle positions are inventoried and converted into contaminant concentrations at grid nodes and sinks.

TWO-DIMENSIONAL CONTAMINANT MIGRATION

In the random walk two-dimensional constant density mass transport model, contaminant particles migrate within a finite-difference grid which is superposed over the plan view map of the aquifer or aquifer layer (Figure 11.2). Grid lines are indexed using the i (column), j (row) notation colinear with the x and y directions, respectively. The i,j plane is in quadrant 4 and i-coordinates increase left to right and j-coordinates increase top to bottom. The origin (1,1) is the upper-left corner of the grid. Particle movement is based on an

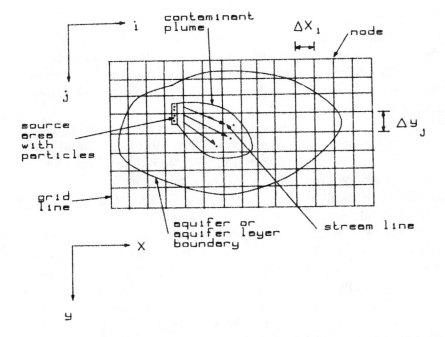

FIGURE 11.2   Plan view of finite-difference grid superposed on contaminant source area and plume

x,y coordinate coordinate system congruent with and on the same scale as the i,j grid. x,y grid borders are no flow boundaries which preclude the flow of particles across the borders. The grid extends beyond the contaminant plume and the area of interest to allow migration of particles to and from that area. If a variable grid is used, the grid density is usually greatest at sources and sinks. Any variable grid, the grid spacing, and the number of columns and rows match those of the flow model from which aquifer or aquifer layer heads were calculated.

At the end of time increments, each particle is assigned to the nearest grid node vector volume or sink. The grid node vector volume is centered at the node and extends the full aquifer or aquifer layer thickness. The number of particles assigned to each node are translated into nodal concentration values with the following equation (Prickett et al., 1981 p. 50):

$$c_{a(i,j)} = np_{i,j}pm/(n_{i,j}m_{i,j}\Delta x_i \Delta y_j) \qquad (11.1)$$

where $c_{a(i,j)}$ is the contaminant concentration at node i,j at time $t + \Delta t$; t is the time at the start of the time increment t; $np_{i,j}$ is the number of particles assigned to node i,j at time $t + \Delta t$; pm is the particle mass; $n_{i,j}$ is the actual aquifer or aquifer layer porosity at node i,j; $m_{i,j}$ is the aquifer or aquifer layer thickness at node i,j; Δx_i is the grid spacing in the x-direction; and Δy_j is the grid spacing in the y-direction.

The general roughness of the calculated concentration distribution is minimized by specifying a small grid spacing and injecting a large number of particles into the aquifer or aquifer layer. The calculation of significant concentration requires at least 20 particles in a grid block (Kinzelbach, 1988, p.229). Useful unit conversion factors for use with equation 11.1 are: 1 mg = 2.205×10^{-6} pounds and 1 ft^3 = 28.32 liters.

Dispersion may be considered as a random walk (Scheidegger, 1954 and De Josselin de Jong, 1958) which from the theory of Brownian motion finally progresses to an equation of the same type as the contaminant advection-dispersion equation. The theory of the Fokker-Planck equation (Ito, 1951) shows that a random walk is equivalent to the two-dimensional contaminant advection-dispersion equation in the following form (Kinzelbach, 1988, pp. 228-229):

$$x_p(t + \Delta t) = x_p(t) + u'_x\Delta t + R_1(2\alpha_L u_{xy}\Delta t)^{0.5}$$

$$(u_x/u_{xy}) + R_2(2\alpha_T u_{xy}\Delta t)^{0.5}(u_y/u_{xy}) \qquad (11.2)$$

$$y_p(t + \Delta t) = y_p(t) + u'_y\Delta t + R_1(2\alpha_L u_{xy}\Delta t)^{0.5}$$

$$(u_y/u_{xy}) - R_2(2\alpha_T u_{xy}\Delta t)^{0.5}(u_x/u_{xy}) \qquad (11.3)$$

with

$$u'_x = u_x + \partial D_{xx}/\partial x + \partial D_{xy}/\partial y \qquad (11.4)$$

$$u'_y = u_y + \partial D_{yx}/\partial x + \partial D_{yy}/\partial y \qquad (11.5)$$

$$u_x = u_x[x_p(t), y_p(t), t] \tag{11.6}$$

$$u_y = u_y[x_p(t), y_p(t), t] \tag{11.7}$$

$$u_{xy} = (u_x^2 + u_y^2)^{0.5} \tag{11.8}$$

$$R_1, R_2 = -6 + \sum_{i=1}^{12} RN_i \tag{11.9}$$

where $x_p(t + \Delta t)$ is the new x-coordinate of a particle at time $t + \Delta t$ (Figure 11.3), $x_p(t)$ is the old (initial) x-coordinate of a particle at time t, α_L is the aquifer or aquifer layer longitudinal dispersivity, α_T is the aquifer or aquifer layer transverse dispersivity, t is the time at the start of the time increment Δt, $y_p(t + \Delta t)$ is the new y-coordinate of a particle at time $t + \Delta t$, $y_p(t)$ is the old (initial) y-coordinate of a particle at time t, u_x is the average pore velocity in the x-direction, u_y is the average pore velocity in the y-direction, R_1 and R_2 are normally distributed random numbers with zero mean and unit standard deviation, and RN_i is an evenly distributed random variable available on most computers.

The average pore velocities in Equations 11.2 to 11.8 are based on the nodal aquifer or aquifer layer heads generated by a flow model. The x and y components of Darcy velocity along grid lines between nodes are calculated with the following equations (see Prickett, et al., 1981, p. 21):

$$v_{x(i,j)} = -P2_{h(i,j)}(h_{i+1,j} - h_{i,j})/[(\Delta x_i + \Delta x_{i+1})/2] \tag{11.10}$$

$$v_{y(i,j)} = -P1_{h(i,j)}(h_{i,j+1} - h_{i,j})/[(\Delta y_j + \Delta y_{j+1})/2] \tag{11.11}$$

where $v_{x(i,j)}$ is the Darcy velocity along a grid line in the x-direction midway between nodes i,j and i + 1,j; $v_{y(i,j)}$ is the Darcy velocity along a grid line in the y-direction midway between nodes i,j and i,j + 1; $P2_{h(i,j)}$ is the aquifer or aquifer layer horizontal hydraulic conductivity of the vector volume between nodes i,j and i + 1,j with horizontal dimensions the same as those of $T2_{i,j}$ defined in Figure 10.2; $Pl_{h(i,j)}$ is the aquifer or aquifer layer horizontal hydraulic conductivity of the vector volume between nodes i,j and i,j + 1 with horizontal dimensions the same as those of $T1_{i,j}$ defined in Figure 10.2; $h_{i,j}$ is the aquifer or aquifer layer head at node i,j at the end of the time increment Δt; Δx_i is the grid spacing in the x-direction; and Δy_j is the grid spacing in the y-direction.

Average pore velocities along a grid line (grid velocities) between nodes are calculated with the following equations:

$$u_{x(ij)} = v_{x(ij)}/n2_{e(ij)} \tag{11.12}$$

$$u_{y(ij)} = v_{y(ij)}/n1_{e(ij)} \tag{11.13}$$

where $n2_{e(i,j)}$ is the aquifer or aquifer layer effective porosity of the vector volume between nodes i,j and i + 1,j with horizontal dimensions the same as those of $T2_{i,j}$ defined in Figure 10.2; $n1_{e(i,j)}$ is the aquifer or aquifer layer effective porosity of the vector volume between nodes i,j and i,j + 1 with horizontal dimensions the same as those of $T1_{i,j}$ defined in Figure 10.2; $u_{x(i,j)}$ is the average pore velocity in the x-direction along a grid line midway between nodes i,j and i + 1,j; and $u_{y(i,j)}$ is the average pore velocity in the y-direction along a grid line midway between nodes i,j and i,j + 1.

Average pore velocity vectors within grid lines (block velocities) are obtained from grid velocities by bilinear interpolation (Konikow and Bredehoeft, 1978, p. 7; Prickett, et al, 1981, pp. 57-61). Bilinear block velocity interpolation is a three step procedure that is point position dependent. The first step is the interpolation of velocity x and y components at the point of interest based on the four closest grid velocities. The second step, if required, is the interpolation of velocity x and y components at the point based on the next closest four grid velocities. The third step, if required, is the final interpolation of velocity x and y components based on the former two interpolations. Relevant grid velocities for bilinear interpolation at points in the four quadrants of interior nodes are shown in Figure 11.3. Along boundaries, the second and third in the direction normal to the boundary are omitted. In grid corners, only the first step is performed. Omitting the second and third steps constitutes linear interpolation. No interpolation is needed if the point is located at a node.

Suppose that the point of interest is in quadrant 1. The closest grid velocities are: $u_x(i,j)$, $u_x(i,j + 1)$, $u_y(i,j)$, and $u_y(i + 1,j)$. The next closest grid velocities are: $u_x(i – 1,j)$, $u_x(i-1,j + 1)$, $u_y(i + 1,j – 1)$, and $u_y(i,j – 1)$. The bilinear interpolation equations for the point are as follows (Kinzelbach, 1986, p. 255):

Step 1

$$u_x^{(1)} = (1 – A)u_x(i,j) + Au_x(i,j + 1) \qquad (11.14)$$

$$u_y^{(1)} = (1 – B)u_y(i,j) + Bu_y(i + 1,j) \qquad (11.15)$$

Step 2

$$u_x^{(2)} = Au_x(i – 1,j + 1) + (1 – A)u_x(i – 1,j) \qquad (11.16)$$

$$u_y^{(2)} = Bu_y(i + 1,j – 1) + (1 – B)u_y(i,j – 1) \qquad (11.17)$$

Step 3

$$u_x(x,y) = u_x^{(1)}(0.5 + B) + u_x^{(2)}(0.5 – B) \qquad (11.18)$$

$$u_y(x,y) = u_x^{(1)}(0.5 + A) + u_y^{(2)}(0.5 – A) \qquad (11.19)$$

with

$$A = (y – y_j)/\Delta y \text{ and } B = (x – x_i)/\Delta x \qquad (11.20)$$

where $u_x(x,y)$ is the block velocity in the x-direction at point x,y; $u_y(x,y)$ is the block velocity in the y-direction at point x,y; $u_x(i,j)$ is the grid velocity in the x-direction; $u_y(i,j,)$ is the grid velocity in the y-direction; Δx is the grid spacing in the x-direction, Δy is the grid spacing in the y-direction, x is the particle coordinate in the x-direction, y is the particle coordinate in the y-direction, x_i is the x-coordinate of the closest grid line, and y_j is the j-coordinate of the closest grid line.

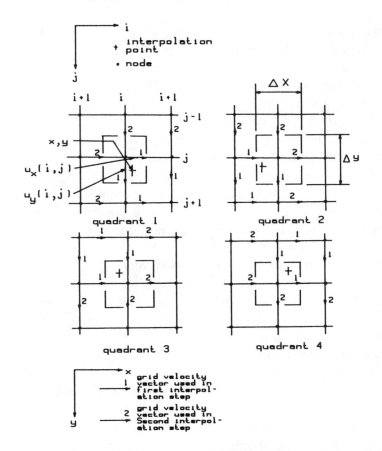

FIGURE 11.3   Grid velocity vectors used in bilinear interpolation of block velocity vectors

Other symbols are defined in Figure 11.4. A bilinear interpolation FORTRAN subroutine is listed by Prickett, et al (1981, pp. 41-42).

The derivative terms in Equations 11.4 and 11.5 require very accurate flow field description and commonly are neglected because of the lack of adequate field data and to reduce computation time (Kinzelbach, 1988, p. 231). However, neglecting these terms has serious consequences in areas with large velocity gradients, near stagnation points, or at layer boundaries.

Time and grid spacing are adjusted so that the Courant-criterion (Kinzelbach, 1986, p. 264) is satisfied and particle density oscillations are avoided. This can be accomplished by limiting successive particle movements to one-fifth of the grid spacing during the time increment Δt (Prickett, et al., 1981, p. 62).

To obtain satisfactory results, the presence of 200 or more particles is required in each grid block (Kinzelbach, 1986, p. 305). Scaling the apparent growth of dispersivities with time or distance may be introduced once at the beginning of the time period or continuously as part of a memory of a particle using the techniques developed by Kinzelbach (1988, pp. 235-242).

Although the random walk model is simple in principle, the programming effort is large because of the extensive book-keeping system required to keep track of particle positions. Particles are injected at sources, inventoried at grid nodes and sinks, and reflected back inside the grid at barrier boundaries. The adding of particles beyond the limits of the grid is prevented. At the end of time increments, particles at positions within 1/2 grid spacing of a grid node are assigned to that grid node. A criterion commonly used to determine if a particle is captured by a sink is a limit radius around the sink. If a particle is closer to the sink than the limit radius it is captured by the sink. In practice, the limit radius is usually 1/2 grid spacing. Advective time steps are kept smaller than the limit radius, otherwise, a particle could move across a sink and not be captured (Kinzelbach, 1986, pp. 305 and 306).

Linear adsorption may be simulated through division of average pore velocities, dispersivities without adsorption, and source masses by the retardation factor R_d (Kinzelbach, 1986, p. 305). For each particle of mass M_p in solution at a particular node there is a shadow mass $M_p(R_d - 1)$ at the same node adsorbed on the aquifer skeleton (Kinzelbach, 1988, p. 234). Slow adsorption or particle exchange between mobile and immobile pore water in an aquifer can be simulated with techniques described by Kinzelbach (1988, pp. 234-235). Radioactive decay may be simulated with equations 7.14-7.15.

Flow streamlines may be traced by setting dispersion terms in equations 11.2 and 11.3 equal to zero and plotting particle tracks from selected initial to secondary positions. Pollock (1988, pp. 743-750) developed a semianalytical 3-dimensional particle tracking model for computation of flow streamlines using velocities generated with block-centered finite-difference flow models. Contaminant plumes may be delineated with Equations 11.2 and 11.3 by plotting many particle tracks from a source area to secondary positions.

The simulation of contaminant migration in fractured rock with random walk methods is described by Schwartz and Smith (1988) and Smith, et al. (1989, pp. 425-440). Uffink (1986) developed a random-walk simulation of dispersion at a fresh-salt water interface.

INITIAL CONTAMINANT CONDITIONS

In the random walk model, particles are injected as a slug at the start of a time increment to simulate aquifer or aquifer layer initial contaminant conditions (see FORTRAN program listed by Walton, 1989b, p. 229), otherwise, zero initial contamination is assumed. Injecting particles means assigning initial coordinates to particles and storing these coordinates in the computer memory (see FORTRAN program listed by Walton, 1989b, pp. 157-246). Initial contaminant conditions are determined from field data or previous simulations. The number of particles to be injected at a grid node is calculated with the following equations (Prickett, et al, 1981, pp. 50):

$$np_{i,j} = M_{i(ij)}/pm \qquad (11.21)$$

with

$$M_{i(i,j)} = c_{i,j}\Delta x_i \Delta y_j m_{i,j} n_{i,j} \qquad (11.22)$$

where $np_{i,j}$ is the number of particles to be injected at initial time t at node i,j; t is the time at the start of the time increment Δt; pm is the particle mass; $M_{i(i,j)}$ is the initial contaminant mass in the aquifer or aquifer layer vector volume centered at node i,j at time t; c_{ij} is the initial contaminant concentration in the aquifer or aquifer layer vector volume centered at node i,j at time t; Δx_i is the grid spacing in x-direction; Δy_j is the grid spacing in y-direction; $m_{i,j}$ is the aquifer or aquifer layer thickness at node i,j; n_{ij} is the aquifer or aquifer layer actual porosity at node i,j.

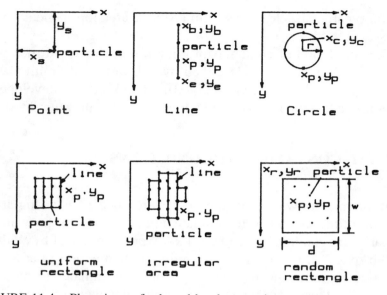

FIGURE 11.4   Plan views of selected local contaminant source areas

SLUG AND CONTINUOUS LOCAL SOURCES

Local contaminant source areas within selected grid blocks are simulated in the random walk model by injecting a known number of particles at a point, uniformly spaced along a line, uniformly spaced around the circumference of a circle, uniformly spaced inside a rectangle, or randomly spaced inside a rectangle (Figure 11.4). Equations for injecting particles as a slug at a point are (Prickett, et al., 1981, p. 26):

$$x_{p(n)} = x_s \tag{11.23}$$

$$y_{p(n)} = y_s \tag{11.24}$$

with

$$n = 1,...,np \tag{11.25}$$

where $x_{p(n)}$ is the initial x-coordinate of particle n at time t, x_s is the x-coordinate of the source point, $y_{p(n)}$ is the initial y-coordinate of particle n at time t, y_s is the y-coordinate of the source point, t is the time at the start of the time increment Δt, and np is the number of particles injected at x_s, y_s.

Equations for injecting 3 or more particles as a slug uniformly along a line are (Prickett, et al. 1981, p. 25):

$$X_{p(1)} = X_b \tag{11.26}$$

$$y_{p(1)} = y_b \tag{11.27}$$

$$X_{p(np)} = X_e \tag{11.28}$$

$$y_{p(np)} = y_e \tag{11.29}$$

$$X_{p(n)} = X_b + (n-1)(x_e - x_b)/(np - 1) \tag{11.30}$$

$$y_{p(n)} = y_b + (n-1)(y_e - y_b)/(np - 1) \tag{11.31}$$

with

$$n = 2, \ldots, np - 1 \tag{11.32}$$

where $x_{p(1)}$ is the initial x-coordinate of particle 1 at time t, $y_{p(1)}$ is the initial y-coordinate of particle 1 at time t, $x_{p(n)}$ is the initial x-coordinate of particle n at time t, $y_{p(n)}$ is the initial y-coordinate of particle n at time t, $x_{p(np)}$ is the initial x-coordinate of particle np at time t, $y_{p(np)}$ is the initial y-coordinate of particle np at time t, x_b is the x-coordinate of the beginning of the line, y_b is the y-coordinate of the beginning of the line, x_e is the x-coordinate of the end of the line, y_e is the y-coordinate of the end of the line, t is the time at the start of the time increment Δt, and np is the number of particles injected uniformly along the line.

Equations for injecting 4 or more particles as a slug uniformly around the circumference of a circle are (Prickett, et al., 1981, p. 27):

$$x_{p(n)} = x_c + r\sin(2\pi n/np) \qquad (11.33)$$

$$y_{p(n)} = y_c + r\cos(2\pi n/np) \qquad (11.34)$$

with

$$n = 1,\ldots,np \qquad (11.35)$$

where $x_{p(n)}$ is the initial x-coordinate of particle n at time t, $y_{p(n)}$ is the initial y-coordinate of particle n at time t, x_c is the x-coordinate of the circle center, y_c is the y-coordinate of the circle center, r is the circle radius, t is the time at the start of the time increment Δt, and np is the number of particles injected uniformly around the circumference of the circle.

Equations for injecting particles as a slug randomly within a rectangular area are (Prickett, et al., 1981, p. 25):

$$x_{p(n)} = x_r + wRN_n \qquad (11.36)$$

$$y_{p(n)} = y_r + dRN_n \qquad (11.37)$$

with

$$n = 1,\ldots,np \qquad (11.38)$$

where $x_{p(n)}$ is the initial x-coordinate of particle n at time t, $y_{p(n)}$ is the initial y-coordinate of particle n at time t, x_r is the x-coordinate of the upper left corner of the rectangular area, y_r is the y-coordinate of the upper left corner of the rectangular area, w is the width of the rectangular area, d is the length of the rectangular area, RN_n is a random number between 0 and 1, t is the time at the start of the time increment Δt, and np is the number of particles injected randomly within the rectangular area.

The size of the rectangle may be adjusted from a point to the grid area and a line or circle (square) may be simulated with Equations 11.36 and 11.37 by appropriate selection of w and d. Irregular source areas may be simulated with multiple line, circle, or rectangle sources.

Particles are injected within a rectangular area at the start of the time increment to simulate slug sources or randomly during the time increment to simulate continuous sources. Random particle continuous injection during the time increment is governed by equations 11.36-11.38 and the following equation (Prickett, et al., 1981, p. 25):

$$\Delta t_n = \Delta t R N_n \tag{11.39}$$

with

$$n = 1,\ldots,np \tag{11.40}$$

where Δt_n is the portion of the time increment Δt the particle n is allowed to migrate, Δt is the time increment, RN_n is a random number between 0 and 1, and np is the number of particles injected randomly during the time increment.

Two types of local contaminant sources (landfills, pits, basins, ditches, and surface water bodies) with known mass or leachate concentration are simulated in the random walk model. In the type 1 local source, the contaminant mass reaching the aquifer or aquifer layer is known. In the type 2 local source, the source leachate concentration is known and there is leachate leakage through a sourcebed and into the aquifer or aquifer layer (Figure 11.5). The number of particles to be injected into the aquifer or aquifer layer with a type 1 local source is calculated with the following equations (Prickett, et al, 1981, pp. 51, 52, 70):

For a slug source

$$np_k(t) = M_{c(k)}/pm \tag{11.41}$$

For a continuous source

$$np_k(\Delta t) = M_{r(k)}\Delta t/pm \tag{11.42}$$

The number of particles to be injected into the aquifer or aquifer layer with a type 2 continuous local source in accordance with Equations 11.39 to 11.40 is calculated with the following equations:

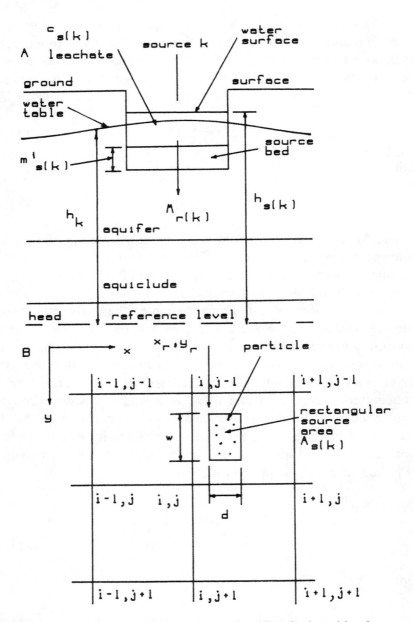

FIGURE 11.5   Cross section (A) and plan view (B) of selected local contaminant source area

$$np_k(\Delta t) = M_{r(k)}/pm \qquad (11.43)$$

with

$$M_{r(k)} = (h_{s(k)} - h_k)R_{r(k)}\Delta tc_{s(k)} \qquad (11.44)$$

$$R_{r(k)} = (P'_{s(k)}/m'_{s(k)})A_{s(k)} \qquad (11.45)$$

where $np_k(\Delta t)$ is the number of particles to be injected into the aquifer or aquifer layer at source k during the time increment Δt, $np_k(t)$ is the number of particles to be injected into the aquifer or aquifer layer at source k at time t, t is the time at the start of the time increment t, pm is the particle mass, $M_{c(k)}$ is the contaminant mass reaching the aquifer or aquifer layer at source k at time t, $M_{r(k)}$ is the contaminant mass leakage rate at source k during time increment Δt, $h_{s(k)}$ is the constant elevation of the source water surface at source k during time increment Δt, h_k is the constant elevation of the aquifer or aquifer layer head at source k during time increment Δt, $P'_{s(k)}$ is the vertical hydraulic conductivity of the sourcebed at source k, $m'_{s(k)}$ is the sourcebed thickness at source k, $A_{s(k)}$ is the sourcebed horizontal area at source k, and $c_{s(k)}$ is the constant leachate concentration at source k during time increment Δt.

Equation 11.44 assumes that h_k is greater than the elevation of the sourcebed base.

INJECTION WELL CONTINUOUS SOURCE

Particles are injected at selected grid nodes to simulate a continuous injection well source in the random walk model (see FORTRAN program listed by Walton, 1989b, pp. 221-222). With the concentration of the injected water and the well injection rate known, the number of particles to be injected at nodes in accordance with Equations 11.39 to 11.40 is calculated with the following equation (Prickett, et al, 1981, p. 70):

$$np_{(i,j)} = M_{w(i,j)}/pm \tag{11.46}$$

with

$$M_{w(ij)} = Q_{w(ij)}\Delta t c_{w(ij)} \tag{11.47}$$

where $np_{(i,j)}$ is the number of particles injected into the aquifer or aquifer layer at node i,j during the time increment Δt; $M_{w(i,j)}$ is the mass injected through a well at node i,j during the time increment Δt; pm is the particle mass; $Q_{w(i,j)}$ is the constant injection rate at node i,j during time increment Δt; and $c_{w(i,j)}$ is the constant concentration of the injected water at node i,j during time increment Δt.

CONTINUOUS AREAL SOURCE

Two types of continuous areal sources centered at nodes are simulated in the random walk model (see FORTRAN program listed by Walton, 1989b, pp. 157-246). In the type 1 areal source (Figure 11.6), there is migration of contaminant mass from a sourcebed through an aquitard into an aquifer or aquifer layer. Contaminant concentration in the sourcebed is known and the contaminant mass in the sourcebed is not depleted as the result of contaminant leakage. In the type 2 areal source, areal sink contaminant mass from another aquifer or aquifer layer constitutes the source. Mass leakage rates into the aquifer or aquifer layer are known.

In the type 1 areal source, the sourcebed head, the aquifer or aquifer layer hydraulic characteristics, and the contaminant concentration are known and it is assumed that the aquifer or aquifer layer head does not decline below the aquitard top. The number of particles to be injected in accordance with Equations 11.39 to 11.40 at each areal source node during the time increment is calculated with the following equation (Prickett, et al, 1981, pp. 51, 52, 70):

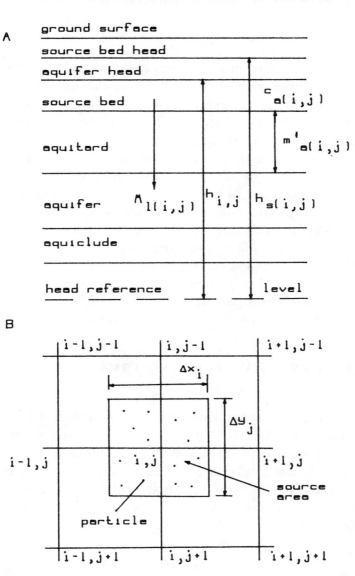

FIGURE 11.6   Cross section (A) and plan view (B) of areal contaminant source

$$np_{i,j} = M_{l(i,j)}/pm \qquad (11.48)$$

with

$$M_{l(i,j)} = (h_{s(i,j)} - h_{i,j})R_{l(i,j)}\Delta tC_{a(i,j)} \qquad (11.49)$$

$$R_{l(i,j)} = (P'_{a(i,j)}/m')_{a(i,j)}\Delta x_i \Delta y_j \qquad (11.50)$$

where $np_{i,j}$ is the number of particles to be injected into the aquifer or aquifer layer at node i,j during the time increment Δt; pm is the particle mass; $M_{l(i,j)}$ is the mass injected during the time increment Δt at node i,j; $h_{s(i,j)}$ is the constant elevation of the head in the sourcebed at node i,j during time increment Δt; $h_{i,j}$ is the constant elevation of the head in the aquifer or aquifer layer at node i,j during time increment Δt; $P'_{a(i,j)}$ is the vertical hydraulic conductivity of the aquitard at node i,j; $m'_{a(i,j)}$ is the aquitard thickness at node i,j; Δx_i is the grid spacing in the x-direction at node i,j; Δy_j is the grid spacing in the y-direction at node i,j; $c_{a(i,j)}$ is the constant concentration in the sourcebed at node i,j during time increment Δt.

In the type 2 areal source, particles captured by areal nodal sinks are regenerated randomly in time and in spaces centered at the areal sink nodes at a later time. The number of particles to be injected at nodes is known from previous areal sink calculations.

WELL, SURFACE WATER, AND AREAL SINKS

Three types of contaminant sinks are simulated in the random walk model (see FORTRAN program listed by Walton, 1989b, pp. 157-246): production well sink-type 1, stream or lake sink-type 2 (Figure 11.7), and areal sink-type 3. In the type 1 sink, the contaminant concentration of the water discharged from a production well located at a grid node is calculated with the following equation (Prickett et al., 1981, p. 70):

$$c_{p(i,j)} = n_{pp(i,j)}pm/(Q_{p(i,j)}\Delta t) \qquad (11.51)$$

where $c_{p(i,j)}$ is the contaminant concentration of the water discharged from a production well sink at node i,j at time $t + \Delta t$; t is the time at the start of the time increment Δt; $n_{pp(i,j)}$ is the number of particles captured by the production well sink at node i,j during the time increment Δt; pm is the particle mass; and $Q_{p(i,j)}$ is the constant production well sink discharge rate at node i,j during time increment Δt.

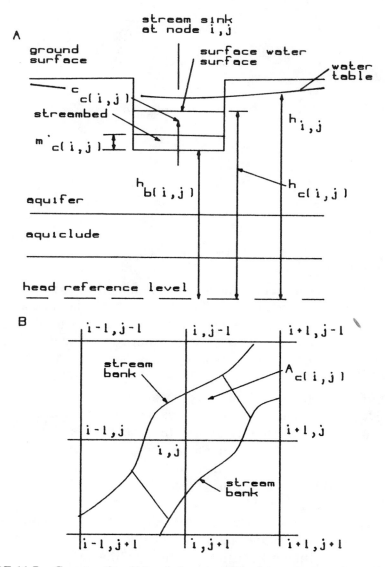

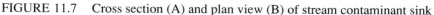

FIGURE 11.7   Cross section (A) and plan view (B) of stream contaminant sink

Useful unit conversion factors for use with Equation 11.51 are: 1 mg = 2.205×10^{-6} pound and 1 gal = 3.785 liter.

In the type 2 sink, the contaminant concentration of the water discharged into a stream or lake at a grid node is calculated with the following equation (Prickett, et al., 1981, p. 70):

$$c_{c(i,j)} = np_{c(i,j)}pm/(\Delta h_{i,j}\Delta t R_{c(i,j)}) \qquad (11.52)$$

with

$$R_{c(ij)} = (P'_{c(ij)}/m'_{c(ij)})A_{c(ij)} \qquad (11.53)$$

For $h_{i,j} > h_{b(i,j)}$

$$\Delta h_{i,j} = h_{c(i,j)} - h_{(i,j)} \qquad (11.54)$$

For $h_{i,j} \leq h_{b(i,j)}$

$$\Delta h_{i,j} = h_{c(i,j)} - h_{b(i,j)} \qquad (11.55)$$

where $c_{c(i,j)}$ is the contaminant concentration of the water discharged to a stream or lake at node i,j at time $t + \Delta t$; t is the time at the start of the time increment Δt; $np_{c(i,j)}$ is the number of particles captured by the stream or lake sink at node i,j during time increment Δt; pm is the particle mass; $R_{c(i,j)}$ is the stream or lake sink simulation factor at node i,j; $P'_{c(i,j)}$ is the vertical hydraulic conductivity of the stream or lake bed at node i,j; $m'_{c(i,j)}$ is the stream or lake bed thickness at node i,j; $A_{c(i,j)}$ is the area of the stream or lake bed assigned to node i,j; $h_{i,j}$ is the constant elevation of the head in the aquifer or aquifer layer at node i,j during time increment Δt; $h_{b(i,j)}$ is the constant elevation of the stream or lake bed base at node i,j during time increment Δt; and $h_{c(i,j)}$ is the constant elevation of the surface of the stream or lake at node i,j during time increment Δt.

In the type 3 sink, there is areal migration of contaminants with or without adsorption and/or radioactive decay from one aquifer through an aquitard and into another aquifer or between adjacent aquifer layers as shown in Figure 11.8. The number of particles leaving the aquifer or aquifer layer at grid node centered areal sinks is calculated with the following equation:

$$np_{s(i,j)} = M_{s(i,j)}/pm \qquad (11.56)$$

where $np_{s(i,j)}$ is the number of particles migrating from the aquifer or aquifer layer at node i,j during the time increment Δt; pm is the particle mass; and $M_{s(i,j)}$ is the contaminant mass migrating from the aquifer or aquifer layer at node i,j during time increment Δt.

$M_{s(i,j)}$ is calculated based on the following mass balance equation provided the contaminant mass migrating from the aquifer or aquifer layer during time increment Δt is less than or equal to the contaminant mass injected into the aquifer or aquifer layer during time increment Δt (Walton, 1989b, pp. 61-64):

$$M_{a(i,j)} = M_{t(i,j)} - M_{s(i,j)} \tag{11.57}$$

with

$$M_{a(i,j)} = c_{e(i,j)}n_{i,j}m_{i,j}\Delta x_i \Delta y_j \tag{11.58}$$

$$M_{t(i,j)} = c_{t(i,j)}n_{i,j}m_{i,j}\Delta x_i \Delta y_j \tag{11.59}$$

For an aquifer-aquitard system (see Prickett, et al, 1981, pp. 51,52)

$$M_{s(i,j)} = c_{s(i,j)}\Delta t(P'_{i,j}/m'_{i,j})$$

$$\Delta x_i \Delta y_j(h_{1(i,j)} - h_{2(i,j)}) \tag{11.60}$$

For an aquifer layer system

$$M_{s(i,j)} = c_{s(i,j)}\Delta t \Delta x_i \Delta y_j(h_{L1} - h_{L2})/$$

$$[(m_{1(i,j)}/P_{v1(i,j)}) + (m_{2(i,j)}/P_{v2(i,j)})] \tag{11.61}$$

with

$$c_{s(i,j)} = (c_{b(i,j)} + c_{e(i,j)})/2 \tag{11.62}$$

where $M_{a(i,j)}$ is the unknown mass remaining in the aquifer or aquifer

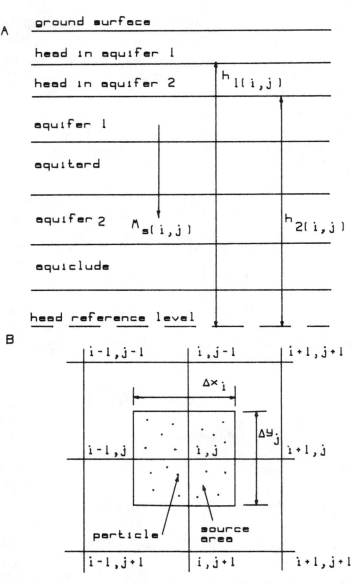

FIGURE 11.8   Cross section (A) and plan view (B) of areal contaminant sink

layer with areal leakage at node i,j at time $t + \Delta t$; t is the time at the start of the time increment Δt; $M_{t(i,j)}$ is the known mass remaining in the aquifer or aquifer layer at time $t + \Delta t$ without areal leakage at node i,j; $M_{s(i,j)}$ is the unknown mass migrating from the aquifer or aquifer layer during the time increment Δt as the result of areal leakage at node i,j; $c_{e(i,j)}$ is the unknown concentration in the aquifer or aquifer layer at time $t + \Delta t$ with areal leakage at node i,j; $c_{t(i,j)}$ is the known concentration in the aquifer or aquifer layer at time $t + \Delta t$ without areal leakage at node i,j; $c_{s(i,j)}$ is the unknown average concentration in the aquifer or aquifer layer during the time increment Δt with areal leakage at node i,j; $c_{b(i,j)}$ is the known concentration in the aquifer or aquifer layer at time t at node i,j; $P'_{i,j}$ is the aquitard vertical hydraulic conductivity at grid node i,j; $m'_{i,j}$ is the aquitard thickness at node i,j; $h_{1(i,j)}$ is the constant aquifer head elevation at node i,j during time increment Δt; $h_{2(i,j)}$ is the constant sourcebed head elevation at node i,j during time increment Δt; Δx_i is the grid spacing in the x-direction at node i,j; Δy_j is the grid spacing in the y-direction at node i,j; $P_{v1(i,j)}$ is the aquifer layer 1 vertical hydraulic conductivity at node i,j; $P_{v2(i,j)}$ is the aquifer layer 2 vertical hydraulic conductivity at node i,j; $m_{1(i,j)}$ is the aquifer layer 1 thickness at node i,j; $m_{2(i,j)}$ is the aquifer layer 2 thickness at node i,j; $h_{L1(i,j)}$ is the constant aquifer layer 1 head elevation at node i,j during time increment Δt; $h_{L2(i,j)}$ is the constant aquifer layer 2 head elevation at node i,j during time increment Δt; $m_{i,j}$ is the aquifer thickness at node i,j; and $n_{i,j}$ is the actual aquifer or aquifer layer porosity at node i,j.

Equation 11.57 assumes that the average concentration in the aquifer or aquifer layer during the time increment is the sum of the concentrations at the start and end of the time increment divided by 2. Concentration at the start of the time increment is affected by initial conditions and any slug contaminant sources. Sink contaminant mass is considered in Equation 11.57 by excluding particles captured by local sinks such as production wells and streams from computations of $M_{t(i,j)}$ (see Prickett, et al, 1981, p. 50). The time increment and particle mass are kept as small as possible so that values of $c_{s(i,j)}$ are reasonably precise.

Substituting Equations 11.58 to 11.62 into Equation 11.57 and solving that equation for $c_{e(i,j)}$ results in the following expression:

$$c_{e(i,j)} = (2c_{t(i,j)} - c_{b(i,j)}F1)/(F1 + 2) \tag{11.63}$$

with

For an aquifer-aquitard system

$$F1 = \Delta t(P'_{i,j}/m'_{i,j})(h_{1(i,j)} - h_{2(i,j)})/(n_{i,j}m_{i,j}) \qquad (11.64)$$

For an aquifer layer system

$$F1 = \Delta t(h_{L1(ij)} - h_{L2(ij)})/\{[(m_{1(ij)}/P_{v1(ij)})$$

$$+ (m_{2(ij)}/P_{v2(ij)})]n_{ij}m_{1(ij)}\} \qquad (11.65)$$

Knowing $c_{e(i,j)}$, the mass to be exported as an areal source mass may be calculated with Equations 11.60 or 11.61 and the contaminant mass remaining in the aquifer or aquifer layer with areal leakage may be calculated with Equation 11.58.

The average contaminant travel time (aquitard thickness or average aquifer layer thickness divided by the average vertical pore velocity) with or without adsorption through the aquitard or between aquifer layers assuming vertical dispersion is negligible may be calculated with the following equations (see Prickett, et al, 1981, p. 7):

For an aquifer-aquitard system

$$t_t = n_{e(i,j)}R_{d1v}\, m'^2_{i,j}/[P'_{i,j}(h_{1(i,j)} - h_{2(i,j)})] \qquad (11.66)$$

For an aquifer layer system

$$tt = n_{e(ij)}R_{d1}m_{av(ij)}^2/[P_{va(ij)}$$

$$(h_{L1(ij)} - h_{L2(ij)})] \qquad (11.67)$$

with

$$P_{va(ij)} = (m_{1(ij)} + m_{2(ij)})/[(m_{1(ij)}/P_{v1(ij)})$$

$$+ (m_{2(ij)}/P_{v2(ij)})] \qquad (11.68)$$

(see Raudkivi and Callander, 1976, p. 118)

where t_t is the average contaminant travel time through the aquitard;

$n_{e(i,j)}$ is the aquifer or aquifer layer effective porosity at node i,j; R_{d1} is the aquifer or aquifer layer retardation factor; R_{d1v} is the aquitard retardation factor; and $m_{av(i,j)} = (m_{1(i,j)} + m_{2(i,j)})/2$.

It is assumed that fairly uniform vertical head losses occur within the area of interest. Vertical head losses reflect average head conditions during the travel time. Heads in the aquifer or aquifer layer receiving contaminant mass from areal sinks reflect average head conditions during the time increment after the travel time specified for the corresponding areal sink. The time increment for the areal source is the same as that used in areal sink calculations.

THREE-DIMENSIONAL CONTAMINANT MIGRATION

A simple algorithm for simulating three-dimensional contaminant migration from one aquifer layer into another aquifer layer (aquifer horizontal stratification) developed by Uffink (1985, pp. 103-114) involves the reflection principle. A fraction $(D_T'^{0.5} \cdot D_T''^{0.5})/(D_T'^{0.5} + D_T''^{0.5})$ of the particles in one aquifer layer are reflected into the other aquifer layer when the velocities in the two aquifer layers are different. In the fraction, D_T' is the dispersion coefficient in aquifer layer 1 and D_T'' is the dispersion coefficient in aquifer layer 2. Another technique for simulating three-dimensional contaminant migration with aquifer stratification is described by Ackerer (1988, pp. 479-480).

Ackerer (1988, pp. 475-488) developed a random walk method for simulating three-dimensional contaminant migration in uniformly porous aquifers or fractured rock. In fractured rock, advection and dispersion are simulated in the fracture and only dispersion is simulated in the rock matrix. At the fracture/matrix interface, the dispersion is divided into two steps: (1) from the initial position to the interface with the fracture dispersivity and (2) from the interface back to the original position (probability = 0.5) with the fracture dispersivity or from the interface to the matrix with the matrix dispersivity.

The theory of the Fokker-Planck equation (Ito, 1951) shows that a random walk is equivalent to the three-dimensional contaminant advection-dispersion equation in the following form (Ackerer, 1988, pp. 476-478; also see Ahlstrom, et al., 1977, pp. II-8, II-9, and III-8):

$$x_p(t + \Delta t) = x_p(t) + u'_x\Delta t + [R_1(2D_{xx}\Delta t)^{0.5}$$

$$+ R_2(2D_{xy}\Delta t)^{0.5} + R_3(2D_{xz}\Delta t)^{0.5}] \tag{11.69}$$

$$y_p(t + \Delta t) = y_p(t) + u'_y\Delta t + [R_1(2D_{yx}\Delta t)^{0.5}$$

$$+ R_2(2D_{yy}\Delta t)^{0.5} + R_3(2D_{yz}\Delta t)^{0.5}] \tag{11.70}$$

$$z_p(t + \Delta t) = z_p(t) + u'_z\Delta t + [R_1(2D_{zx}\Delta t)^{0.5}$$

$$+ R_2(2D_{zy}\Delta t)^{0.5} + R_3(2D_{zz}\Delta t)^{0.5}] \tag{11.71}$$

with

$$D_{xx} = D_L(u_x^2/u_{xyz}^2) + D_T[(u_y^2/u_{xy}^2)$$

$$+ (u_z^2/u_{xyz}^2)(u_x^2/u_{xy}^2)] \tag{11.72}$$

$$D_{xy} = (D_L - D_T)(u_xu_y/u_{xyz}^2) \tag{11.73}$$

$$D_{xz} = (D_L - D_T)(u_xu_z/u_{xyz}^2) \tag{11.74}$$

$$D_{yx} = D_{xy} \tag{11.75}$$

$$D_{yy} = D_L(u_y^2/u_{xyz}^2) + D_T[(u_x^2/u_{xy}^2)$$

$$+ (u_z^2/u_{xyz}^2)(u_y^2/u_{xy}^2)] \tag{11.76}$$

$$D_{yz} = (D_L - D_T)(u_yu_z/u_{xyz}^2) \tag{11.77}$$

$$D_{zx} = D_{xz} \tag{11.78}$$

$$D_{zy} = D_{yz} \tag{11.79}$$

$$D_{zz} = D_L(u_z^2/u_{xyz}^2) + D_T(u_{xy}^2/u_{xyz}^2) \tag{11.80}$$

$$D_L = \alpha_L u_{xyz} \tag{11.81}$$

$$D_T = \alpha_T u_{xyz} \tag{11.82}$$

$$u_{xy} = (u_x^2 + u_y^2)^{0.5} \tag{11.83}$$

$$u_{xyz} = (u_{xy}^2 + u_z^2)^{0.5} \tag{11.84}$$

$$u'_x = u_x + \partial D_{xx}/\partial x + \partial D_{xy}/\partial y + \partial D_{xz}/\partial z \tag{11.85}$$

$$u'_y = u_y + \partial D_{yx}/\partial x + \partial D_{yy}/\partial y + \partial D_{yz}/\partial z \tag{11.86}$$

$$u'_z = u_z + \partial D_{zx}/\partial x + \partial D_{zy}/\partial y + \partial D_{zz}/\partial z \tag{11.87}$$

$$R_1, R_2, R_3 = -6 + \sum_{i=1}^{12} RN_i \tag{11.88}$$

where $x_p(t + \Delta t)$ is the new x-coordinate of a particle at time $t + \Delta t$; $x_p(t)$ is the old (initial) x-coordinate of a particle at time t; α_L is the aquifer or aquifer layer longitudinal dispersivity; α_T is the aquifer or aquifer layer transverse dispersivity; t is the time at the start of the time increment Δt; $y_p(t + \Delta t)$ is the new y-coordinate of a particle at time $t + \Delta t$; $y_p(t)$ is the old (initial) y-coordinate of a particle at time t; $z_p(t + \Delta t)$ is the new z-coordinate of a particle at time $t + \Delta t$; $z_p(t)$ is the old (initial) coordinate of a particle at time t; u_x is the average pore velocity in the x-direction; u_y is the average pore velocity in the y-direction; u_z is the average pore velocity in the z-direction; R_1, R_2, and R_3 are normally distributed random numbers with zero mean and unit standard deviation, and RN_i is an evenly distributed random variable available on most computers.

The derivative terms in Equations 11.85 to 11.87 require very accurate flow field description and commonly are neglected because of the lack of adequate field data and to reduce computation time. However, neglecting these terms has serious consequences in areas with large velocity gradients, near stagnation points, or at layer boundaries.

The extensive two-dimensional contaminant migration model book keeping system is extended as follows in a three-dimensional contaminant migration model: the aquifer, aquifer layer, or aquitard is divided into selected depth intervals; sinks are assigned depth intervals; particle z-coordinates are tracked in addition to particle x- and y-coordinates; particles leaving or entering an aquifer, aquifer layer, or aquitard are tabulated; particles migrating beyond vertical barrier boundaries are reflected back inside the aquifer or aquifer layer; the criterion used to determine if a particle is assigned to a grid node aquifer , aquifer layer, or aquitard depth interval includes the particle's z-coordinate and the aquifer, aquifer layer, or aquitard depth interval z-coordinates; and the criterion used to determine if a particle is captured by a sink includes the particle z-coordinate and the sink's depth interval z-coordinates.

Nodal concentration values are calculated by extending Equation 11.1 to cover aquifer, aquifer layer, or aquitard depth intervals. Initial contaminant conditions, slug and continuous local sources, and injection well continuous sources are simulated by extending Equations 11.21 to 11.47. Each injected particle is assigned a z-coordinate in addition to i- and j- or x- and y-coordinates. Well and surface water sinks are simulated by extending Equations 11.51 to 11.55. Each sink is assigned depth interval z-coordinates in addition to i- and j- or x- and y-coordinates.

The i and j components of grid velocities and the x and y components of block velocities are calculated with Equations 11.10 to 11.20. The z components of grid and block velocities are calculated with the following equations:

$$v_{z(i,j)} = -P_{v(i,j)}(h_{1(i,j)} - h_{2(i,j)})/m_{v(i,j)} \qquad (11.89)$$

$$u_{z(i,j)} = v_{z(i,j)}/n_{e(i,j)} \qquad (11.90)$$

$$v_z(x,y) = -P_v(x,y)(h_1(x,y) - h_2(x,y))/m_x(x,y) \qquad (11.91)$$

$$u_z(x,y) = v_z(x,y)/n_e(x,y) \qquad (11.92)$$

where $v_{z(i,j)}$ is the Darcy grid velocity in the z-direction at node i,j;

$P_{v(i,j)}$ is the vertical hydraulic conductivity of the aquifer, aquifer layer, or aquitard at node i,j; $h_{1(i,j)}$ is the head at the aquifer top, the head at the aquifer layer top, or the head in the source bed above or below an aquifer or aquifer layer at node i,j; $h_{2(i,j)}$ is the head at the aquifer base, the head at the aquifer layer base, or the head in the aquifer above or below an aquitard at node i,j; $m_{v(i,j)}$ is the aquifer, aquifer layer, or aquitard thickness at node i,j; $u_{z(i,j)}$ is the average pore grid velocity at node i,j: $n_{e(i,j)}$ is the aquifer, aquifer layer, or aquitard effective porosity at node i,j; v_z (x,y) is the Darcy block velocity in the z-direction at a point x,y; $P_v(x,y)$ is the vertical hydraulic conductivity of the aquifer, aquifer layer, or aquitard at a point x,y; $h_1(x,y)$ is the head at the aquifer top, the head at the aquifer layer top, or the head in the source bed above or below an aquifer or aquifer layer at a point x,y; $h_2(x,y)$ is the head at the aquifer base, the head at the aquifer layer base, or the head in the aquifer above or below an aquitard at a point x,y; $m_v(x,y)$ is the aquifer, aquifer layer, or aquitard thickness at a point x,y; $u_z(x,y)$ is the average pore block velocity at a point x,y: and $n_e(x,y)$ is the aquifer, aquifer layer, or aquitard effective porosity at a point x,y.

Values of $P_v(x,y)$, $h_1(x,y)$, $h_2(x,y)$, $m_v(x,y)$, and $n_e(x,y)$ are commonly obtained from corresponding grid nodal values by Lagrangian interpolation using Equation 10.21. FORTRAN random walk method programs (3-D RANDOM WALK and INTERTRANS) simulating three-dimensional contaminant migration, developed by T.A Prickett and M.L. Voorhees, respectively, are distributed by T.A. Prickett and Associates and the Scientific Software Group.

References

Abu-Zied, M.A. and V.H. Scott. 1963. Nonsteady flow for wells with decreasing discharge. Hydraulics division. Journal of American Society of Civil Engineers. Proceedings. Vol. 89, No. H-13. pp. 119-132.

Abramowitz, M. and I.A. Stegun, eds. 1964. Handbook of mathematical functions with formulas, graphs, and mathematical tables. U.S. Department of Commerce, National Bureau of Standards. Applied Mathematics Series. Vol. 55. 1046 pp.

Ackerer, P. 1988. Random-walk method to simulate pollutant transport in alluvial aquifers or fractured rocks. In Groundwater flow and quality modelling. eds. E. Custodio, A. Gurgui, and J.P Lobo Ferreira. D. Reidel Publishing Company. Boston. pp. 475-486.

Ackroyd, E.A., W.C. Walton, and D.L. Hills. 1967. Groundwater contribution to streamflow and its relation to basin characteristics in Minnesota. Minnesota Geological Survey. Report of Investigation. No. 6.

Adrion, W.R., M.A. Branstad, and J.C. Cherniqvsky. 1982. Validation, verification, and testing of computer software. Computing Surveys. Vol. 14, No. 2. pp. 334-363.

Andriole, S.J. 1986. Software validation, verification, testing, and documentation. Petrocelli Books. Princeton, N.J.

Ahlstrom, S.W., H.P. Foote, R.C. Arnett, C.R. Cole, and R.J. Serne. 1977. Multicomponent mass transport model. Theory and numerical implementation. Battelle, Pacific Northwest Laboratories Report. BNWL 2127. Richland, Washington.

Ahrens, T.P. 1957. Well design criteria. Water Well Journal. Sept. and Nov. National Water Well Association.

Aller, L., T. Bennett, J.H. Lehr, R.J. Petty, and G. Hackett. 1987. DRASTIC: A standardized system for evaluating ground water pollution potential using hydrogeological settings. U.S. Environmental Protection Agency. EPA-600/2-87-035. 455 pp.

American Water Works Association. 1971. Water quality and treatment. McGraw-Hill. New York. 654 pp.

Anderson, M.P. 1979. Using models to simulate the movement of contaminants through groundwater flow systems. Critical Reviews on Environmental Controls. Vol. 9, No. 2. pp. 97-156.

Anderson, M.P. 1984. Movement of contaminants in groundwater: groundwater transport-advection and dispersion. In: Groundwater Contamination. National Academy Press. pp. 37-45.

Anderson, M.P. 1986. Field validation of ground water models. In Evaluation of pesticides in ground water. Eds. Garner, W.Y., R.C. Honeycutt, and H.N. Nigg. American Chemical Society Symposium Series 315. 573 pp.

Anon. 1977. Ground water manual. U.S. Department of the Interior, Bureau of Reclamation. Water Resources Technical Publication. 480 pp.

Aral, M.M. 1990a. Ground water modeling in multilayered aquifers-steady flow. Lewis Publishers, Inc. 192 pp.

Aral, M.M. 1990b. Ground water modeling in multilayered aquifers-unsteady flow. Lewis Publishers, Inc. 240 pp.

Aron, G. and V.H. Scott. 1965. Simplified solutions for decreasing flow in wells. Proceedings of American Society of Civil Engineers. Vol. 91, HY 5. pp. 1-12.

Back, W., J.S. Rosenshein, and P.R. Seaber. Eds. 1989. Hydrogeology, the geology of North America. The Geological Society of America. 534 pp.

Baetsle, L.H. 1967. Computational methods for the prediction of underground movement of radio-nuclides. Journal of Nuclear Safety. Vol. 8. pp. 576-588.

Baetsle, L.H. 1969. Migration of radionuclides in porous media. In Duhamel, A.M.F., Ed. Progress in nuclear energy, Series 12: Health Physics. Vol. 2, Pt. I.

Barcelona, M.J., J.P. Gibb, and R.A. Miller. 1983. A guide to the selection of materials for monitoring well construction and groundwater sampling. Illinois State Water Survey. Contract Report 327. EPA-600/52-84-024. 78 pp.

Barcelona, M.J. and others. 1985. Practical guide for ground-water monitoring. Illinois State Water Survey. Contract Report 374. 94 pp.

Bear, J. 1972. Dynamics of fluids in porous media. American Elsevier Publishing Company, Inc. 764 pp.

Bear, J. 1979. Hydraulics of groundwater. McGraw-Hill Inc. 567 pp.

Bear, J. and M.Y. Corapcioglu. 1981a. Mathematical model for regional land subsidence due to pumping. 1. Integrated aquifer subsidence equations based on vertical displacement only. Water Resources Research. Vol. 17. pp. 937-946.

Bear, J. and M.Y. Corapcioglu. 1981b. Mathematical model for regional land subsidence due to pumping. 2. Integrated aquifer subsidence equations for vertical and horizontal displacements. Water Resources Research. Vol. 17. pp. 947-958.

Bear, J. and A. Verruijt. 1987. Modeling groundwater flow and pollution. D. Reidel Publishing Company. Boston. 414 pp.

Beljin, M.S. 1985. SOLUTE-a program package of ten analytical models for solute transport in groundwater. BAS-15. International Ground Water Modeling Center. Holcomb Reasearch Institute. Butler University. Indianapolis, Indiana.

Beljin, M.S. 1987. Representation of individual wells in two-dimensional groundwater modeling. In Proceedings of the NWWA Conference on solving groundwater problems with models. National Water Well Association. pp. 340-351.

Bennett, G.D., A.L. Kontis, and S.P. Larson. 1982. Representation of multiaquifer well effects in three-dimensional ground-water flow simulation. Ground Water. Vol. 20, No. 3. pp. 334-341.

Bentall, R. Comp. 1963. Shortcuts and special problems in aquifer tests. U.S. Geological Survey. Water-Supply Paper 1545-C. 117 pp.

Bierschenk, W.H. 1964. Determining well efficiency by multiple step-drawdown tests. International Association of Scientific Hydrology. Publication 64. pp. 493-506.

Bonn, B.A. and S.T. Rounds. 1989. DREAM-Analytical ground water flow programs. Lewis Publishers, Inc. pp. 350.

Boonstra, J. and Boehmer, W.K. 1989. Analysis of data from well tests in dikes and fractures. In: Groundwater contamination: use of models in decision-making. Eds. Jousma, G., J. Bear, Y.Y. Haimes, and F. Walter. Kluwer Academic Publishers. Boston, MA. pp. 171-180.

Borg, I.Y., R. Stone, H.B. Levy, and L.D. Ramspott. 1976. Information pertinent to the migration of radionuclides in groundwater at the Nevada Test Site. Part 1: Review and analysis of existing information. Lawrence Livermore Laboratory. Report UCRL-52078.

Boulton, N.S. 1954. The drawdown of the water table under non-steady conditions near a pumped well in an unconfined formation. Proceedings (British) of Institute of Civil Engineers. Vol. 3, Pt. 3.

Boulton, N.S. and T.D. Streltsova. 1976. The drawdown near an abstraction well of large diameter under non-steady conditions in an unconfined aquifer. Journal of Hydrology. Vol. 30. pp. 29-45.

Boulton, N.S. and T.D. Streltsova. 1977. Unsteady flow to a pumped well in a fissured water-bearing formation. Journal of Hydrology. Vol. 35. pp. 257-269.

Boulton, N.S. and T.D. Streltsova. 1978. Unsteady flow to a pumped well in a fissured aquifer with a free surface level maintained constant. Water Resources Research. Vol. 14, No. 3. pp. 527-532.

Bourke, P.D. 1987. A contouring subroutine. BYTE Publications, Inc. June. pp. 143-160.

Boutwell, S.H., S.M. Brown, B.R. Roberts, and D.F. Atwood. 1986. Modeling remedial actions at uncontrolled hazardous waste sites. Pollution Technology Review No. 130. Noyes Publications. Park Ridge, New Jersey. 440 pp.

Bouwer, H. 1964. Unsaturated flow in ground-water hydraulics. Journal of the Hydraulics Division, Proceedings of the American Society of Civil Engineers. Vol. 90, No. HY5. pp. 121-144.

Bouwer, H. and R.C. Rice. 1976. A slug test for determining conductivity of unconfined aquifers with completely or partially penetrating wells. Water Resources Research. Vol. 12, No.3. pp. 423-428.

Bouwer, H. 1978. Groundwater Hydrology. McGraw-Hill Book Company. New York. 480 pp.

Bredehoeft, J.D. and I.S. Papadopulos. 1965. Rates of vertical groundwater movement estimated from the earth's thermal profile. Water Resources Research. Vol.1, No. 2. pp. 325-328.

Bredehoeft, J.D. 1967. Response of well-aquifer systems to earth tides. Journal of the Geophysical Union. Vol. 72. pp. 3075-3087.

Bredehoeft, J.D. and G.F. Pinder. 1970. Digital analysis of areal flow in multiaquifer groundwater systems: A quasi three-dimensional model. Water Resources Research. Vol. 6, No. 3. pp. 883-888.

Bredehoeft, J.D. 1976. Status of quantitative groundwater hydrology. In: Advances In Groundwater Hydrology. Ed. Z. A. Saleem. American Water Resources Association. pp. 8-14.

Bredehoeft, J.D. and S.S. Papadopulos. 1980. A method for determining the hydraulic properties of tight formations. Water Resources Research. Vol. 16, No. 1. pp. 233-238.

Brody, P. and T.H. Illangasebare. 1989. A laboratory study and model for the separate phase transport of an immiscible contaminant floating on the water table. In: Solving ground water problems with models. Proceedings of the fourth international conference on the use of models to analyze and find working solutions to ground water problems. National Water Well Association and International Ground Water Modeling Center. pp. 431-452.

Broten, M., L. Fenstermaker, and J. Shafer. 1987. An automated GIS for groundwater contamination investigation. In: Proceedings of the solving groundwater problems with models conference, Feb., 1987. National Water Well Association. pp. 1143-1159.

Bruch, J.C. 1967. Two-dimensional dispersion. American Society of Civil Engineers. Journal of the Sanitary Engineering Division. Vol. 93, No. SA6. pp. 17-39.

Brutsaert, W. and M.Y. Corapcioglu. 1976. Pumping of aquifer with viscoelastic properties. Journal of the Hydraulics Division, American Society of Civil Engineers. Vol. 102 (HY11). pp. 1663-1675.

Burns, W.A. 1969. New single-well test for determining vertical permeability. Transactions of American Institute of Mining Engineers. Vol. 246. pp. 743-752.

Butler, S.S. 1957. Engineering hydrology. Prentice-Hall. Englewood Cliffs, New Jersey.

Buxton, B.E. ed. 1989. Geostatistical, sensitivity, and uncertainty methods for ground-water flow and radionuclide transport modeling. Proceedings of Conference sponsored by U.S Department of Energy and Atomic Energy of Canada Limited. Conf-870971. Batelle Press, Columbus, Ohio. 670 pp.

California Department of Health. 1975. Hazardous waste management-law regulations-guidelines for the handling of hazardous waste.

Callahan, M.A., M.W. Slimak, N.W. Gable, I.P. May, C.F. Fowler, J.R. Freed, P. Jennings, R.L. Durfee, F.C. Whitmore, B. Maestri, W.R. Mabey, B.R. Holt, and C. Gould. 1979. Water related fate of 129 priority pollutants. Environmental Protection Agency, Office of Water Planning and Standards, Washington, D.C. Vol. 1 and 2. EPA 600/3-82-023ab.

Campbell, M.D. and J.H Lehr. 1973. Water Well Technology. McGraw-Hill Book Company. New York. 681 pp.

Canter, L.W. and R.C. Knox. 1985. Septic tank system effects on ground water quality. Lewis Publishers, Inc. 336 pp.

Canter, L.W. and R.C. Knox. 1985. Ground Water Pollution Control. Lewis Publishers, Inc. 526 pp.

Carr, J.R., D.E. Myers, and C.E. Glass. 1985. Co-kriging-a computer program. Computers and Geosciences, Vol. 11, No. 2. pp. 111-127.

Carroll, D. 1962. Rainwater as a chemical agent of geologic processes-A review. U.S. Geological Survey Water-Supply Paper 1535-G. 18 pp.

Carslaw, H.S. and J.C. Jaeger. 1959. Conduction of heat in solids. Oxford University Press, London. 496 p.

Cedergren, H.R. 1977. Seepage, drainage, and flow nets. 2nd ed. John Wiley & Sons. 334 p.

Ceroici, W. 1980. International system (SI) of units in hydrogeology. Alberta Research Council. Information Series 91. 13 pp.

Chapuis, R.P. 1989. Shape factors for permeability tests in boreholes and piezometers. Ground Water. Vol. 27, No. 5. pp. 647-654.

Chin, C.S. and D.L. Reddell. 1983. Temperature distribution around a well during thermal injection and a graphical technique for evaluating aquifer thermal properties. Water Resources Research. Vol. 19, No. 2. pp 351-363.

Cherry, J.A., R.W. Gillham, and J.F. Barker. 1984. Contaminants in groundwater: chemical processes. In Groundwater contamination. Studies in geophysics. National Academy Press. pp. 46-66.

Clark, D. 1987. Microcomputer programs for groundwater studies. Developments In Water Science, 30. Elsevier Science Publishing Co., Inc.

Clark, D. 1988. Groundwater discharge tests: simulation and analysis. Developments In Water Science, 37. Elsevier Science Publishing Co., Inc. 376 pp.

Clark, I., K.L. Basinger, and W.V. Harper. 1989. MUCK-a novel approach to co-kriging. In: Proceedings of the conference on geostatistical, sensitivity, and uncertainty methods for ground water flow and radionuclide transport modeling. Battelle Memorial Institute. Columbus, Ohio. 670 pp.

Cleary, R.W. 1978. Analytical models for groundwater pollution and hydrology. Princeton University. Water Resources Program Report No. 78-WR-15.

Codell, R.B. and D.L. Schreiber. 1979. NRC models for evaluating the transport of radio-nuclides in groundwater. In Carter, M.W., Ed. Management of low-level radioactivity. Vol. 2. Pergamon Press. New York. pp. 1193-1212.

Cohen, R.M. and W.J. Miller. 1983. Use of analytical models for evaluating corrective actions at hazardous waste sites. In: Proceedings of the Third National Symposium On Aquifer Restoration And Ground-Water Monitoring, May 25-27, 1983, Columbus, Ohio. National Water Well Association. pp. 85-97.

Cook, N.G.W. 1982. Ground-water problems in open-pit and underground mines. In Recent Trends In Hydrogeology. Geological Society of America. Special Paper 189. pp. 397-405.

Cooper, H.H. and C.E. Jacob. 1946. A generalized graphical method for evaluating formation constants and summarizing well-field history. Transactions of the American Geophysical Union. Vol. 27, No. 4.

Cooper, H.H. and M.I. Rorabaugh. 1963. Groundwater movements and bank storage due to flood stages in surface streams. U.S. Geological Survey. Water Supply Paper 1536-J.

Cooper, H.H., J.D. Bredehoeft, and I.S. Papadopulos. 1967. Response of a finite-diameter well to an instantaneous charge of water. Water Resources Research. Vol. 3, No. 1. pp. 263-269.

Corapcioglu, M.Y. and J. Bear. 1983. A mathematical model for regional land subsidence due to pumping. 3. Integrated equations for a phreatic aquifer. Water Resources Research. Vol. 19. pp. 895-908.

Corapcioglu, M.Y. and A.L. Baehr. 1987. Groundwater contamination by petroleum products: 1. Theoretical considerations. Water Resources Research. Vol.23, No. 1. pp. 191-200.

Cuellar, G. 1984. Fancy programming in IBM PC BASIC. Reston Publishing Company, Inc. 269 pp.

Dansby, D.A. 1987. Graphical well package- a graphical, computer-assisted, curve matching approach to well test analysis. In: Proceedings of the Solving Ground Water Problems with Models Conference and Exposition. Feb. 10-12. Denver. pp. 1523-1534.

Darcy, H. 1856. Les fontaines publiques de la villa de Dijon, V. Dalmont, Paris. 647 pp.

Davis, S.N. 1969. Porosity and permeability in natural materials. In: Flow through porous media. ed. R.J.M. DeWeist. Academic Press. pp. 53-89.

Davis, P.R. and W.C. Walton. 1982. Factors involved in evaluating ground water impacts of deep coal mine drainage. American Water Resources Association. Water Resources Bulletin. Vol. 18, No. 5. pp. 841-848.

Davis, A.D. 1986. Deterministic modeling of dispersion in heterogeneous permeable media. Ground Water Vol. 24, No. 5.

Davis, J.C. 1973. Statistics and data analysis in geology. John Wiley & Sons, Inc. 550 p.

Davis, J.C. 1986. Statistics and data analysis in geology. 2nd edition. John Wiley & Sons, Inc. 646 p.

Davis, S.N. and R.J.M. DeWeist. 1966. Hydrogeology. John Wiley & Sons, Inc. 202 pp.

Davis, S.N., D.J.Campbell, H.W. Bentley, and T.J. Flynn. 1985. Ground water traces. National Water Well Association. 200 pp.

Dawson, G.W., C. J. English, and S.E. Petty. 1980. Physical chemical properties of hazardous waste constituents. U.S. Environmental Protection Agency, Environmental Research Laboratory, Athens, Georgia.

Dean, J.D., P.P. Jowise, and A.S. Donigian. 1984. Leaching evaluation of agricultural chemicals (LEACH) handbook. Prepared for the U.S. Environmental Protection Agency, Environmental Research Laboratory, Athens, GA. Anderson-Nichols and Co., Inc. EPA-600/3-84-068.

De Jossselin de Jong, G. 1958. Longitudinal and transverse diffusion in granular deposits. Transactions of American Geophysical Union. Vol. 39. pp. 67-74.

D'Itri, F.M. and L.G. Wolfson. 1987. Rural groundwater contamination. Lewis Publishers, Inc. 416 pp.

Domenico, P.A. and M.D. Mifflin. 1965. Water from low-permeability sediments and land subsidence. Water Resources Research. Vol. 1, No. 4. pp. 563-576.

Domenico, P.A. 1972. Concepts and models in groundwater hydrology. McGraw-Hill Book Company. 405 pp.

Domenico, P.A. 1977. Transport phenomena in chemical rate processes in sediments. Earth and Planetary Science. Vol. 5. pp. 287-317.

Domenico, P.A. and V.V. Palciauskas. 1982. Alternative boundaries in solid waste management. Ground Water. Vol. 20, No. 3. pp. 303-311.

Domenico, P.A. and G.A. Robbins. 1985. A new method of contaminant plume analysis. Ground Water. Vol. 23, No. 4. pp. 476-485.

Dougherty, D.E. and D.K. Babu. 1984. Flow to a partially penetrating well in a double-porosity reservoir. Water Resources Research. Vol. 20., No. 8. pp. 1116-1122.

Dougherty, D. E. 1989. Computing well hydraulics solutions. Ground Water. Vol. 27, No.4. pp. 564-569.

Drever, J.I. 1982. The geochemistry of natural waters. Prentice-Hall. 388 pp.

Driscoll, F.G. 1986. Groundwater and wells. 2nd edition. Johnson Division. St. Paul, Minn. 1089 pp.

Dunlap, W.J. and others. 1977. Sampling for organic chemicals and microorganisms in the subsurface. U.S. Environmental Protection Agency. EPA-600/2-77-176. 27 pp.

El-Kadi, Aly I. 1989. A semi-analytical model for pollution by light hydrocarbons. In: Solving ground water problems with models. Proceedings of the fourth international conference on the use of models to analyze and find working solutions to ground water problems held February 7-9, 1989. National Water Well Association and International Ground Water Modeling Center. pp. 453-471.

Ericson, R. and A. Moskol. 1986. Mastering Reflex. SYBEX Inc. Berkeley, Ca. 336 pp.

Etter, D.M. 1987. Structured FORTRAN 77 for engineers and scientists. The Benjamin/Cummings Publishing Company, Inc. Menlo Park, California. 519 pp.

Everett, L.G. 1989. Groundwater monitoring. Genium Publishing Corp., Schenectady, NY. 440 pp.

Fairchild, D.M. 1987. Ground water quality and agriculture. Lewis Publishers, Inc. 402 pp.

Faust, C.R. and J.W. Mercer. 1979a. Geothermal reservoir simulation, 1. Mathematical models for liquid and vapor-dominated hydrothermal systems. Water Resources Research. Vol. 15, No. 1. pp. 23-30.

Faust, C.R. and J.W. Mercer. 1979b. Geothermal reservoir simulation, 2. numerical solution techniques for liquid and vapor-dominated hydrothermal systems. Water Resources Research. Vol. 15, No. 1. pp. 31-46.

Faust, C.R. and J.W. Mercer. 1981. Geother, a finite-difference model of the three-dimensional, single- and two-phase heat transport in a porous medium. Geo Trans Report 200-00K-01. Reston, Va.

Faust, C.R. and J.W. Mercer. 1983. GEOTHER: a two-phase fluid flow and heat transfer code. ONWI-434. Battelle Memorial Institute. Columbus, Ohio.

Faust, C.R. and J.W. Mercer. 1984. Evaluation of slug tests in wells containing a finite-thickness skin. Water Resources Research. Vol. 20, No. 4. pp. 504-506.

Fenn, D. and others. 1977. Procedures manual for ground water monitoring at solid waste disposal facilities. U.S. Environmental Protection Agency. SW-611. 269 pp.

Fenn, D.G., K.J. Hanley, and T.V. De Geare. 1975. Use of the water balance method for predicting leachate generation from solid waste disposal sites. U.S. Environmental Protection Agency. Cincinnati, Ohio. Report SW-168.

Ferris, J.G., D.B. Knowles, R.H. Brown, and R.W.Stallman. 1962. Theory of aquifer tests. U.S. Geological Survey. Water-Supply Paper 1536-E. 174 pp.

Freeman, R.C. 1989. Using Generic CADD. Osborne McGraw-Hill, Inc. Berkeley, California. 465 pp.

Fetter, C.W. 1988. Applied hydrogeology. Merrill Publishing Company. Columbus, OH. 592 pp.

Ford, P.A., P.J. Turina, and D.E. Seely. 1983. Characterization of hazardous waste sites- A methods manual, Vol. II, Available sampling methods. U.S. Environmental Protection Agency. EPA-600/4-83-040. 215 pp.

Fowler, J. 1984. IBM PC/XT graphics book. Prentice-Hall, Inc. 351 pp.

Franke, O.L., T.E. Reilly, and G.D. Bennett. 1985. Definition of boundary and initial conditions in the analysis of saturated ground-water flow systems-An introduction. U.S. Geological Survey. Open-File Report 84-4302. 26 pp.

Freeze, R.A. and P.A. Witherspoon. 1967. Theoretical analysis of regional groundwater flow: 2. Effect of water-table configuration and subsurface permeability variation. Water Resources Research. Vol. 3. pp. 623-634.

Freeze, R. and J.A Cherry. 1979. Groundwater. Prentice-Hall, Inc. 604 pp.

Fried, J.J. 1975. Groundwater pollution. Elsevier Scientific Publishing Company. 330 pp.

Fuori, W. M., S. Gaughran, L. Gioia, and M. Fuori. 1986. FORTRAN 77 elements of programming style. Hayden Book Company. Hasbrouck Heights, New Jersey. 358 pp.

Gardner, W.R. 1958. Some steady-state solutions of the unsaturated moisture flow equation with application to evaporation from a water table. Soil Science. Vol. 85. pp. 228-233.

Garrison, P. 1982. Programming the TI-59 and the HP-41 calculators. TAB Books, Inc. Blue Ridge Summit, PA. 294 pp.

Gelhar, L.W., A. Mantoglou, C. Welty, and K.R. Rehfeldt. 1985. A review of field scale physical solute transport processes in saturated and unsaturated media. Electric Power Research Institute. Pal Alto, CA. Report No. EPRI EA-4190. 116 pp.

GeoTrans, Inc. 1987. FTWORK: Groundwater flow and solute transport in three dimensions, version 1.6 documentation. 93 pp.

Gibb, J.P., R.M. Schuller, and R.A. Griffin. 1981. Procedures for the collection of representative water quality data from monitoring wells. Illinois State Geological Survey and Illinois State Water Survey Cooperative Ground Water Report 7. 61. pp.

Gillham, R.W. and J.A. Cherry. 1982. Contaminant migration in saturated unconsolidated geologic deposits. In: Recent trends in hydrogeology. The Geological Society of America. Special Paper 189. pp. 31-62.

Gilham, R.W. 1982. Applicability of solute transport models to problems of aquifer restoration. In Proceedings of Second National Symposium on Aquifer Restoration and Ground Water Monitoring. National Water Well Association.

Glover, K.C. 1987. A dual-porosity model for simulating solute transport in oil shale. U.S. Geological Survey. Water-Resources Investigations Report 86-4047.

Goldfarb, W. 1988. Water Law. Lewis Publishers, Inc. 290 pp.

Goldstein, L.J. 1984. Advanced BASIC and beyond for the IBM PC. Robert J. Brady Company. 360 pp.

Goode, A.J., S.M. Neuder, R.A. Pennifill, and T. Ginn. 1986. Onsite disposal of radioactive waste: estimating potential groundwater contamination. U.S. Nuclear Regulatory Commission, Division of Waste Management, Office of Nuclear Material Safety and Safeguards. Washington, D.C. NUREG-1101, Vol.3.

Goodman, R.E., D.G. Moye, A. van Schalkwyk, and I. Javandel. 1965. Ground water flows during tunnel driving. Engineering Geology. Vol. 2. pp. 39-56.

Gringarden, A.C. and J.P. Sauty. 1975. A theoretical study of heat extraction from aquifers with uniform regional flow. Journal of Geophysical Research. Vol. 80, No. 35. pp. 4956-4962.

Gringarten, A.C. 1984. Interpretation of tests in fissured and multilayered reservoirs with double-porosity behavior: Theory and practice. Journal of Petroleum Technology. Vol. 36. pp. 549-564.

Grove, D.B. and K.L. Kipp. 1980. Modeling contaminant transport inporous media in relation to nuclear waste disposal, a review. In: Modeling and low-level waste management, an interagency workshop, ORD-821. Oak Ridge National Laboratory, Oak Ridge, TN.

Gupta, S.K., C.R. Cole, and F.W. Bond. 1979. Finite element three-dimensional ground-water (FE3DGW) flow model-formulation, program listing, and user's manual. Report PNL-2939. Battelle Pacific NW Laboratories, Richland, WA.

Gupta, S.K., C.R. Cole, F.W. Bond, and A.M. Monti. 1984. Finite-element three-dimensional ground-water (FE3DGW) flow model: computer source listings and user's manual. Battelle Memorial Institute Report BMI/ONWI-548. Columbus, Ohio.

Gupta, S.K., C.T. Kincaid, P.R. Meyer, C.A. Newbill, and C.R. Cole. 1986. a multi-dimensional finite-element code for analysis of coupled fluid, energy, and solute transport (CFEST). Pacific Northwest Laboratory Report PNL-4260. Richland, Wash.

Gupta, S.K., C.R. Cole, C.T. Kincaid, and A.M. Monti. 1987. Coupled fluid, energy, and solute transport (CFEST) model: formulation and user's manual. Office of Nuclear Waste Isolation. Battelle Memorial Institute. BMI/ONWI-660. Columbus, Ohio.

Haji-Djafari, S, P.E. Antommaria, and H.L. Crouse. 1981. Attenuation of radionuclides and toxic elements by In Situ soils at a uranium tailings pond in central Wyoming. In: Permeability and groundwater contaminant transport. American Society For Testing And Materials. Special Technical Publication 746. pp. 221-242.

Hantush, M.S. and C.E. Jacob. 1955. Nonsteady radial flow in an infinite leaky aquifer. Transaction American Geophysical Union. Vol. 36. pp. 95-100.

Hantush, M.S. 1959a. Nonsteady flow to flowing wells in leaky artesian aquifers. Journal of Geophysical Research. Vol. 64, No. 8. pp. 1043-1052.

Hantush, M.S. 1959b. Analysis of data from pumping wells near a river. Journal of Geophysical Research. Vol. 64, No. 11. pp. 1921-1932.

Hantush, M.S. 1960. Modification of the theory of leaky aquifers. Journal of Geophysical Research. Vol. 65, No. 11. pp. 3713-3725.

Hantush, M.S. 1961a. Drawdown around a partially penetrating well. Hydraulics Division Journal of American Society of Civil Engineers. Proceedings. pp. 83-98.

Hantush, M.S. 1961b. Economical spacing of interfering wells. International Association of Scientific Hydrology. Publication No. 56. pp. 350-364.

Hantush, M.S. 1962. Flow of ground water in sands of nonuniform thickness. Journal of Geophysical Research. Vol. 67, No. 4.

Hantush, M.S. 1964. Hydraulics of wells. In Advances In Hydroscience. Vol. 1. Academic Press Inc. pp. 281-432.

Hantush, M.S. 1965. Wells near streams with semipervious beds. Journal of Geophysical Research. Vol. 70, No. 12. pp. 2829-2838.

Hantush, M.S. 1966. Analysis of data from pumping tests in anisotropic aquifers. Journal of Geophysical Research. Vol. 71. pp. 421-426.

Hantush, M.S. 1967a. Flow of groundwater in relatively thick leaky aquifer. Water Resources Research. Vol. 3, No. 2. pp. 583-590.

Hantush, M.S. 1967b. Flow to wells in aquifers separated by a semipervious layer. Journal of Geophysical Research. Vol. 72, No. 6. pp. 1709-1720.

Hantush, M.S. 1967c. Growth and decay of groundwater-mounds in response to uniform percolation. Water Resources Research. Vol. 3, No. 4. pp. 227-234.

Hantush, M.S. 1967d. Depletion of flow in right-angle stream bends by steady wells. Water Resources Research. Vol. 3, No. 1. pp. 235-240.

Hantush, M.S. 1978. Unsteady movement of fresh water in thick unconfined saline aquifers. Bulletin of the International Association of Scientific Hydrology. Vol. 13, No. 2. pp. 40-60.

Harada, M. P.L. Chambe, M. Foglia, K. Higashi, F. Iwamoto, D. Leung, D.H. Pigford, and D. Ting. 1980. Migration of radionuclides through sorbing media. Analytical solutions. I. Report LBL-10500. Lawrence Berkeley Laboratory. Berkeley, CA.

Harr, M.E. 1962. Groundwater and Seepage. McGraw-Hill Book Company. New York. 315 pp.

Harris, J., S. Gupta, M. Bitner, and M. Broten. 1989. San Gabriel basin ground water flow analyses-integrated use of GIS and three-dimensional finite-element modeling. In: Solving ground water problems with models. Proceedings of the fourth international conference on the use of models to analyze and find working solutions to ground water problems. National Water Well Association and International Ground Water Modeling Center. pp. 371-384.

Hazen, A. 1893. Some physical properties of sand and gravels with special reference to their use in filtration. 24th Annual Report. Mass. State Bd. Boston.

Hazen, A. 1911. Discussion of dams on sand foundations, by A.C. Koenig. Transactions of the American Society of Civil Engineers. Vol. 73.

Hearn, D. and M.P. Baker. 1983. Computer graphics for the IBM personal computer. Prentice-Hall, Inc. 330 pp.

Heath, R.C. 1982. Classification of ground-water systems of the United States. Ground Water. Vol. 20, No. 4.

Helfferich, F. 1962. Ion exchange. McGraw-Hill. New York.

Helm, D.C. 1975. One-dimensional simulation of aquifer system compaction near Pixley, California, 1. Constant parameters. Water Resources Research. Vol. 11, No. 3. pp. 465-478.

Helm, D.C. 1982. Conceptual aspects of subsidence due to fluid withdrawal. In Recent Trends In: Hydrogeology. The Geological Society of America, Inc. Special Paper 189. pp. 103-142.

Helm, D.C. 1984. Analysis of sedimentary skeletal deformation in a confined aquifer and the resulting drawdown. In: Groundwater Hydraulics. American Geophysical Union. Water Resources Monograph 9. pp. 29-82.

Hem, J.D. 1970. Study and interpretation of the chemical characteristics of natural water. U.S. Geological Survey Water-Supply Paper 1473. 363 pp.

Hirasaki, G.J. 1974. Pulse tests and other early transient pressure analyses for in-situ estimation of vertical permeability. Transactions of American Institute of Mining Engineers (AIME). Vol. 257, pp. 75-90.

Holzer, T.L. Ed. 1984. Man-induced subsidence. The Geological Society of America, Inc. Reviews in Engineering. 232 pp.

Hoopes, J.A. and D.R.F. Harleman. 1967. Dispersion in radial flow from a recharging well. Journal of Geophysical Research. Vol. 72, No. 14. pp. 3595-3607.

Hubbert, M.K. 1940. The theory of groundwater motion. Journal of Geology. Vol. 48. pp. 785-944.

Huisman, L. 1972. Groundwater Recovery. Winchester Press, New, New York. 336 pp.

Huisman, L. and T.N. Olsthoorn. 1983. Artificial groundwater recharge. Pitman Advanced Publishing Program. Boston. 320 pp.

Hund-Der Yeh and Hund-Yuang Han. 1989. Numerical identification of parameters in leaky aquifers. Ground Water. Vol. 27, No. 5. pp. 655-663.

Hunt, B.W. 1973. Dispersion from pit in uniform seepage. American Society of Civil Engineers. Journal of the Hydraulics Division. Vol. 99, No. HY1. pp. 13-21.

Hunt, B.W. 1978. Dispersive sources in uniform ground-water flow. American Society of Civil Engineers. Journal of the Hydraulics Division. Vol. 104, No. HY1. pp. 75-85.

Huntoon, P.W. 1974. Finite-difference methods as applied to the solution of groundwater flow problems. Wyoming Water Resources Research Institute. Laramie Wyoming. 108 pp.

Hunt, B. 1983. Mathematical analysis of groundwater resources. Butterworth & C. Boston. 271 pp.

Hutchinson, W.R. and J.L. Hoffman. 1983. A ground water pollution priority system: New Jersey Geological Survey. Open-file Report No. 83-4. Trenton, New Jersey. 32 pp.

Huyakorn, P.S. and G.F. Pinder. 1983. Computational methods in subsurface flow. Academic Press, Inc. New York. 473 pp.

Huyakorn, P.S., S.D. Thomas, and B.M. Thompson. 1984. Techniques for making finite elements competitive in modeling flow in variably saturated porous media. Water Researches Research. Vol. 20, No. 8. pp. 1099-1115.

Huyakorn, P.S. et al. 1984. SEFTRAN:a simple and efficient flow and transport code. GeoTrans, Inc. Reston, Virginia.

Huyakorn, P.S., M.J. Ungs, L.A. Mulkey, and E.D. Sudicky. 1987. A three-dimensional analytical method for predicting leachate migration. Ground Water. Vol. 25, No. 5. pp. 588-598.

Hvorslev, M.J. 1951. Time lag and soil permeability in ground water observations. U.S. Army Corps. of Engineers Water Ways Experimental Station. Bulletin 36. 50 pp.

Ingersol, L.R., O.J. Zobel, and A.C. Ingersol. 1948. Heat conduction with engineering, geological, and other applications. McGraw-Hill Book Co. New York.

Intera, Inc. 1979. Revision of the documentation for a model for calculating effects of liquid waste disposal in deep saline aquifers. Water Resources Investigations 79-96. U.S. Geological Survey. Washington, D.C.

Intercomp, Inc. 1976. A model for calculating effects of liquid waste disposal in deep saline aquifer. Parts I and II. National Technical Information Service. PB-256903.

Ito, K. 1951. On stochastic differential equations. American Mathematics Society. New York.

Jacob, C.E. 1944. Notes on determining permeability by pumping tests under water table conditions. U.S. Geological Survey. Mimeo. Report.

Jacob, C.E. 1947. Drawdown test to determine effective radius of artesian well. Transactions of the American Society of Civil Engineers. Vol. 112. pp. 1047-1070.

Jacob, C.E. and S.W. Lohman. 1952. Nonsteady flow to a well of constant drawdown in an extensive aquifer. American Geophysical Union Transactions. Vol. 33, No. 4. pp. 559-569.

Javandel, I., C. Doughty, and C.F. Tsang. 1984. Groundwater transport: handbook of mathematical models. American Geophysical Union. Water Resources Monograph 10. 228 pp.

Jenkins, C.T. 1968. Computation of rate and volume of stream depletion by wells. U.S. Geological Survey. Techniques of Water-Resources Investigations. Book 4, Chapter D1. 17 pp.

Johnson, A.I. 1967. Specific yield-compilation of specific yields for various materials. U.S. Geological Survey. Water Supply Paper 1662-D. 74 pp.

Jones, T.A., D.E. Hamilton, and C.R. Johnson. 1986. Contouring geologic surfaces with the computer. Van Nostrand Reinhold Company. 314 pp.

Jousma, G., J. Bear, Y.Y. Haimes, and F. Walter. Eds. 1989. Groundwater contamination: Use of models in decision making. Kluwer Academic Publishers. Norwell, Maine. 656 pp.

JRB Associates, Inc. 1982. Handbook, remedial actions at waste disposal sites. U.S. Environmental Protection Agency. EPA-625/6-82-006. 497 pp.

Kanwar, R.S., H.P. Johnson, and H.S. Chauhan. 1979. Nonsteady flow to nonpenetrating well. American Society of Civil Engineers, Journal of Irrigation and Drainage Division. Vol. 105, No. IR1.

Kaplan, O. B. 1987. Septic systems handbook. Lewis Publishers, Inc. 290 pp.

Karickhoff, S.W., D.S. Brown, and T.A. Scott. 1979. Sorption of hydrophobic pollutants on natural sediments. Water Resources Research. Vol. 13. pp. 241-248.

Karplus, W.J. 1958. Analog simulation. McGraw Hil. New York.

Kashef, A.I. 1986. Groundwater Engineering. McGraw-Hill, Inc. New York.

Kazmann, R.G. 1948. River infiltration as a source of ground-water supply. Transactions of the American Society of Civil Engineers. Vol. 113. pp. 404-424.

Kazmann, R.G. and W.R. Whitehead. 1980. The spacing of heat pump supply and discharge wells. National Water Well Association. Ground Water Heat Pump Journal. Summer. pp. 28-31.

Keely, J.F. 1987. The use of models in managing ground-water protection programs. Office of Research and Development, U.S. Environmental Protection Agency. EPA/600/8-87/003.

Kimbler, O.S., R.G. Kazmann, and W.R. Whitehead. 1975. Cyclic storage of fresh water in saline aquifers. Louisiana Water Resources Research Institute. Louisiana State University. Bulletin 10. 78 pp.

Kinzelbach, W. 1986. Groundwater modelling-an introduction with sample programs in BASIC. Elsevier Science Publishers. Amsterdam. 334 p.

Kinzelbach, W. 1988. The random walk method in pollutant transport simulation. In Groundwater Flow and Quality Modelling. eds. Custodio, E., A. Gurgui, and J.P. Lobo Ferreira. D. Reidel Publishing Company. Boston. pp. 227-245.

Kipp, K.L. 1987. HST3D: A computer code for simulation of heat and solute transport in three-dimensional ground-water flow systems. U.S. Geological Survey. Water Resources Investigations Report 86-4095. 517 pp.

Kirkham, D. and S.B. Affleck. 1977. Solute travel times to wells. Ground Water. Vol. 15, No. 3. pp. 231-242.

Klaer, F.H. 1953. Providing large industrial water supplies by induced infiltration. Mining Engineering. Vol. 5. pp. 620-624.

Konikow, L.F. and Bredehoeft, J.D. 1978. Computer model of two-dimensional solute transport and dispersion in ground water. U.S. Geological Survey Techniques of Water-Resources Investigations. Book 7, Chapter C2. 90 pp.

Konikow, L.F. 1988. Present limitations and perspectives on modelling pollution problems in aquifers. In Groundwater flow and quality modelling. eds. Custodio, e., A. Gurgui, and J.P. Lobo Ferreira. D. Reidel Publishing Company. Boston. pp. 643-664.

Kontis, A.L. and R.J. Mandle. 1988. Modifications of a three-dimensional ground-water flow model to account for variable water density and effects of multiaquifer wells. U.S. Geological Survey Water Resources Investigations Report 87-4265. 78 pp.

Korites, B.J. 1982. Data plotting software for micros. Kern Publications.

Kruseman, G.P. and N.A De Ridder. 1976. Analysis and evaluation of pumping test data. International Institute for Land Reclamation and Improvement Wageningen The Netherlands. Bulletin 11. 200 pp.

Kuiper, L.K. 1986. A comparison of methods for the solution of the inverse problem in two-dimensional steady-state groundwater flow modeling. Water Resources Research. Vol. 22. pp. 705-714.

Kuppusany, T.J., J. Sheng, J.C. Parker, and R.J. Lenhard. 1987. Finite-element analysis of multiphase immiscible flow through soils. Water Resources Research. Vol. 23, No. 4. pp. 625-631.

Lauwerier, H.A. 1955. The transport of heat in an oil layer caused by the injection of hot fluid. Applied Science Research. Section A., 5. pp. 145-150.

LeGrand, H.E. 1983. A standard system for evaluating waste-disposal sites. National Water Well Association. 49 pp.

Lenau, C.W. 1972. Dispersion from recharge well. American Society of Civil Engineers. Journal of the Engineering Mechanics Division. Vol. 98, No. EM2. pp. 331-344.

Lenau, C.W. 1973. Contamination of discharge well from recharge well. American Society of Civil Engineers. Journal of the Hydraulics Division. Vol. 99, No. HY8. pp. 1247-1263.

Lenk, F. 1988. Inside AutoSketch. New Riders Publishing, Thousand Oaks, Ca.

Lennox, D.H. 1966. Analysis and application of step-drawdown test. Journal of the Hydraulics Division, Proceedings of the American Society of Civil Engineers. Vol. 92, No. HY6. pp. 25-48.

Linsley, R.K., and M.A. Kohler, and J.L. Paulhus. 1958. Hydrology for engineers. McGraw-Hill Book Company. New York.

Lohman, S.W. 1972. Ground-water hydraulics. U.S. Geological Survey Professional Paper 708. 70 pp.

Longsine, D.E., E.J. Bonano, and C.P. Harlan. 1987. User's manual for the NEFTRAN computer code. U.S. Nuclear Regulatory Commission, Office of Nuclear Materials Safety And Safeguards. NU-REG/CR-4766, SAND86-2405-RW.

Luthin, J.N. 1966. Drainage engineering. John Wiley & Sons, Inc. New York.

Lyman, W.J., W.F. Reehl, and D.H. Rosenblatt. 1982. Handbook of Chemical property estimation methods. McGraw-Hill. New York, New York.

Mackay, D.M., P.V. Roberts, and J.A. Cherry. 1985. Transport of organic contaminants in groundwater. Environmental Science Technology. Vol. 19, No. 5. pp. 384-392.

Masch, F.D. and K.J. Denny. 1966. Grain size distribution and its effect on the permeability of unconsolidated sands. Water Resources Research. Vol. 2, No. 4.

Mercado, A. 1967. The spreading patterns of injected water in a permeable stratified aquifer. In: Proceedings of International Association of Scientific Hydrology. Symposium at Haifa. Publication No. 72. pp. 23-26.

Mercer, J., S. Larson, and C. Faust. 1980. Finite-difference model to simulate the areal flow of saltwater and freshwater separated by an interface. U.S. Geological Survey. Open-File Report. 80-407. 88 pp.

Marino, M.A. and J.N. Luthin. 1982. Seepage and groundwater. Elsevier Scientific Publishing Co. New York. 489 pp.

Marsily, Ghislain de. 1986. Quantitative hydrogeology. Academic Press. 440 p.

Matthess, G. 1982. The properties of groundwater. John Wiley & Sons, Inc., New York. 406 p.

Matthess, G. and A. Pekdeger. 1985. Survival and transport of pathogenic bacteria and viruses in ground water. In Ground Water Quality. John Wiley & Sons, Inc.

McCarty, P.L., M. Reinhard, and B.E. Rittman. 1981. Trace organics in ground water. Environmental Science and Technology. Vol. 15, No. 1.

McDonald, M.G. and A.W. Harbaugh. 1984. A modular three-dimensional finite-difference ground-water flow model. U.S. Geological Survey. Open-file Report 83-875. 528 pp.

McNellis, M. and C.O. Morgan. 1969. Modified Piper diagrams by the digital computer. Kansas State Geological Survey. University of Kansas. Special Distribution Publication No. 43. 36 pp.

McWorter, D.B. 1981. Predicting groundwater response to disturbance by mining-selected problems. Proceedings 1981 Symposium on Surface Mining Hydrology, Sedimentology, and Reclamation. University of Kentucky. UKY BU126. pp. 89-95.

Mercer,J.W. and G.F. Pinder. 1975. A finite element model of two-dimensional single-phase heat transport in a porous medium. U.S. Geological Survey. Open File Report 75-574. 115 pp.

Mercer, J.W., S.P. Larson, and C.R. Faust. 1980. Finite-difference model to simulate the areal flow of saltwater and freshwater separated by an interface. U.S. Geological Survey. Open File Report 80-407. 88 pp.

Mercer, J.W. and Faust, C.R. 1981. Ground-water modeling. National Water Well Association. 60 pp.

Michigan Department of Natural Resources. 1983. Site assessment system (SAS) for the Michigan priority ranking system under the Michigan Environmental Response Act. Michigan Department of Natural Resources. 91 pp.

Microsoft. 1987. Microsoft QuickBASIC BASIC language reference. 533 pp.

Mikels, F.C. and F.H. Klaer. 1956. Application of ground water hydraulics to the development of water supplies by induced infiltration. International Association of Hydrology. Publication No. 41. pp. 232-24

Miller, D.W. ed. 1980. Waste disposal effects on ground water. Premier Press, Berkeley, Cal. 512 pp.

Miller, A.R. 1981. BASIC programs for scientists and engineers. SYBEX, Inc. 318 p.

Mills, W., J. Dean, D. Porcella, S. Gherini, R. Hudson, W. Frick, G. Rupp, and G. Bowie. 1982. Water quality assessment: a screening procedure for toxic and conventional pollutants, U.S. Environmental Protection Agency, Athens, Georgia. Vol. 1 and 2. EPA 600/6082-004ab.

Milnes, A.G. 1985. Geology and radwaste. Academic Press Inc. 328 pp.

Moench, A.F. and T.A. Prickett. 1972. Radial flow in an infinite aquifer undergoing conversion from artesian to water table conditions. Water Resources Research. Vol. 8, No. 2. pp. 494-499.

Moench, A.F. and A. Ogata. 1981. A numerical inversion of the Laplace transform solution to radial dispersion in a porous medium. Water Resources Research. Vol. 17, No. 1. pp. 250-252.

Moench, A.F. and A. Ogata. 1984. Analysis of constant discharge wells by numerical inversion of Laplace transform solutions. In: Groundwater Hydraulics. American Geophysical Union. Water Resources Monograph 9. pp. 146-170.

Moench, A.F. 1984. Double-porosity models for a fissured groundwater reservoir with fracture skin. Water Resources Research. Vol. 20. pp. 831-846.

Montgomery, J.H. and L.M. Welkom. 1989. Ground water chemicals desk reference. Lewis Publishers, Inc. 600 pp.

Moody, W.T. 1966. Nonlinear differential equation for drain spacing. Proceedings of the American Society of Civil Engineers. Vol. 92.

Moore, J.E. 1979. Contribution of ground-water modeling to planning. Journal of Hydrology. Vol. 43. pp. 121-128.

Morgan, C.O. and M. McNellis. 1969. Stiff diagrams of water-quality data programmed for the digital-computer. Kansas State Geological. University of Kansas. Special Distribution Publication No. 43. 27 pp.

Morris, D.A. and A.I. Johnson. 1967. Summary of hydrologic and physical properties of rock and soil materials as analyzed by the Hydrologic Laboratory of the U.S. Geological Survey. Water Survey Paper 1839-D. 42. pp.

Muskat, M. 1937. The flow of homogeneous fluids through porous medium. McGraw-Hill. 763 pp.

Narasimhan, T.N. 1980. Program TERZAGHI user's manual. Report LBL-10908. Lawrence Berkeley Laboratory. University of California. Berkeley.

National Research Council. 1927. International critical tables of numerical data, physics, chemistry, and technology. Vol.2. McGraw-Hill Book Co. New York.

National Academy of Sciences, National Academy of Engineering. 1972. Water quality criteria. 594 pp.

Naval Facilities Engineering Command (NAVFAC). 1971. Chapter 4. Field tests and measurements. Design manual-Soil mechanics, foundations, and earth structures, Dept.of the Navy. Alexandria, Va.

Neuman, S.P. and P.A. Witherspoon. 1969. Applicability of current theories of flow in leaky aquifers. Water Resources Research. Vol. 5, No. 4. pp. 817-829.

Neuman, S.P. 1972. Theory of flow in unconfined aquifers considering delayed response of the water table. Water Resources Research. Vol. 8, No. 4. pp. 1031-1045.

Neuman, S.P. and P.A. Witherspoon. 1972. Field determination of the hydraulic properties of leaky multiple aquifer systems. Water Resources Research. Vol. 8, No. 5. pp. 1284-1298.

Neuman, S.P. 1974. Effect of partial penetration on flow in unconfined aquifers considering delayed gravity response. Water Resources Research. Vol. 10, No. 2. pp. 303-312.

Neuman, S.P. 1975a. Analysis of pumping test data from anisotropic unconfined aquifers considering delayed gravity response. Water Resources Research. Vol. 11, No. 2. pp. 329-342.

Neuman, S.P. 1975b. A computer program to calculate drawdown in an aniso-tropic unconfined aquifer with a partially penetrating well. Unpublished manuscript. Department of Hydrology and Water Resources, University of Arizona.

Neuman, S.P. 1979. Perspective on "delayed yield." Water Resources Research. Vol. 15, No. 4. pp. 899-908.

Neuzil, C.E. 1982. On conducting the modified 'slug' test in tight formations. Water Resources Research. Vol. 18, No. 2. pp. 439-441.

Nguyen, V. and G.F. Pinder. 1984. Direct calculation of aquifer parameters in slug test analysis. In: Groundwater Hydraulics. Water Resources Monograph Series 9. American Geophysical Union. pp. 222-239.

Nowak, S.F. 1984. Getting the most from your pocket computer. TAB Books, Inc. Blue Ridge Summit, PA. 232 pp.

Olea, R.A. 1975. Optimal mapping techniques using regionalized theory. Kansas Geological Survey Series on Spatial Analysis. No.2 137 p.

Onishi, Y., R.J. Serne, E.M. Arnold, C.E. Cowan, and F.L. Thompson. 1981. Critical review: radioactive transport, sediment transport, and water quality modeling; and radioactive adsorption/desorption mechanisms. U.S. Nuclear Regulatory Commission. NUREG/CR-1322. Washington, D.C.

ORD. 1981. Treatability manual. Vol. 1. Treatability Data. U.S. Environmental Protection Agency. Office of Research and Development. EPA-600/2-82-001a.

OSM. 1981. Ground water model handbook. U.S. Office of Surface Mining. Denver. H-D3004-021-81-1062D.

OST. 1984. Protecting the nation's groundwater from contamination. Office of Technology Assessment. Vol. 1. U.S. Congress, Washington, D.C. 503 pp.

Oudijk, G. and K. Mujica. 1989. Handbook for the identification, location and investigation of pollution sources affecting ground water. National Water Well Association. 185 pp.

Page, A.L., T.J. Logan, and J.A. Ryan. 1987. Land application of sludge. Lewis Publishers, Inc. 168 pp.

Papadopulos, I.S. 1966. Nonsteady flow to multiaquifer wells. Journal of Geophysical Research. Vol. 71, No. 20. pp. 4791-4797.

Papadopulos, I.S. 1967. Drawdown distribution around a large-diameter well. National Symposium on Ground-water Hydrology. San Francisco, California. November 6-8, 1967. Proceedings. pp. 157-168.

Papadopulos, I.S. and H.H. Cooper. 1967. Drawdown in a well of large diameter. Water Resources Research. Vol.3, No. 1. pp. 241-244.

Papadopulos, S.S, J.D. Bredehoeft, and H.H. Cooper. 1973. On the analysis of slug test data. Water Resources Research. Vol. 9, No. 4. pp. 1087-1089.

Parker, J.C. and J.J. Kaluarachchi. 1989. A numerical model for design of free product recovery systems at hydrocarbon spill sites. In: Solving ground water problems with models. Proceedings of the fourth international conference on the use of models to analyze and find working solutions to ground water problems. National Water Well Association and International Ground Water Modeling Center. pp. 271- 281.

Parker, J.C., T. Kuppusamy, and B.H. Lien. 1989. Modelling immiscible organic chemical transport in soils and groundwater. In: Groundwater contamination: use of models in decision-making.Eds. Jousma, G., J. Bear, Y.Y. Haimes, and F.Walter. Kluwer Academic Publishers. Boston, MA. pp. 301-312.

Patchick, P.F. 1967. Predicting well yields-two case histories. Ground Water. Vol.5, No. 2.

Peaceman, D.W. and Rachford, H.H. 1955. The numerical solution of parabolic and elliptical difference equations. Journal of Society of Industrial and Applied Mathematics. Vol. 3, No. 11. pp. 28-41.

Peaceman, D.W. 1983. Interpretation of well-block pressures in numerical reservoir simulation with nonsquare grid blocks and anisotropic permeability. Society of petroleum Engineers Journal. Vol. 23, No. 3. pp. 531-543.

Pedrosa, O.A. and Aziz, K. 1986. Use of a hybrid grid in reservoir simulation. Society of Petroleum Engineers. Reservoir Engineering. Vol.1, No. 6.

Peterson, D. M. 1989. Modeling the effects of variably saturated flow on stream losses. In Solving ground water problems with models. Proceedings of the fourth international conference on the use of models to analyze and find working solutions to ground water problems held February 7-9, 1989. National Water Well Association and International Ground Water Modeling Center. pp. 899-928.

Pinder, G.F. and L.M. Abriola. 1986. On the simulation of nonaqueous phase organic compounds in the subsurface. Water Resources Rearch. Vol. 22. No. 9. Supplement. pp. 109s-1119s.

Pinder, G.F. 1988. An overview of groundwater modelling. In Groundwater flow and quality modelling. eds. Custodio, E., A. Gurgui, and J.P. Lobo Ferreira. D. Reidel Publishing Company. Boston. pp. 119-134.

Pickens, J.F. and G.E. Grisak. 1981a. Scale-dependent dispersion in a stratified aquifer. Water Resources Research. Vol. 17. pp. 1191-1212.

Pickens, J.F. and G.E. Grisak. 1981b. Modeling of scale-dependent dispersion in hydrogeologic systems. Water Resources Research. Vol. 17. pp. 1701-1711.

Pickens, J.F., R.E. Jackson, K.J. Inch, and W.F. Merritt. 1981. Measurement of distribution coefficients using a radial injection dual-tracer test. Water Resources Research, Vol. 17, No. 3. pp. 529-544.

Pigford, T.H., P.L. Chambre, M. Albert, M. Foglia, M. Harada, F. Iwamoto, T. Kanki, D. Leung, S. Masuda, S. Muraoka, and D. Ting. 1980. Migration of radionuclides through sorbing media. Analytical solutions. II. Report LBL-11616, 2 vol. Lawrence Berkeley Laboratory. Berkeley, CA.

Piper, A.M. 1944. A graphical procedure in the geochemical interpretation of water-analyses. Transactions of the American Geophysical Union. Vol. 25. pp. 914-928.

Penman, H.L. 1948. Natural evaporation from openwater, bare soil, and grass. Proceedings of the Royal Society (London) A. Vol.193. pp. 120-145.

Plomb, D.J. 1989. A 3-D finite element model to predict drawdown caused by infiltration into a 32-foot diameter tunnel. In: Solving ground water problems with models. Proceedings of the fourth international conference on the use of models to analyze and find working solutions to ground water problems. National Water Well association and International Ground Water Modeling Center. pp. 955-978.

Plummer, L.N. 1984. Geochemical modeling: a comparison of forward and inverse methods. In: Proceedings of first Canadian/American conference on hydrogeology-Practical applications of ground water geochemistry. National Water Well Association. pp. 149-177.

Poland, J.F. 1961. The coefficient of storage in a region of major subsidence caused by compaction of an aquifer system. U.S. Geological Survey. Professional Paper 424-B. pp. B52-B54.

Poland, J.P. and G.H. Davis. 1969. Land subsidence due to withdrawal of fluids. In Reviews in Engineering Geology. Vol. 2. The Geological Society of America, Inc. pp. 1878-269.

Pollock, D.W. 1988. Semianalytical computation of path lines for finite-difference models. Ground Water. Vol. 26, No. 6. pp. 743-750.

Polubarinova-Kochina, P.YA. 1962. Theory of ground water movement. Princeton University Press. 613 pp.

Poole, L., M. Borchers, and K. Koessel. 1981. Some common BASIC programs. Osborne/McGraw-Hill, Inc. 193 p.

Potter, S.T. and W.J. Gburek. 1987. Seepage face simulation using PLASM. Ground Water. Vol. 25, No. 6. pp.722-732.

Powers, J.P. 1981. Construction dewatering: A guide to theory and practice. John Wiley & Sons, Inc. New York. 484 pp.

Prakash, A. 1984. Ground-water contamination due to transient sources of pollution. American Society of Civil Engineers. Journal of the Hydraulics Division. Vol. 110, No. 11. pp. 1642-1658.

Prats, M. 1970. A method for determining the net vertical permeability near a well from in-situ measurements. Transactions of American Institute of Mining Engineers (AIME). Vol. 249, pp. 637-643.

Press, W.H., B.P. Flannery, S.A. Teukolsky, and W.T. Vetterling. 1986. Numerical recipes-the art of scientific computing. Cambridge Univ. Press. 818 p.

Prickett, T.A. and C.G. Lonnquist. 1971. Selected digital computer techniques for groundwater resource evaluation. Illinois State Water Survey. Bulletin 55. 60 pp.

Prickett, T.A. 1975. Modeling techniques for groundwater evaluation. In Advances In Hydroscience. Ed. Ven Te Chow. Vol. 10. pp. 1-143.

Prickett, T.A. 1980. Oral communication at Champaign, Illinois.

Prickett, T.A. 1981. Oral communication at Champaign, Illinois.

Prickett, T.A. 1983. Oral communication at Champaign, Illinois.

Prickett, T.A. 1987. Oral communication at Champaign, Illinois.

Prickett, T.A., Naymik, T.S., and Lonnquist, C.G. 1981. A "random walk" solute transport model for selected groundwater quality evaluations. Illinois State Water Survey. Bulletin 65. 103 p.

Rai, D., J.M. Zachara, R.A. Schmidt, and A.P. Schwab. 1984. Chemical attentuation rates, coefficients and constants in leachate migration. Electric Power Research Institute. Palo Alto, California. Vol. 1: A critical review. EA-3356.

Ramey, H.J. 1982. Well-loss function and the skin effect: a review. In Recent Trends In Hydrogeology. The Geological Society of America. Special Paper 189. pp. 265-272.

Rasmussen, W.C. 1964. Permeability and storage of heterogeneous aquifers in the U.S. International Association of Scientific Hydrology. Publication No. 64. pp. 317-325.

Raudkivi, A.J. and R.A. Callander. 1976. Analysis of groundwater Flow. John Wiley & Sons, Inc. New York. 214 pp.

Reed, J.E. 1980. Type curves for selected problems of flow to wells in confined aquifers. Techniques of Water Resources Investigations of the U.S. Geological Survey. Book 3. Chapter B3. 106 p.

Reeves, M. and Cranwell, R.M. 1981. User's manual for the Sandia waste-isolation flow and transport model (SWIFT). Release 4.81: NUREG/CR-2324 USNRC. Division of Risk Analysis. Office of Nuclear Regulatory Research. Washington, D.C. 145 pp.

Reeves, M., D.S. Ward, N.D. Johns, and R.M. Cranwell. 1986a. Theory and implementation for SWIFT II, the Sandia waste-isolation flow and transport model. Release 4.84. Sandia National Laboratories. NUREG/CR-3328 and Sand83-1159. 189 pp.

Reeves, M., D.S. Ward, N.D. Johns, and R.M. Cranwell. 1986b. Data input guide for SWIFT II, the Sandia waste-isolation flow and transport model. Release 4.84. Sandia National Laboratories. NUREG/CR-3162 and Sand83-0242. 144 pp.

Reeves, M., D.S. Ward, P.A. Davis, and E.J. Bonano. 1986c. SWIFT II self-teaching curriculum: illustrative problems for the Sandia waste-isolation flow and transport model for fractured media. Sandia National Laboratories. NUREG/CR-3925 and Sand83-1586. 96 pp.

Reily, T.E., O.L. Franke, H.T. Buxton, and G.D. Bennett. 1987. A conceptual framework for ground-water solute-transport studies with emphasis on physical mechanisms of solute movement. U.S. Geological Survey Water-Resources Investigation Report 87-4191. 44 pp.

Reisenauer, A.E., K.T. Key, T.N. Narasimhan, and R.W. Nelson. 1982. TRUST: a computer program for variable saturated flow in multidimensional, deformable media. Report NUREG/CR-2360, PNL-3975. Battelle Pacific NW Laboratories, Richland, WA.

Remson, I., G.M. Hornberger, and F.J. Molz. 1971. Numerical methods in sub-surface hydrology. Wiley-Interscience a Division of John Wiley & Sons, Inc. New York. 389 pp.

Rice, R.E. ed. 1988. Chemistry of ground water. Lewis Publishers, Inc. 300 pp.

Rifai, H.S., P.B. Bedient, R.C. Borden, and J. F. Hassbeek. 1987. BIOPLUME II-computer model of two-dimensional transport under the influence of oxygen limited biodegradation in ground water. User's manual, version 1.0, Rice University, Houston, Texas.

Rorabaugh, M.I. 1953. Graphical and theoretical analysis of step-drawdown test of artesian well. Proceedings Separate. No. 362. American Society of Civil Engineers. Vol. 79.

Rorabaugh, M.I. 1956. Ground water in Northeastern Louisville and Kentucky with reference to induced infiltration. U.S. Geological Survey. Water Supply Paper 1360-B.

Rounds, S.A. and B.A. Bonn. 1989. DREAM a menu-driven program that calculates drawdown, streamlines, velocity and water level elevation. In Solving ground water problems with models. Proceedings of the fourth international conference on the use of models to analyze and find working solutions to ground water problems. National Water Well Association and International Ground Water Modeling Center. pp. 329-350.

Rovey, C.E.K. 1975. Numerical model of flow in a stream-aquifer system. Hydrology Paper No. 74. Colorado State University.

Roy, W.R. and R.A. Griffin. 1985. Mobility of organic solvents in water-saturated soil materials. Environmental Geology and Water Sciences. Vol. 7. pp. 241-247.

Rumbaugh, J. O. 1989. Increasing the efficiency and accuracy of applied modeling using a database approach. In: Solving ground water problems with models. Proceedings of the fourth international conference on the use of models to analyze and find working solutions to ground water problems. National Water Well Association and International Ground Water Modeling Center. pp. 683-698.

Rushton, K.R. and S.C. Redshaw. 1979. Seepage and groundwater flow. John Wiley and Sons. New York. 339 pp.

Safai, N.M. and G.F. Pinder. 1977. Numerical model of land subsidence due to pumpage from fully and partially penetrating wells. Civil Engineering Department, Princeton, University. Tech. Rept. 78-WR-1.

Sagar, B. 1982. Dispersion in three dimensions: approximate analytical solutions. American Society of Civil Engineers. Journal of the Hydraulics Division. Vol.108., No. HY1. pp. 47-62.

Sageev, A. 1986. Slug test analysis. Water Resources Research. Vol. 22, No. 8. pp. 1323-1333.

SAI. 1981. Tabulation of waste isolation computer models. Prepared by Science Applications, Inc. for the Office of Nuclear Waste, Battelle, Memorial Instaitute. Columbus, ohio. ONWI-78.

Salinity Laboratory. 1954. Diagnosis and improvement od saline and alkaline soil. U.S. Department of Agriculture Handbook No. 6. 160 pp.

Sandberg, R., R.B. Scheiback, D. Koch, and T.A. Prickett. 1981. Selected hand-held calculator codes for the evaluation of the probable cumulative hydrologic impacts of mining. U.S. Dept. of the Interior, Office of Surface Mining. H-D3004/030-81-1029F.

Sandler, C. 1984. Desktop graphics for the IBM PC. Creative Computing Press. Morris Plains, New Jersey. 190 pp.

Sanford, W.E. and L.F. Konikow. 1985. A two-constituent solute transport model for ground water having variable density. U.S. Geological Survey. Water Resources Investigations. 85-4279. 88 pp.

Sapik, D.B. 1988. Documentation of a steady-state saltwater-intrusion model for three-dimensional ground-water flow, and user's guide. U.S. Geological Survey Open File Report 87-526. 174 pp.

Sauty, J.P. 1977. Interpretation of tracer tests by means of type curves application to uniform and radial flow. In: Invitational Well-Testing Symposium Proceedings. Lawrence Berkeley Laboratory. Report LBL-7027. pp. 82-90.

Sauty, J.P. 1980. An analysis of hydrodispersive transfer in aquifers. Water Resources Research. Vol. 16, No. 1. pp. 145-158.

Sauty, J.P. and W. Kinzelbach. 1988. On the identification of the parameters of groundwater mass transport. In Groundwater flow and quality modelling. Eds. E. Custodio, A. Gurgui, and J.P. Lobo Ferreira. D. Reidel Publishing Company. Boston. pp. 33-56.

Sauveplane, C. 1984a. Pumping test analysis in fractured aquifer formations: state of the art and some perspectives. In: Groundwater Hydraulics. American Geophysical Union Water Resources Monograph 9. pp. 171-206.

Sauveplane, C.M. 1984b. On the use of approximate analytical inversion of Laplace transform for radial flow problems. In International Groundwater Symposium on Groundwater Resources Utilization and Contaminant Hydrogeology. Vol. 1. Atomic Energy of Canada Ltd.. pp. 197-215.

Sawyer, C.N. and P.L. McCarty. 1967. Chemistry for sanitary engineers. McGraw-Hill. New York. 518 pp.

Schaetzle, W.F., C.E. Brett, D.M. Grubbs, and M.S. Seppanen. 1980. Thermal energy storage in aquifers. Pergamon Press, Inc. New York. 177 pp.

Scalf, M.R. and others. 1981. Manual of ground water quality sampling procedures. National Water Well Association. 1981.

Schapery, R.A. 1961. Approximate methods of transform inversion for viscoelastic stress analysis. Proceedings of U.S. National Congress for Applied Mechanics. 4th. pp. 1075-1085.

Scheidegger, A.E. 1954. Statistical hydrodynamics in porous media. Journal of Geophysical Research. Vol. 66, No. 10. pp. 3273-3278.

Schicht, R.J. and W.C. Walton. 1961. Hydrologic budgets for three small watersheds in Illinois. Illinois State Water Survey. Report of Investigation. No. 40.

Schmorak, S. and A. Mercado. 1969. Upconing of fresh water-sea water interface below pumping wells, field study. Water Resources Research. Vol. 5, No. 6. pp. 1290-1310.

SCS Engineers. 1982. Costs of remedial response actions at uncontrolled hazardous waste sites. U.S. Environmental Protection Agency, Municipal Environmental Research Laboratory, Cincinnati, Ohio. EPA 600/2-82-035.

Schwartz, F.W. and A. Crowe. 1980. A deterministic probabilistic model for contaminant transport. NUREG/CR-1609, CGS Inc. Prepared for Nuclear Regulatory Commission.

Schwartz, F.W. and L. Smith. 1988. A continuum approach for modeling mass transport in fractured media. Water Resources Research. Vol. 24, No. 8.

Schwille, F. 1988. Dense chlorinated solvents in porous and fractured media: Model experiments. Lewis Publishers, Inc. 142 pp.

Seller, L.E. and L.W. Canter. 1980. Summary of selected ground-water quality impact assessment methods. National Center for Ground Water Research. Report No. NCGWR 80-3. Norman, Oklahoma. 142 pp.

Shafer, J.M. 1987. GWPATH: interactive ground-water flow path analysis. Illinois State Water Survey. Bulletin 69. 42 pp.

Shen, H.T. 1976. Transient dispersion in uniform porous media flow. American Society of Civil Engineers. Journal of the Hydraulics Division. Vol. 102, No. HY6. pp. 707-716.

Shepherd, R.G. 1989. Correlations of permeability and grain size. Ground Water. Vol. 27, No. 5. pp. 633-638.

Simons, S.L. 1983. Make fast and simple contour plots on a microcomputer. BYTE Publications, Inc. Nov. pp. 487-492.

Sims, P.N., P.F. Andersen, D.E. Stephenson, and C.R. Faust. 1989. Testing and benchmarking of a three-dimensional groundwater flow and solute transport model. In: Solving ground water problems with models. Proceedings of the fourth international conference on the use of models to analyze and find working solutions to ground water problems. National Water Well Association and International Ground Water Modeling Center. pp. 821-841.

Skrivan, J.A. and M.R. Karlinger. 1980. Semi-variogram estimation and universal kriging program. U.S. Geological Survey Computer Contribution. Tacoma, Washington. PB81-120560. 98 p.

Smith, H.F. 1954. Gravel packing water well. Illinois State Water Survey. Circular 44.

Smith, J.W. 1970. Chemical engineering kinetics. McGraw-Hill. New York.

Smith, L. and F.W. Schwartz. 1980. Mass transport 1. A stochastic analysis of macroscopic dispersion. Water Resources Research. Vol. 16, No. 2. pp. 303-313.

Smith, L., F. Schwartz, and C. Mase. 1989. Application of stochastic methods for the simulation of solute transport in discrete and continuum models of fractures rock systems. In: Proceedings of the conference on geostatistical, sensitivity, and uncertainty methods for ground-water flow and radionuclide transport modeling. CONF-870971. Battelle Press. Columbus, Ohio. 670 pp.

Souza, W.R. 1987. Documentation of a graphical display program for the Saturated-Unsaturated Transport (SUTRA) finite-element simulation model. U.S. Geological Survey. Water-Resources Investigations Report 87-4245.

Spillette, A.G. 1972. Heat transfer during hot fluid injection into an oil reservoir. In Thermal recovery techniques. Society of Petroleum Engineers of A.I.M.E., Reprint Series. N0. 10. pp. 21-25.

Spiridonoff, S.V. 1964. Design and use of radial collector wells. Journal of the American Water Works Association. June.

Spitzer, F.L. 1964. Principles of random walk. Princeton University Press.

Stallman, R.W. 1962. Variable discharge without vertical leakage. In Theory of aquifer tests. U.S. Geological Survey. Water-Supply Paper 1536-E. pp. 118-122.

Stallman, R.W. 1963. Type curves for the solution of single-boundary problems. In Shortcuts and Special Problems in Aquifer Tests. U.S. Geological Survey Water-Supply Paper 1545-C. pp. 45-47.

Stallman, R.W. 1971. Aquifer-test design, observation and data analysis. U.S. Geological Survey. Techniques of Water-Resources Investigations. Book 3, Chapter B1.

Stehfest, H. 1970. Algorithm 368. Numerical inversion of Laplace transforms. Commun. ACM. Vol. 13, No. 1. pp. 47-49.

Sternberg, Y.M. 1967. Transmissivity determination from variable discharge pumping tests. Ground Water. Vol. 5, No. 4. pp 27-29.

Sternberg, Y.M. 1968. Simplified solution for variable rate pumping test. Proceedings of American Society of Civil Engineers. Vol. 94, HY 1. pp. 177-180.

Strack, O.D.L. 1976. A single-potential solution for regional interface problems in coastal aquifers. Water Resources Research. Vol. 12, No. 6. pp. 1165-1174.

Strack, O.D.L. 1989. Groundwater Mechanics. Prentice- Hall, Inc. 732 pp.

Streltsova, T.D. 1972. Unsteady radial flow in an unconfined aquifer. Water Resources Research. Vol. 8, No. 4. pp. 1059-1066.

Streltsova, T.D. 1978. Well hydraulics in heterogeneous aquifer formations. In Advances in Hydrosciences. Vol. 11 pp. 357-423. Academic Press, Inc.

Streltsova, T.D. 1988. Well testing in heterogeneous formations. John Wiley and Sons, Inc. 413 p.

Terzaghi, K. 1925. Principles of soil mechanics: IV, settlement and consolidation of clay. Engineering News- Record. pp. 874-878.

Theis, C.V. 1935. The relation between the lowering of the piezometric surface and the rate and duration of discharge of a well using ground-water storage. American Geophysical Union Transactions. Vol.16. pp. 519-524.

Theis, C.V. 1963. Drawdowns caused by a well discharging under equilibrium conditions from an aquifer bounded by a straight-line source. In Shortcuts and Special Problems in Aquifer Tests. U.S. Geological Survey Water-Supply Paper 1545-C. pp. 101-105.

Thompson, D.B. 1987. A microcomputer program for interpreting time-lag permeability tests. Ground Water. Vol. 25, No. 2. pp. 212-218.

Thompson, S.D., B. Ross, and J.W. Mercer. 1982. A summary of repository siting models. U.S. Nuclear Regulatory Commission. NUREG/CR-2782. Washington, D.C.

Thornthwaite, C.W. 1954. The loss of water to the air. Meteorological Monographs. Vol. 6. pp. 165-180.

Todd, D.K. 1980. Groundwater Hydrology. John Wiley and Sons, Inc. New York. 535 pp.

Toth, J. 1962. A theory of groundwater motion in small drainage basins in central Alberta. Journal of Geophysical Research. Vol. 67. pp. 4375-4387.

Toth, J. 1963. A theoretical analysis of groundwater flow in small drainage basins. Journal of Geophysical Research. pp. 4795-4812.

Townley, L.R. and J.L. Wilson. 1980. Description of and user's manual for a finite element aquifer flow model. AQUIFEM-1. Ralph M. Parsons Laboratory for Water Resources and Hydrodynamics. Massachusetts Institute of Technology. Report No. 252.

Tracy, J.V. 1982. Users guide and documentation for adsorption and decay modifications to the U.S.G.S. solute transport model. U.S. Nuclear Regulatory Commission. NUREG/CR-2502. U.S. Government Printing Office. Washington, D.C.

Trescott, P.C., G.F. Pinder, and S.P. Larson. 1976. Finite-difference model for aquifer simulation in two dimensions with results of numerical experiments. Techniques of Water-Resources Investigations, U.S. Geological Survey. Chapter C1, Book 7. 116 pp.

Tsang. C.T., T. Buscheck, and C. Doughty. 1981. Aquifer thermal energy storage: numerical simulation of Auburn University field experiments. Water Resources Research. Vol. 17, N0. 3. pp. 647-658.

Tsang, C.F. and D.L. Hopkins. 1982. Aquifer thermal energy storage: a survey. In Recent trends in hydrogeology. Geological Society of America. Special Paper 189. pp. 427-441.

Tyson, H.N. and E.M. Weber. 1964. Groundwater management for the nation's future-computer simulation of groundwater basins. Journal of the Hydraulics Division, Proceedings of American Society of Civil Engineers. Vol. 90, No. HY4. pp. 59-77.

Uffink, G.J.M. 1985. A random walk method for the simulation of macrodispersion in a stratified aquifer. In: Relation of Groundwater Quantity and Quality. Proceedings of the Hamburg Symposium, August 1983. IAHS Publication No. 146. pp. 103-114.

Uffink, G.J.M. 1986. A random-walk simulation of dispersion at an interface between fresh and saline groundwater. Proceedings of the 9 th Salt Water Intrusion Meeting. Delft.

van der Heijde, P.K.M. and P. Srinivasan. 1983. Aspects of the use of graphic techniques in ground water modeling. International Ground Water Modeling Center. Report No. GWMI 83-11.

van der Heijde, P., Y. Bachmat, J. Bredehoeft, B. Andrews, D. Holtz, and S. Sebastian. 1985. Groundwater management: the use of numerical models. American Geophysical Union. Water Resources Monograph 5. 180 pp.

van der Heijde. 1987. Quality assurance in computer simulations of groundwater contamination. Environmental Software. Vol. 2, No. 1. pp. 19-25.

Van Voast, W.A. and Hedges, R.B. 1975. Hydrogeologic aspects of existing and proposed strip coal mines near Decker, southeastern Montana. State of Montana Bureau of Mines and Geology. Bulletin 97.

Virdee and Kottegoda. 1984. A brief review of kriging and its application to optimal interpolation and observation well selection. Journal of Hydrological Sciences. Vol. 29, No. 4. pp. 367-387.

Voisinet, D.D. 1986. Introduction to computer-aided drafting. McGraw-Hill Book Company. 274 pp.

Voss, C.I. 1984. A finite-element simulation model for saturated-unsaturated, fluid-density-dependent ground-water flow with energy transport or chemically-reactive single-species solute transport. U.S. Geological Survey. Water Resources Investigations Report 84-4369. 409 pp.

Wagener, J.L. 1980. FORTRAN77-principles of programming. John Wiley & Sons, Inc. New York. 370 pp.

Wang, H.F. and M.P. Anderson. 1982. Introduction to groundwater modeling-Finite difference and finite element methods. W.H. Freeman and Company. San Francisco. 237 pp.

Walton, W.C. 1962. Selected analytical methods for well and aquifer evaluation. Illinois State Water Survey. Bulletin 49. 81 pp.

Walton, W.C. 1963. Estimating the infiltration rate of a streambed by aquifer-test analysis. International Association of Scientific Hydrology. General Assembly, Berkeley.

Walton, W.C. 1965. Groundwater recharge and runoff in Illinois. Illinois State Water Survey. Report of Investigation. No. 48.

Walton, W.C. 1970. Groundwater Resource Evaluation. McGraw-Hill, Inc. 664 pp.

Walton, W.C. 1979. Progress in analytical groundwater modeling. Journal of Hydrology. Vol. 43. pp. 149-159.

Walton, W.C. 1984a. Practical aspects of groundwater modeling. National Water Well Association. 587 pp.

Walton, W.C. 1984b. Thirty-five BASIC groundwater programs for desktop microcomputers. International Ground Water Modeling Center, Holcomb Research Institute, Butler University. WALTON84-BASIC.

Walton, W.C. 1984c. Analytical groundwater modeling with programmable calculators and hand-held computers. In: Groundwater Hydraulics, eds. Rosenshein, J and G.D. Bennett. American Geophysical Union. Water Resources Monograph 9. pp. 298-312.

Walton, W.C. 1987. Groundwater Pumping Tests. Lewis Publishers, Inc. 201 pp.

Walton, W.C. 1989a. Analytical groundwater modeling. Lewis Publishers, Inc. 173 pp.

Walton, W.C. 1989b. Numerical groundwater modeling. Lewis Publishers, Inc. 272 pp.

Ward, D.S., D.R. Buss, J.W. Mercer, and S.S. Hughes. 1987. Evaluation of groundwater corrective action at the Chem-Dyne hazardous waste site using telescopic mesh refinement modeling approach. Water Resources Research. Vol. 23, No. 4.

Warner, D.L. and J.H. Lehr. 1981. Subsurface wastewater injection. Premier Press, Berkeley, Cal. 344 pp.

Way, Shao-Chih and C.R. McKee. 1982. In-situ determination of three-dimensional aquifer permeabilities. Ground Water. Vol. 20, No. 5. pp. 594-603.

Weeks, E.P. 1969. Determining the ratio of horizontal to vertical permeability by aquifer-test analysis. Water Resources Research. Vol.5, No. 1. pp. 196-214.

Weinman, D.G. and B.L. Kurshan. 1985. IBM PC BASIC for scientists and engineers. Reston Publishing Company, Inc. 344 pp.

Wenzel, L.K. 1942. Methods for determining permeability of water-bearing materials with special reference to discharging-well methods. U.S. Geological Survey. Water-Supply Paper 887. 192 pp.

Werner, P.W. 1946. Notes on flow-time effects in the great artesian aquifers of the earth. American Geophysical Union Transactions. Vol. 27, No. 5. pp. 687-708.

Wilcox, L.V. 1955. Classification and use of irrigation waters. U.S. Department of Agriculture Circular 969. 19 pp.

Wilson, J.L. and D.A. Hamilton. 1978. Influence of strip mines on regional ground-water flow. Journal of the Hydraulics Division, American Society of Civil Engineers. Vol. 104, No. HY9. pp. 1213-1223.

Wilson, J.L. and P.J. Miller. 1978. Two-dimensional plume in uniform ground-water flow. American Society of Civil Engineers. Journal of the Hydraulics Division. Vol. 104, No. HY4. pp. 503-514.

Winter, T.C. 1976. Numerical simulation analysis of the interaction of lakes and groundwaters. U.S. Geological Survey Professional Paper 1001. 45. pp.

Winter, T.C. 1978. Numerical simulation of steady-state three-dimensional groundwater flow near lakes. Water Resources Research. Vol. 14. pp. 245-254.

Winter, T.C. 1981. Effects of water-table configuration on seepage through lake beds. Limnology and Oceanography. Vol. 26. pp. 925-934.

Winter, T.C. 1983. The interaction of lakes with variable saturated porous media. Water Resources Research. Vol. 19. pp. 1203-1218.

Witherspoon, P.A. and S.P. Neuman. 1967. Evaluating a slightly permeable caprock in aquifer gas storage: I. Caprock of infinite thickness. Journal of Petroleum Technology. pp. 949-955.

Wood, E.F., R.A. Ferrara, W.G. Gray, and G.F. Pinder. 1984. Groundwater contamination from hazardous wastes. Prentice-Hall, Inc. Englewood Cliffs, New Jersey. 163 pp.

Yeh, G.T. and D.S. Ward. 1980. FEMWATER: a finite-element model of water flow through saturated-unsaturated porous media. Oak Ridge National Laboratory. ORNL-5567.

Yeh, G.T. and D.S. Ward. 1981. FEMWASTE: a finite-element model of a waste transport through porous media. Oak Ridge National Laboratory. ORNL-5601.

Yeh, G.T. 1981. AT123D: analytical transient one-, two-, and three-dimensional simulation of waste transport in the aquifer system. Oak Ridge National Laboratory. Environmental Sciences Division Publication No. 1439. ORNL-5602. 83 pp.

Appendix A
Widely Used Textbooks and Journals

Table A.1 . Textbooks

Bear, J. 1979. *Hydraulics of Groundwater*. McGraw-Hill Inc. 567 pp.

Bouwer, H. 1978.*Groundwater Hydrology.*. McGraw-Hill Inc. 480 pp.

Davis, S.N. andR.J.M. De Wiest. 1966. *Hydrogeology*. John Wiley & Sons Inc. 463 pp.

De Wiest, R.J.M. 1965. Geohydrology. John Wiley & Sons Inc. 366 pp.

Strack, O.D.L. 1989. *Groundwater Mechanics*. Prentice-Hall, Inc. 732 pp.

Fetter, C.W. 1988. *Applied Hydrogeology*. Merrill Publishing Co. 591 pp.

Freeze, R.A. and J.A. Cherry. 1979. *Groundwater*. Prentice-Hall Inc. 604 pp.

Marsily, Ghislain de.1986. Quantitative Hydrogeology. Academic Press, Inc. 440 pp.

McWhorter, D.B and Sunada, D.K. 1977. *Ground-Water Hydrology and Hydraulics*. Water Resources Publications. Fort Collins, Colorado. 292 pp.

Todd, D.K. 1980. *Groundwater Hydrology*. John Wiley & Sons, Inc. 535 pp.

Hunt, B. 1983. *Mathematical Analysis of Groundwater Resources*. Butterworth & Co. 271 pp.

Table A.2 . Journals

Advances In Water Resources. Quarterly by C.M.L. Publications Ashurst Lodge, Ashurst, South-ampton SO4 2AA, England.

American Society of Agricultural Engineers Transactions. Bimonthly by ASAE, 2950 Niles Rd. St. Joseph, Michigan 49085.

Journal of Hydraulic Engineering. Monthly by American Society of Civil Engineers, 345 E. 47th Street, New York, New York 10017

Journal of Irrigation and Drainage Engineering. Quarterly by Association American Society of Civil Engineers.

American Water Works Association Journal. Monthly by AWWA, 6666 W. Quincy Ave., Denver, Colorado. 80235.

Ground Water. Bimonthly. by Water Well Journal Publishing Co., 6375 Riverside Drive., Dublin, Ohio 43017.

Ground Water Monitoring Review. Quarterly by Water Well Journal Publishing Co.

Hydrological Sciences Journal. Quarterly by the International Association of Hydrological Sciences. Blackwell Scientific Publications Ltd., Osney Mead, Oxford OX2 OEL, England.

Journal of Contaminant Hydrology. Quarterly by Elsevier Scientific Publishing Co. Journal Div., P.O. Box 211, 1000 AE Amsterdam. The Netherlands.

Journal of Hydraulic Research. Quarterly by the International Association for Hydraulic Research. Rotterdamseweg 185, P.O. Box No. 177, 2600 MH Delft, The Netherlands.

Journal of Hydrology. Four issues per volume and four volumes per year by Elsevier Scientific Publishing Co.

Water Resources Bulletin. Bimonthly by the American Water Resources Association, 5410 Grosvenor Ln., Suite 220, Bethesda, Maryland 20814

Table A.2 .   Journals (continued)

Journal of Water Resources Planning & Management. Quarterly by American Society of

Water Resources Research. Bimonthly by the American Geophysical Union, 2000 Florida Ave., N.W.,

Appendix B
Representative Aquifer System Characteristic Values

Table B.1. Porosity

Deposit	Porosity (dimensionless)
Volcanic, pumice	0.80 — 0.90
Peat	0.60 — 0.80
Silt	0.35 — 0.60
Clay	0.35 — 0.55
Loess	0.40 — 0.55
Sand, dune	0.35 — 0.45
Sand, fine	0.25 — 0.55
Sand, coarse	0.30 — 0.45
Gravel, coarse	0.25 — 0.35
Sand and gravel	0.20 — 0.35
Till	0.25 — 0.45
Siltstone	0.25 — 0.40
Sandstone	0.25 — 0.50
Volcanic, vesicular	0.10 — 0.50
Volcanic, tuff	0.10 — 0.40
Limestone	0.05 — 0.55
Schist	0.05 — 0.50
Basalt	0.05 — 0.35
Shale	0.01 — 0.10
Volcanic, dense	0.01 — 0.10
Igneous, dense	0.01 — 0.05
Salt bed	0.005— 0.03

(After Davis, 1969; Johnson, 1967; Morris and Johnson, 1967; Davis and DeWiest, 1966; Marsily, 1986, pp. 68,79; Rasmusussen, 1964, pp. 317-325; Polubarinova-Kochina, 1962; Freeze and Cherry, 1979, p. 404-410; Walton, 1988, pp. 19-23, 58)

Table B.2 Specific Yield

Deposit	Specific Yield (dimensionless)
Peat	0.30 — 0.50
Sand, dune	0.30 — 0.40
Sand, coarse	0.20 — 0.35
Sand, gravelly	0.20 — 0.35
Gravel, fine	0.20 — 0.35
Gravel, coarse	0.10 — 0.25
Gravel, medium	0.15 — 0.25
Loess	0.15 — 0.35
Sand, medium	0.15 — 0.30
Sand, fine	0.10 — 0.30
Igneous, weathered	0.20 — 0.30
Sandstone	0.10 — 0.40
Sand and gravel	0.15 — 0.30
Silt	0.01 — 0.30
Clay, sandy	0.03 — 0.20
Clay	0.01 — 0.20
Volcanic, tuff	0.02 — 0.35
Siltstone	0.01 — 0.35
Limestone	0.01 — 0.25
Till	0.05 — 0.20

(After Davis, 1969; Johnson, 1967; Morris and Johnson, 1967; Davis and DeWiest, 1966; Marsily, 1986, pp. 68,79; Rasmusussen, 1964, pp. 317-325; Polubarinova-Kochina, 1962; Freeze and Cherry, 1979, p. 404-410; Walton, 1988, pp. 19-23, 58)

Table B.3 Artesian Storativity

Deposit	Artesian Storativity (dimensionless)
Clay, plastic	$6.2 \times 10^{-3} — 7.8 \times 10^{-4}$m
Clay, stiff	$7.8 \times 10^{-3} — 3.9 \times 10^{-4}$m
Clay, medium hard	$3.9 \times 10^{-4} — 2.8 \times 10^{-4}$m
Sand, loose	$3.1 \times 10^{-4} — 1.5 \times 10^{-5}$m
Sand, dense	$6.2 \times 10^{-5} — 3.9 \times 10^{-5}$m
Sand and gravel, dense	$3.1 \times 10^{-5} — 1.5 \times 10^{-5}$m

Table B.3 Artesian Storativity (continued)

Rock, fissured and jointed	2.1×10^{-5} — 1.0×10^{-6}m
Rock, sound	$<1.0 \times 10^{-6}$m

m = aquifer or aquitard thickness (ft)

(After Domenico, 1972, p. 231)

Table B.4. Horizontal Hydraulic Conductivity

Deposit	Horizontal Hydraulic Conductivity (gpd/sq ft)
Gravel	1×10^3 — 3×10^4
Basalt	1×10^{-6} — 2×10^4
Limestone	2×10^{-2} — 2×10^4
Sand and Gravel	2×10^2 — 5×10^3
Sand	2×10^1 — 3×10^3
Sand, quick	50 — 8×10^3
Sand, dune	1×10^2 — 3×10^2
Peat, little decomposed	80 — 300
Peat, moderately decomposed	8 — 40
Peat, young sphagum	8 — 80
Peat, old sphagum	6 — 8
Sandstone	1×10^{-1} — 50
Loess	2×10^{-3} — 20
Clay	2×10^{-4} — 2
Soil bentonite	1×10^{-3} — 2×10^{-1}
Cement bentonite	2×10^{-2}
Till	5×10^{-4} — 1
Shale	1×10^{-5} — 1×10^{-1}
Quartzite	4×10^{-3} — 8
Greenstone	1×10^{-1} — 14
Rhyolite	1 — 20
Schist	1×10^{-2} — 2
Coal	1 — 1×10^3

(After Davis, 1969; Johnson, 1967; Morris and Johnson, 1967; Davis and DeWiest, 1966; Marsily, 1986, pp. 68,79; Rasmusussen, 1964, pp. 317-325; Polubarinova-Kochina, 1962; Freeze and Cherry, 1979, p. 404-410; Boutwell, et al., 1986, p. 294; Walton, 1988, pp. 19-23, 58)

Table B.5. Aquitard Vertical Hydraulic Conductivity

Deposit	Aquitard Vertical Hydraulic Conductivity (gpd/sq ft)
Sand, gravel, and clay	1×10^{-1} — 1×10^{0}
Clay, sand, and gravel	1×10^{-2} — 6×10^{-2}
Clay	5×10^{-4} — 1×10^{-2}
Shale	1×10^{-7} — 1×10^{-3}

(after Walton, 1970, p. 239)

Table B.6. Aquifer Stratification

Degree Of Stratification	P_V/P_H Ratio
Low	1/2
Medium	1/10
High	1/100
Very High	1/1000

(after Walton, 1988, p. 164)

Table B.7. Transmissivity And Hydraulic Conductivity Of Major Aquifers In United States

Region	Geologic Situation	Transmissivity (ft²/d)	Hydraulic Conductivity (ft/d)
Western Mountain Ranges	Mountains with thin soils over fractured rocks alternating with narrow alluvial and partly glaciated valleys	5—1000	0.001—50
Alluvial Basins	Thick alluvial (locally gla-		

**Table B.7.   Transmissivity And Hydraulic Conductivity Of Major
Aquifers In United States (continued)**

Region	Geologic Situation	Transmissivity (ft²/d)	Hydraulic Conductivity (ft/d)
	cial) deposits in basins and valleys bordered by mountains	2,000— 200,000	100—2,000
Columbia Lava Plateau	Thick lava sequence interbedded with unconsolidated deposits and overlain by thin soils	20,000— 5,000,000	500— 10,000
Colorado Plateau and Wyoming Basin	Thin soils over fractured sedimentary rocks	5—1,000	0.01—5
High Plains	Thick alluvial deposits over fractured sedimentary rocks	10,000— 100,000	100—1,000
Nonglaciated Central Region	Thin regolith over fractured sedimentary rocks	3,000— 100,000	10—1,000
Glaciated Central Region	Thick glacial deposits over fractured sedimentary rocks	1,000— 20,000	5—1,000
Piedmont and Blue Ridge	Thick regolith fractured crystalline and metamorphosed sedimentary rocks	100—2,000	0.003—3
Northeast and Superior Uplands	Thick glacial deposits over fractured crystalline rocks	500—5,000	5—100
Atlantic and Gulf	Complexly inter-		

Table B.7. Transmissivity And Hydraulic Conductivity Of Major Aquifers In United States (continued)

Region	Geologic Situation	Transmissivity (ft²/d)	Hydraulic Conductivity (ft/d)
Coastal Plain	bedded sands, silts, and clays	5,000— 100,000	10—400
Southeast Coastal Plain	Thick layers of sand and clay over semi-con-solidated car-bonated rocks	10,000— 1,000,000	100— 10,000
Alluvial Valleys	Thick sand and gravel deposits beneath flood-plains and terraces of streams	2000— 500,000	100—5000
Hawaiian Islands	Lava flows aug-mented by dikes, inter-bedded with ash deposits, and partly overlain by alluvium	100,000— 1,000,000	500— 10,000
Alaska	Glacial and alluvial depos-its in part per-ennially frozen and overlying crystalline,met-amorphic, and sedimentary soils	1,000— 100,000	100—2,000

(after Heath, 1983, p.13)

Table B.8 Fractured Rock Characteristics

Characteristic	Value
Fracture horizontal hydraulic conductivity	0.01 — 100 gpd/sq ft
Matrix rock vertical hydraulic conductivity	1×10^{-6} — 1×10^{-3} gpd/sq ft
Fracture porosity	0.001 — 0.01
Block porosity	0.01 — 0.30
Fracture width	0.001 — 1.0 in
Fracture spacing	0.1 — 10 ft
Block storativity	10^{-6} multiplied by rock thickness
Fracture storativity	1/10 — 1/100 of block storativity

(after Streltsova-Adams, 1978, pp.360-361)

Table B.9 Deposit Elasticity

Deposit	Bulk Modulus Of Elasticity (kg/cm^2)
Igneous, sandstone, limestone, fissured, jointed	1500—30000
Dense gravel and sand	2000—10000
Dense sands	500—2000
Loose sands	100—200
Dense clays and silts	100—1000
Medium clays and silts	50—100
Loose clays	10—50
Peat	1—5

(after Bouwer, 1978, p. 321)

Table B.10 Capillary Fringe Thickness

Deposit	Capillary Fringe Thickness (cm)
Fine gravel	2.5
Very coarse sand	6.5
Coarse sand	13.5
Medium sand	24.6
Fine sand	42.8
Silt	105.5

(after Lohman, 1972)

Table B.11 Induced Streambed Infiltration

Location	Induced Streambed Infiltration Rate (gpd/acre/ft)	Temperature Of Surface Water (°F)
Satsop River, Satsop, WASH.	1.0×10^7	51
Mad River-Springfield, OH	1.0×10^6	39
Sandy Creek-Canton, OH	7.2×10^5	82
Mississippi River-St.Louis, ILL	3.1×10^5	54
White River-Anderson, IND	2.2×10^5	69
Miami River-Cincinnati, OH	1.7×10^5	35
Mississippi River,St.Louis, ILL	9.1×10^4	33
White River-Anderson, IND	4.0×10^4	38
Mississippi River-St.Louis, ILL	3.5×10^4	83

(after Walton, 1970, p. 265)

Table B.12 Water Dynamic Viscosity

Water Temperature (°F)	Dynamic Viscosity (10^{-3} pascal second)
32.0	1.7921
33.8	1.7313
35.6	1.6728
37.4	1.6191
39.2	1.5674
41.0	1.5188
42.8	1.4728
44.6	1.4284
46.4	1.3860
48.2	1.3462
50.0	1.3077
51.8	1.2713
53.6	1.2363
55.4	1.2028
57.2	1.1709
59.0	1.1404
60.8	1.1111
62.4	1.0828
64.4	1.0559
66.2	1.0299
68.0	1.0050
77.0	0.8937
86.0	0.8007
104.0	0.6560

(after Matthess, 1982; p.14)

Table B.13. Dispersivity

Type of Aquifer	Location	Measurement Method	Longitudinal Dispersivity (meters)
Alluvial, full	Chalk River, Ontario	Single well	0.034
Alluvial,			

Table B.13. Dispersivity (continued)

Type of Aquifer	Location	Measurement Method	Longitudinal Dispersivity (meters)
plane of high velocity			0.034—0.1
Alluvial, full		Two well	0.5
Alluvial, plane of high velocity			0.1
Alluvial, stratum scale	Lyons, France	Single well	0.1-0.5
Alluvial, full			5.0
Alluvial			12.0
			8.0
			5.0
			7.0
Alluvial, sediments	Alsace, France		12.0
Fractured dolomite	Carsbad,NM	Two well	38.1
Fractured Schistgneiss	Savannah River, SC		134.1
Alluvial sediments	Barstow, CA		15.2
Chalk, fractured	Dorset, England		3.1
Chalk, intact			1.0
Sand/gravel	Berkeley, CA	Multi-well	2.0—3.0
Limestone	Mississippi	Single well	11.6
Alluvial, sediments	Tucson, Ariz.	Two well	79.2
	Rocky Mtn. Col.	Model calib.	30.5
	Arkansas River Valley, Col.		30.5
	California		30.5
Glacial deposits	Long Island, NY		21.3

Table B.13. Dispersivity (continued)

Type of Aquifer	Location	Measurement Method	Longitudinal Dispersivity (meters)
Limestone	Brunswick, GA		61.0
Basalt, fractured	Snake River, ID		91
Basalt, fractured	Idaho		91
Basalt, fractured	Hanford site, WA		30.5
Alluvial deposits	Barstow, CA		61.0
Limestone	Roswell Basin, NM		21.3
Lava flows and sediments	Idaho Falls, ID		91.0
Alluvial sediments	Barstow, CA		61.0
	Alsace, France		15.0
Limestone	Florida (SE)		6.7
Alluvial	Sutter Basin, CA		80.0—200.0

(after Borg, et al., 1976; Anderson, 1979)

Table B.14 Dry Deposit Density

Deposit	Density (g/cm^3)
Shale	1.54—3.17
Silt	1.01—1.79
Basalt	1.99—2.89
Limestone	1.21—2.69
Dolomite	1.83—2.20
Granite	1.21—1.78
Schist	1.42—2.69
Fine Sand	1.13—1.99
Medium Sand	1.27—1.93

Table B.14 Dry Deposit Density (continued)

Deposit	Density (g/cm^3)
Coarse Sand	1.42—1.94
Medium Gravel	1.47—2.09
Coarse Gravel	1.69—2.08
Loess	0.75—1.62
Eolian Sand	1.33—1.70
Soil	1.13—2.00
Rock Salt	1.68—2.14
Sandstone	1.60—2.68
Siltstone	1.35—2.12
Claystone	1.37—1.60
Clay	1.00—2.40
Till	1.61—2.12
Glacial Drift	1.11—1.83
Coal	0.70—1.50

(after Carslaw and Jaeger, 1959; Oudijk and Mujica, 1989, p. 149)

Table B.15 Dry Deposit Thermal Conductivity

Deposit	Thermal Conductivity (Cal/cm sec °C)
Gneiss	$5 \times 10^{-3} - 6 \times 10^{-3}$
Shale	$4 \times 10^{-3} - 8 \times 10^{-3}$
Salt	$8 \times 10^{-3} - 1 \times 10^{-2}$
Marl	$5 \times 10^{-3} - 7 \times 10^{-3}$
Sandstone	$2 \times 10^{-3} - 9 \times 10^{-3}$
Marble	7×10^{-3}
Limestone	$5 \times 10^{-3} - 8 \times 10^{-3}$
Chalk	2×10^{-3}
Dolomite	$4 \times 10^{-3} - 1 \times 10^{-2}$
Basalt	5×10^{-3}
Granite	$4 \times 10^{-3} - 8 \times 10^{-3}$
Gypsum	3×10^{-3}
Clay	$2 \times 10^{-3} - 3 \times 10^{-3}$
Sand	$3 \times 10^{-3} - 6 \times 10^{-3}$

Table B.15 Dry Deposit Thermal Conductivity (continued)

Deposit	Thermal Conductivity (Cal/cm sec °C)
Soil	2×10^{-3}
Coal	$3 \times 10^{-4} - 7 \times 10^{-4}$

(after Ingersol,et al., 1948; Carslaw and Jaeger, 1959; Bear, 1972, p. 649)

Table B.16 Dry Deposit Thermal Diffusivity

Deposit	Thermal Diffusivity (cm²/sec)
Sandstone	1.1×10^{-2} - 2.3×10^{-2}
Dolomite	8.0×10^{-3}
Granite	6.0×10^{-3} - 2.0×10^{-2}
Limestone	5.0×10^{-3} - 1.0×10^{-2}
Shale	4.0×10^{-3}
Soil	2.0×10^{-3} - 8.0×10^{-3}
Gravel	5.7×10^{-3} - 6.2×10^{-3}
Sandy Clay	3.3×10^{-3}
Quartz Sand	9.0×10^{-3} - 1.3×10^{-2}
Peat	1.0×10^{-2} - 2.0×10^{-2}

(after Ingersol, et al., 1954; National Research Council, 1927; Matthess, 1982, p. 199)

Table B.17 Dry Deposit Specific Heat

Deposit	Specific Heat (Cal/g °C)
Sandstone	0.23
Limestone	0.22
Clay	0.22
Salt	0.22

Table B.17 Dry Deposit Specific Heat (continued)

Deposit	Specific Heat (Cal/g °C)
Chalk	0.21
Basalt	0.20
Soil	0.19
Granite	0.19
Quartz	0.19

(after Carslaw and Jaeger, 1959; Bear, 1972, p. 648)

Appendix C
Unit Conversions And Abbreviations

Table C.1 Conversions

(mega = 10^6, kilo = 10^3, hecto = 10^2, deca = 10^1, deci = 10^{-1}, centi = 10^{-2}, milli = 10^{-3}, and micro = 10^{-6})

Dimension	Unit Conversion	
Area	English-Metric	
	$1\ \text{in}^2$	$= 6.4516\ \text{cm}^2$
		$= 6.4516 \times 10^{-4}\ \text{m}^2$
	$1\ \text{ft}^2$	$= 9.2903 \times 10^{-2}\ \text{m}^2$
		$= 9.2903 \times 10^2\ \text{cm}^2$
	$1\ \text{mi}^2$	$= 2.590\ \text{km}^2$
		$= 2.590 \times 10^6\ \text{m}^2$
	Metric-English	
	$1\ \text{cm}^2$	$= 1.550 \times 10^{-1}\ \text{in}^2$
		$= 1.0764 \times 10^{-3}\ \text{ft}^2$
	$1\ \text{m}^2$	$= 1.550 \times 10^3\ \text{in}^2$
		$= 10.7637\ \text{ft}^2$
		$= 1 \times 10^{-4}\ \text{ha}$
		$= 2.4711 \times 10^{-4}\ \text{acres}$
	$1\ \text{km}^2$	$= 1.0764 \times 10^7\ \text{ft}^2$
		$= 1 \times 10^2\ \text{ha}$
		$= 2.4711 \times 10^2\ \text{acres}$
	Other	
	$1\ \text{km}^2$	$= 1 \times 10^6\ \text{m}^2$
	$1\ \text{m}^2$	$= 1 \times 10^{-6}\ \text{km}^2$
		$= 1 \times 10^4\ \text{cm}^2$

Table C.1   Conversions (continued)

Dimension	Unit Conversion	
	1 cm^2	$= 1 \times 10^{-4}$ m^2
	1 acre	$= 4.047 \times 10^1$ ha
		$= 1.5625 \times 10^{-3}$ mi^2
		$= 4.3560 \times 10^4$ ft^2
	1 ha	$= 2.4711$ acres
		$= 3.8610 \times 10^{-3}$ mi^2
		$= 1.0764 \times 10^5$ ft^2
	1 mi^2	$= 2.7878 \times 10^7$ ft^2
		$= 2.590 \times 10^2$ ha
		$= 6.40 \times 10^2$ acres
	ft^2	$= 144$ in^2
	in^2	$= 6.944 \times 10^{-3}$ ft^2
Density	English-Metric	
	1 lb/in^3	$= 27.680$ gm/cm^3
		$= 2.768 \times 10^4$ kg/m^3
		$= 27.680$ kg/L
	1 lb/ft^3	$= 1.602 \times 10^{-2}$ gm/cm^3
		$= 16.018$ kg/m^3
		$= 1.602 \times 10^{-2}$ kg/L
	1 lb/U.S. gal	$= 1.20 \times 10^{-1}$ gm/cm^3
		$= 1.198 \times 10^2$ kg/m^3
		$= 1.20 \times 10^{-1}$ kg/L
	1 lb/imp. gal	$= 9.983 \times 10^{-2}$ gm/cm^3
		$= 99.827$ kg/m^3
		$= 9.983 \times 10^{-2}$ kg/L
	Metric-English	
	1 gm/cm^3	$= 3.613 \times 10^{-2}$ lb/in^3
		$= 62.429$ lb/ft^3
		$= 8.345$ lb/U.S. gal
		$= 10.017$ lb/imp. gal
	1kg/m^3	$= 3.613 \times 10^{-5}$ lb/in^3
		$= 6.243 \times 10^{-2}$ lb/ft^3
		$= 8.345 \times 10^{-3}$ lb/U.S. gal
		$= 1.002 \times 10^{-2}$ lb/imp. gal

Table C.1   Conversions (continued)

Dimension	Unit Conversion	
	1 kg/L	$= 3.613 \times 10^{-2}$ lb/in^3
		$= 62.429$ lb/ft^3
		$= 8.345$ lb/U.S. gal
		$= 10.017$ lb/imp. gal
	Other	
	1 lb/in^3	$= 1.728 \times 10^3$ lb/ft^3
		$= 2.310 \times 10^2$ lb/U.S. gal
		$= 2.773 \times 10^2$ lb/imp. gal
	1 lb/ft^3	$= 5.787 \times 10^{-4}$ lb/in^3
		$= 1.34 \times 10^{-1}$ lb/U.S. gal
		$= 1.60 \times 10^{-1}$ lb/imp. gal
	1 lb/U.S. gal	$= 4.329 \times 10^{-3}$ lb/in^3
		$= 7.463$ lb/ft^3
		$= 1.201$ lb/imp. gal
	1 lb/imp. gal	$= 3.606 \times 10^{-3}$ lb/in^3
		$= 6.250$ lb/ft^3
		$= 8.33 \times 10^{-1}$ lb/U.S. gal
	1 gm/cm^3	$= 1 \times 10^3$ kg/m^3
		$= 1$ kg/L
	1 kg/m^3	$= 1 \times 10^{-3}$ gm/cm^3
		$= 1 \times 10^{-3}$ kg/L
	1kg/L	$= 1$ gm/cm^3
		$= 1 \times 10^3$ kg/m^3
Discharge Rate	English-Metric	
	1 U.S. gal/min	$= 6.308 \times 10^{-2}$ L/sec
		$= 6.308 \times 10^{-5}$ m^3/sec
		$= 5.450$ m^3/day
	1 imp. gal/min	$= 7.576 \times 10^{-2}$ L/sec
		$= 7.576 \times 10^{-5}$ m^3/sec

Table C.1   Conversions (continued)

Dimension	Unit Conversion
	1 ft³/sec $= 6.546$ m³/day
	$= 28.321$ L/sec
	$= 2.832 \times 10^{-2}$ m³/sec
	$= 2.446 \times 10^3$ m³/day
	1 acre-ft/day $= 14.276$ L/sec
	$= 1.428 \times 10^{-2}$ m³/sec
	$= 1.234 \times 10^3$ m³/day

Metric-English

1 L/sec	$= 13.201$ imp. gal /min
	$= 15.852$ U.S. gal /min
	$= 3.531 \times 10^{-2}$ ft³ /sec
	$= 7.005 \times 10^{-2}$ acre-ft /day
1 m³/sec	$= 1.320 \times 10^4$ imp. gal /min
	$= 1.585 \times 10^4$ U.S. gal /min
	$= 35.313$ ft³/sec
	$= 70.045$ acre-ft/day
1 m³/day	$= 1.53 \times 10^{-1}$ imp. gal /min
	$= 1.84 \times 10^{-1}$ U.S. gal /min
	$= 4.088 \times 10^{-4}$ ft³/sec
	$= 8.107 \times 10^{-4}$ acre-ft /day

Other

1 imp. gal/min	$= 1.201$ U.S. gal/min
	$= 2.675 \times 10^{-3}$ ft³/sec
	$= 5.307 \times 10^{-3}$ acre-ft /day
1 U.S. gal/min	$= 8.33 \times 10^{-1}$ imp. gal /min

Table C.1 Conversions (continued)

Dimension	Unit Conversion	
	1 ft³/sec	$= 2.228 \times 10^{-3}$ ft³/sec $= 4.421 \times 10^{-3}$ acre-ft /day $= 3.738 \times 10^{2}$ imp. gal /min $= 4.488 \times 10^{2}$ U.S. gal /min
	1 acre-ft/day	$= 1.984$ acre-ft/day $= 1.884 \times 10^{2}$ imp. gal /min $= 2.262 \times 10^{2}$ U.S. gal /min $= 5.04 \times 10^{-1}$ ft³/sec
	1 L/sec	$= 1 \times 10^{-3}$ m³/sec $= 86.40$ m³/day
	1 m³/sec	$= 1 \times 10^{3}$ L/sec $= 8.640 \times 10^{4}$ m³/day
	1 m³/day	$= 1.157 \times 10^{-2}$ L/sec $= 1.157 \times 10^{-5}$ m³/sec
	1 million U.S. gal/day	$= 6.944 \times 10^{2}$ U.S. gal/min $= 1.547$ ft³/sec
	1 U.S. miner's in	$= 1.5$ ft³/min
Length	English-Metric	
	1 in	$= 2.540 \times 10^{-5}$ km $= 2.540 \times 10^{-2}$ m $= 2.540$ cm $= 25.40$ mm
	1 ft	$= 3.0480 \times 10^{-4}$ km $= 3.048 \times 10^{-1}$ m $= 30.480$ cm $= 3.0480 \times 10^{2}$ mm
	1 yd	$= 9.1440 \times 10^{-4}$ km

Table C.1 Conversions (continued)

Dimension	Unit Conversion
1 mi	$= 9.1440 \times 10^{-1}$ m
	$= 91.440$ cm
	$= 1.6093$ km
	$= 1.6093 \times 10^3$ m
	$= 1.6093 \times 10^5$ cm
Metric-English	
1 mm	$= 3.2808 \times 10^{-3}$ ft
	$= 3.937 \times 10^{-2}$ in
1 cm	$= 1.0936 \times 10^{-2}$ yd
	$= 3.2808 \times 10^{-2}$ ft
	$= 3.937 \times 10^{-1}$ in
1 m	$= 6.2137 \times 10^{-4}$ mi
	$= 1.0936$ yd
	$= 39.3701$ in
1 km	$= 6.214 \times 10^{-1}$ mi
	$= 1.0936 \times 10^3$ yd
	$= 3.2808 \times 10^3$ ft
	$= 3.9370 \times 10^4$ in
Other	
1 in	$= 1.5783 \times 10^{-5}$ mi
	$= 2.7778 \times 10^{-2}$ yd
	$= 8.3333 \times 10^{-2}$ ft
1 ft	$= 1.8939 \times 10^{-4}$ mi
	$= 6.061 \times 10^{-2}$ rd
	$= 3.333 \times 10^{-1}$ yd
	$= 12$ in
1 yd	$= 5.6818 \times 10^{-4}$ mi
	$= 3$ ft
	$= 36$ in
1 rd	$= 16.50$ ft
1 mi	$= 1.760 \times 10^3$ yd
	$= 5.280 \times 10^3$ ft
	$= 6.336 \times 10^4$ in
1 cm	$= 1 \times 10^{-5}$ km
	$= 1 \times 10^{-2}$ m
	$= 10$ mm

Table C.1   Conversions (continued)

Dimension	Unit Conversion	
	1 m	$= 1 \times 10^{-3}$ km
		$= 1 \times 10^{2}$ cm
	1 km	$= 1 \times 10^{3}$ m
		$= 1 \times 10^{5}$ cm
Mass	English-Metric	
	1 oz	$= 2.835 \times 10^{-2}$ kg
		$= 28.35$ gm
	1 lb	$= 4.536 \times 10^{-1}$ kg
		$= 4.536 \times 10^{2}$ gm
	1 s. ton	$= 9.072 \times 10^{2}$ kg
	1 l. ton	$= 1.016 \times 10^{3}$ kg
	Metric-English	
	1 gm	$= 3.527 \times 10^{-2}$ oz
		$= 2.205 \times 10^{-3}$ lb
	1 kg	$= 35.27$ oz
		$= 2.205$ lb
		$= 1.102 \times 10^{-3}$ s. ton
		$= 9.843 \times 10^{-4}$ l. ton
	Other	
	1 lb	$= 16$ oz
	1 kg	$= 1 \times 10^{3}$ gm
Mass Per Unit Volume	English-Metric	
	1 lb/ft^3	$= 16.02$ kg/m^3
		$= 16.02$ gm/L
	Metric-English	
	1 kg/m^3	$= 6.242 \times 10^{-2}$ lb/ft^3
		$= 1$ gm/L

Table C.1 Conversions (continued)

Dimension	Unit Conversion
Hydraulic Conductivity	English-Metric

1 U.S. gal
/day/ft^2
$= 4.716 \times 10^{-5}$ cm/sec
$= 4.075 \times 10^{-2}$ m/day
$= 4.720 \times 10^{-7}$ m/sec
$= 2.830 \times 10^{-5}$ m/min
$= 4.714 \times 10^{-4}$ L/sec-m

1 imp. gal
/day/ft^2
$= 5.663 \times 10^{-5}$ cm/sec
$= 4.893 \times 10^{-2}$ m/day
$= 3.398 \times 10^{-5}$ m/min
$= 5.663 \times 10^{-7}$ m/sec

1 ft/sec $= 3.048 \times 10^{-1}$ m/sec
$= 2.704 \times 10^{2}$ L/sec-m

1 ft/day $= 3.528 \times 10^{-6}$ m/sec

Metric-English

1 cm/sec $= 1.766 \times 10^{4}$ imp. gal
/day/ft^2
$= 2.121 \times 10^{4}$ U.S. gal
/day/ft^2
$= 3.281 \times 10^{-2}$ ft/sec
$= 9.985$ L/sec-m
$= 864$ m/day

1 m/sec $= 3.281$ ft/sec
$= 2.119 \times 10^{6}$ U.S.
gal/day/ft^2
$= 2.835 \times 10^{5}$ ft/day

1 m/day $= 20.44$ imp. gal
/day/ft^2
$= 24.54$ U.S. gal
/day/ft^2

Other

1 U.S. gal
/day/ft^2
$= 8.327 \times 10^{-1}$ imp.

Table C.1   Conversions (continued)

Dimension	Unit Conversion
	= gal/day/ft^2
	= 1.3368×10^{-1} ft /day
	= 1.547×10^{-6} ft/sec
1 imp. gal /day/ft^2	= 1.201 U.S. gal /day/ft^2
1 m/day	= 1.157×10^{-3} cm/sec

Intrinsic Permeability (for water at 20°C)   English-Metric

1 darcy	=8.347×10^{-1} m/day
	= 9.8697×10^{-12} m^2
	= 9.8697×10^{-9} cm^2
	= 1.062×10^{-11} ft^2
	= 9.613×10^{-4} cm/sec
	= 2.725 ft/day

Metric-English

1 m^2	= 1.01325×10^{12} darcy
1 cm/sec	= 1.035×10^{3} darcy
	= 1.035×10^{6} milli-darcy
1 m/day	= 1.198 darcy

Other

1 darcy	= 18.202 U.S. gal /day/ft^2
1 ft^2	= 9.412×10^{10} darcy
1 U.S. gal /day/ft^2	= 5.494×10^{-2} darcy

Pressure   English-Metric

1 lb/in^2	= 51.711 mm of

Table C.1   Conversions (continued)

Dimension	Unit Conversion
	mercury
	= 6.895 kPa
	= 7.027×10^{-6} kg-force/m^2
	= 7.027×10^{-2} kg-force/cm^2
1 lb/ft^2	= 47.88 Pa
1 atm	= 1.013×10^5 Pa
1 bar	= 1×10^5 Pa
1 ft of water	= 22.40 mm of mercury
	= 2.986 kPa
	= 3.046×10^{-6} kg-force/m^2
	= 3.046×10^{-2} kg-force/cm^2
1 in of mercury	= 25.40 mm of mercury
	= 3.385 kPa
	= 3.452×10^{-6} kg-force/m^2
	= 3.452×10^{-2} kg-force/cm^2
Metric-English	
1 mm of mercury	= 1.934×10^{-2} lb/in^2
	= 4.46×10^{-2} ft of water
	= 3.937×10^{-2} in of mercury
1 Pa	= 2.089×10^{-2} lb/ft^2
	= 1×10^{-5} bars
1 kPa	= 1.45×10^{-1} lb/in^2
	= 3.35×10^{-1} ft of water
	= 2.95×10^{-1} in of mercury

Table C.1   Conversions (continued)

Dimension	Unit Conversion	
	1 kg-force/m²	$= 1.423 \times 10^5$ lb/in²
		$= 3.284 \times 10^5$ ft of water
		$= 2.896 \times 10^5$ in of mercury
	1kg-force/cm²	$= 14.23$ lb/in²
		$= 32.84$ ft of water
		$= 28.96$ in of mercury
	Other	
	1 lb/in²	$= 2.31$ ft of water
		$= 2.036$ in of mercury
	1 ft of water	$= 4.35 \times 10^{-1}$ lb /in²
		$= 8.83 \times 10^{-1}$ in of mercury
	1 in of mercury	$= 4.91 \times 10^{-1}$ lb/in²
		$= 1.133$ ft of water
	1 mm of mercury	$= 1.33 \times 10^{-1}$ kPa
		$= 1.359 \times 10^{-7}$ kg-force/m²
		$= 1.359 \times 10^{-3}$ kg-force/cm²
	1 kg-force/m²	$= 7.356 \times 10^6$ mm of mercury
		$= 9.807 \times 10^5$ kPa
		$= 1 \times 10^4$ kg-force/cm²
	1 kg-force/cm²	$= 7.356 \times 10^2$ mm of mercury
		$= 98.067$ kPa
		$= 1 \times 10^{-4}$ kg-force/m²

Table C.1   Conversions (continued)

Dimension	Unit Conversion

Time

1 day	$= 8.640 \times 10^4$ sec
	$= 1.440 \times 10^3$ min
	$= 24$ hours
1 hour	$= 3.60 \times 10^3$ sec
	$= 60$ min
1 min	$= 60$ sec
1 sec	$= 1.157 \times 10^{-5}$ day
	$= 2.778 \times 10^{-4}$ hour
	$= 1.667 \times 10^{-2}$ min
1 min	$= 6.944 \times 10^{-4}$ day
	$= 1.667 \times 10^{-2}$ hour
1 hour	$= 4.167 \times 10^{-2}$ day

Temperature

$^{\circ}F = 9/5\,^{\circ}C + 32$
$^{\circ}C = 5/9(^{\circ}F - 32)$
$^{\circ}F = {^{\circ}K} - 459.69$
$^{\circ}C = {^{\circ}K} - 273.15$
$^{\circ}K = {^{\circ}F} + 459.69$
$^{\circ}K = {^{\circ}C} + 273.15$

Transmissivity   English-Metric

1 U.S.gal/day/ft	$= 1.242 \times 10^{-2}$ m²/day
	$= 1.438 \times 10^{-7}$ m²/sec
1 imp.gal/day/ft	$= 1.491 \times 10^{-2}$ m²/day
	$= 1.726 \times 10^{-7}$ m²/sec
1 ft²/sec	$= 9.290 \times 10^{-2}$ m²/sec
1 ft²/day	$= 1.075 \times 10^{-6}$ m²/sec
1 ft²/sec	$= 8.027 \times 10^3$ m²/day
1 ft²/day	$= 9.290 \times 10^{-2}$ m²/day

Metric-English

1 m²/day	$= 80.52$ U.S. gal

Table C.1   Conversions (continued)

Dimension	Unit Conversion	
		/day/ft
		= 67.05 imp. gal
		/day/ft
		= 10.74 ft^2/day
	1 m^2/sec	= 10.76 ft^2/sec
		= 6.954 × 10^6 U.S.
		gal/day/ft
	Other	
	1 U.S.gal/day/ft	= 8.326 × 10^{-1} imp.
		gal/day/ft
		= 1.3368 × 10^{-1}
		ft^2/day
	1 imp.gal/day/ft	= 1.201 U.S. gal
		/day/ft
		= 1.6046 × 10^{-1}
		ft^2/day
Velocity	English-Metric	
	1 ft/sec	= 30.48 cm/sec
		= 3.048 × 10^{-1} m/sec
		= 1.097 km/hour
		= 6.817 × 10^{-1} mi/hour
	1 ft/day	= 30.48 cm/day
		= 3.048 × 10^{-1} m/day
	Metric-English	
	1 cm/sec	= 3.281 × 10^{-2} ft/sec
	1 m/sec	= 3.281 ft/sec
	1 cm/day	= 3.281 × 10^{-2} ft/day
	1 m/day	= 3.281 ft/day
Viscosity (dynamic)	English-Metric	
	1 poundal sec	

Table C.1   Conversions (continued)

Dimension	Unit Conversion	
	$/\text{ft}^2$	$= 1.488$ Pa-sec
		$= 14.88$ poise
	1 lb-force sec	
	$/\text{ft}^2$	$= 47.619$ Pa-sec
		$= 4.762 \times 10^2$ poise
	Metric-English	
	1 Pa-sec	$= 6.72 \times 10^{-1}$ poundal sec/ft^2
		$= 2.1 \times 10^{-2}$ pound-force sec/ft^2
	1 poise	$= 6.72 \times 10^{-2}$ poundal sec/ft^2
		$= 2.1 \times 10^{-3}$ pound-force sec/ft^2
	Other	
	1 poundal sec $/\text{ft}^2$	$= 3.125 \times 10^{-2}$ pound-force sec/ft^2
	1 pound-force sec/ft^2	$= 32.002$ poundal sec/ft^2
Viscosity (kinematic)	English-Metric	
	1 ft^2/sec	$= 9.290 \times 10^{-2}$ m^2/sec
		$= 9.29 \times 10^4$ mm^2/sec
	1 in^2/sec	$= 6.452 \times 10^{-4}$ m^2/sec
		$= 6.452 \times 10^2$ mm^2/sec
	1 stoke	$= 1 \times 10^{-4}$ m^2/sec
		$= 1 \times 10^2$ mm^2/sec
	Metric-English	
	1 m^2/sec	$= 10.764$ ft^2/sec

Table C.1　Conversions (continued)

Dimension	Unit Conversion	
	1 mm^2/sec	$= 1.55 \times 10^3$ in^2/sec $= 1 \times 10^4$ stoke $= 1.076 \times 10^{-5}$ ft^2/sec $= 1.55 \times 10^{-3}$ in^2/sec $= 1 \times 10^{-2}$ stoke
	Other	
	1 ft^2/sec	$= 1.44 \times 10^2$ in^2/sec $= 9.29 \times 10^2$ stoke
	1 in^2/sec	$= 6.944 \times 10^{-3}$ ft^2/sec $= 6.452$ stoke
	1 stoke	$= 1.076 \times 10^3$ ft^2/sec $= 1.55 \times 10^{-1}$ in^2/sec
	1 m^2/sec	$= 1 \times 10^6$ mm^2/sec
	1 mm^2/sec	$= 1 \times 10^{-6}$ m^2/sec
Volume	English-Metric	
	1 ft^3	$= 2.8317 \times 10^{-2}$ m^3 $= 2.8317 \times 10^4$ cm^3 $= 28.317$ L
	1 imp. gal	$= 4.5437 \times 10^{-3}$ m^3 $= 4.5437 \times 10^3$ cm^3 $= 4.5437$ L
	1 U.S. gal	$= 3.7854 \times 10^{-3}$ m^3 $= 3.7854 \times 10^3$ cm^3 $= 3.7854$ L
	1 acre-ft	$= 1.2335 \times 10^3$ m^3 $= 1.2335 \times 10^9$ cm^3 $= 1.2335 \times 10^6$ L
	Metric-English	
	1 m^3	$= 35.3137$ ft^3 $= 2.2008 \times 10^2$ imp. gal $= 2.6417 \times 10^2$ U.S. gal

Table C.1   Conversions (continued)

Dimension	Unit Conversion
1 cm^3	= 8.1071 × 10^{-4} acre -ft = 3.5314 × 10^{-5} ft^3 = 2.2008 × 10^{-4} imp. gal = 2.6417 × 10^{-4} U.S. gal = 8.1071 × 10^{-10} acre -ft
1 L	= 3.5314 × 10^{-2} ft^3 = 2.201 × 10^{-1} imp. gal = 2.642 × 10^{-1} U.S. gal = 8.1071 × 10^{-7} acre -ft
Other	
1 ft^3	= 6.2321 imp. gal = 7.4805 U.S. gal = 2.2957 × 10^{-5} acre -ft
1 imp. gal	= 1.605 × 10^{-1} ft^3 = 1.2003 U.S.gal = 3.6846 × 10^{-6} acre -ft
1 U.S. gal	= 1.337 × 10^{-1} ft^3 = 8.331 × 10^{-1} imp. gal = 3.0697 × 10^{-6} acre -ft
1 acre-ft	= 4.3560 × 10^4 ft^3 = 2.7140 × 10^5 imp. gal = 3.2576 × 10^5 U.S gal

Table C.1   Conversions (continued)

Dimension	Unit Conversion	
Work or Energy	English-Metric	
	1 ft lb	$= 1.356$ J
	1 cal	$= 4.185$ J
	1 BTU	$= 1.055 \times 10^3$ J
	Metric-English	
	1 J	$= 7.374 \times 10^{-1}$ ft lb
	1 J	$= 2.389 \times 10^{-1}$ cal
	1 J	$= 9.479 \times 10^{-4}$ BTU

Chemical Constituent	Conversion Factor (mg/L to meq/L)
Aluminum(Al^{+3})	0.11119
Ammonium(NH_4^+)	0.05544
Barium(Ba^{+2})	0.01456
Beryllium(Be^{+3})	0.33288
Bicarbonate(HCO_3^-)	0.01639
Bromide($Br-$)	0.01251
Cadmium(Cd^{+2})	0.01779
Calcium(Ca^{+2})	0.04990
Carbonate(CO_3^{2-})	0.03333
Chloride(Cl^-)	0.02821
Cobalt(Co^{+2})	0.03394
Copper(Cu^{+2})	0.03148
Fluoride(F^-)	0.05264
Hydrogen(H^+)	0.99209
Hydroxide(OH^-)	0.05880
Iodide($I-$)	0.00788
Iron(Fe^{+2})	0.03581
Iron(Fe^{+3})	0.05372
Lithium(Li^+)	0.14411
Magnesium(Mg^{+2})	0.08226
Manganese(Mn^{+2})	0.03640
Nitrate(NO_3^-)	0.01613
Nitrite(NO_2^-)	0.02174
Phosphate(PO_4^{3-})	0.03159

Table C.1   Conversions (continued)

Phosphate(HPO_4^{2-})	0.02084
Potassium(K^+)	0.02557
Rubidium(Rb^+)	0.01170
Sodium(Na^+)	0.04350
Strontium(Sr^{+2})	0.02283
Sulfate(SO_4^{2-})	0.02082
Sulfide(S^{-2})	0.06238
Zinc(Zn^{+2})	0.03060

(after Ceroici, 1980; Hem, 1970)

able C.2   Abbreviations

Unit	Dimension	Abbreviation
hectare	area	ha
square feet	area	ft^2
British Thermal Unit	energy	BTU
joule	energy	J
calorie	energy	cal
meter	length	m
centimeter	length	cm
kilometer	length	km
inch	length	in
feet	length	ft
yard	length	yd
rod	length	rd
mile	length	mi
millimeter	length	mm
micrograms	mass	g
milligrams	mass	mg
grams	mass	gm
ounch	mass	oz
kilograms	mass	kg

Table C.2 Abbreviations (continued)

Unit	Dimension	Abbreviation
tons	mass	ton
pound	mass	lb
pascal	pressure	Pa
kilopascal	pressure (barometric)	kPa
atmosphere	pressure (barometric)	atm
Fahrenheit	temperature	F
Celsius	temperature	C
Kelvin	temperature	K
second	time	sec
minute	time	min
day	time	day or d
milliliters	volume	mL
liter	volume	L
cubic feet	volume	ft^3
gallon	volume	gal or g
U.S. gallon	volume	U.S.gal
U.K. gallon	volume	imp.gal

Table C.3 Greek Characters

α	alpha
β	beta
γ	gamma
δ	delta
ϵ	epsilon
ζ	zeta
ν	eta
θ	theta
ι	iota
κ	kappa
λ	lambda
μ	mu
ϖ	nu
ξ	xi
o	omicron

Table C.3 Greek Characters

π	pi
ρ	rho
σ	sigma
τ	tau
υ	upsilon
φ	phi
ξ	chi
ψ	psi
ω	omega

Appendix D
Natural Recharge And Discharge

Table D.1 Hydrologic Budget Factors

(for 3 drainage basins in central Illinois during a year of near-normal precipitation; after Schicht and Walton, 1961)

Budget Factor	Basin 1 (inch)	Basin 2 (inch)	Basin 3 (inch)
Precipitation	32.62	37.18	39.73
Streamflow	9.82	9.48	13.93
Surface runoff	2.66	5.68	12.04
Groundwater runoff	7.16	3.80	1.89
Evapotranspiration	23.94	24.30	24.68
Surface and soil evapotranspiration	21.93	21.10	23.80
Groundwater evapotranspiration	2.01	3.20	0.88
Groundwater recharge	8.03	10.40	3.89

Table D.2 Natural Recharge Rates

(for major groundwater regions in United States; after Heath, 1982)

Region	Recharge Rate (inch/ year)
Western mountain ranges	0.1—2
Alluvial basins	0.0001—1
Columbia lava plateau	0.2—10
Colorado plateau and	

Table D.2 Natural Recharge Rates (continued)

Region	Recharge Rate (inch/ year)
Wyoming basin	0.01—2
High plains	0.2—3
Nonglaciated central region	0.2—20
Glaciated central region	0.2—10
Piedmont and Blue Ridge	1—10
Northeast and Superior Uplands	1—10
Atlantic and Gulf Coastal Plain	2—20
Southeast Coastal plain	1—20
Alluvial valleys	2—20
Hawaiian Islands	1—40
Alaska	0.1—10

Appendix E
Water Quality Classifications

Table E.1 Saline Groundwater

	Total Dissolved Solids (mg/L)
Fresh water	0—1000
Brackish water	1000—10000
Saline water	10000—100000
Brine	>100000

(after Carroll, 1962)

Table E.2 Groundwater Hardness

Hardness (mg/L as $CaCO_3$)	Water Class
0-75	Soft
75-150	Moderately hard
150-300	Hard
> 300	Very hard

(after Sawyer and McCarty, 1967)

Table E.3 National Interim Drinking Water Standards

Maximum contaminant levels for inorganic chemicals

Contaminant	Level, mg/L (μg/L)
Arsenic	0.05 (50)
Barium	1.0 (1000)
Cadmium	0.010 (10)
Chromium	0.05 (50)
Fluoride	2.2
Lead	0.05 (50)
Mercury	0.002 (2)
Nitrate (as N)	10.0
Selenium	0.01 (10)
Silver	0.05 (50)

Secondary inorganic chemical and physical standard

Standard	Level, mg/L (μg/L)
Chloride	250
Color	15 units
Copper	1.0 (1000)
Corrosivity	Noncorrosive
Foaming agents MBAS (methylene blue active substances)	0.5
Hydrogen sulfide	not detectable
Iron	0.3
Manganese	0.05 (50)
Odor	3 (Threshold No.)
Sulfate	250
Total residue	500
Zinc	5 (5000)

Table E.3 National Interim Drinking Water Standards (continued)

Maximum contaminant levels for organic chemicals

Contaminant	Level, mg/L
Chlorinated hydrocarbons: Endrin (1,2,3,4,10, 10-hexachloro- 6,7-expoxy-1,4,4a,5,6,7,8,8a-octahydro-1,4-endo, endo-5,8-dimethano naphthalene)	0.0002
Lindane (1,2,3,4,5,6-hexachloro-cyclohexane, gamma isomer)	0.004
Methoxychlor (1,1,1-trichloro-2,2-bis (p-methoxyphenyl ethane)	0.1
Toxaphene ($C_{10}H_{10}Cl_8$- Technical chlorinated camphene,67—69 % chlorine)	0.005
Chlorophenoxys: 2,4-D(2,2-dichlorophen-oxyacetic acid)	0.1
2,4,5-TP Silvex(2,4,5-trichlorophenoxy-propionic acid)	0.01
Total trihalomethanes [sum of the concentrations of bromodichoromethane, dibro-mochloromethane, tribromo-methane (bromoform), and trichloromethane (chloroform)]	0.10

Table E.3 National Interim Drinking Water Standards (continued)

Maximum contaminants levels for radium-226, radium-228, and gross alpha
particle radioactivity

Combined radium-226 and radium-228 5pCi/L

Gross alpha particle activity (including radium-226 but excluding
radon and uranium) 15pCi/L

Maximum contaminant levels for beta particle and photon radioactivity from
manmade radionuclides in community water systems

The average annual concentration of beta particle and photon
radioactivity from manmade radionuclides in drinking water shall not
produce an annual dose equivalent to the total body or any internal
organ greater than 4 millirem/year

Except for the radionuclides listed below, the concentration of
manmade radionuclides causing 4 mrem total body or organ dose
equivalents shall be calculated on the basis of a 2-liter per day
drinking water intake. If 2 or more radionuclides are present, the
sum of their annual dose equivalent to the total body or to any
organ shall not exceed 4 millirem/year

Average annual concentrations assumed to produce a total body or organ dose
of 4 mrem/year

Radionuclide	Critical Organ	pCi per liter
Tritium	Total body	20000
Strontium-90	Bone marrow	8

(from Federal Register, Feb. 1978, No. 266)

Table E.4 Contaminants Regulated Under Safe Drinking Water Act. 1986 Amendments

(some were included in EPA proposed and final rules published in Federal Register, Nov. 13, 1985)

Volatile Organic Chemicals

> Trichloroethylene
> Tetrachloroethylene
> Carbon tetrachloride
> 1,1,1-Trichloroethane
> 1,2-Dichloroethane
> Vinyl chloride
> Methylene chloride
> Benzene
> Chlorobenzene
> Dichlorobenzene(s)
> Trichlorobenzene(s)
> 1,1-Dichloroethylene
> trans-1,2-Dichloroethylene
> cis-1,2-Dichloroethylene

Microbiology and Turbitity

> Total coliforms
> Turbity
> Giardia lamblia
> Viruses
> Standard plate count
> Legionella

Inorganics

> Arsenic
> Barium
> Cadmium
> Chromium
> Lead
> Mercury
> Nitrate
> Selenium
> Silver
> Fluoride

Table E.4 Contaminants Regulated Under Safe Drinking Water Act. 1986 Amendments (continued)

Aluminum
Antimony
Molybdenum
Asbestos
Sulfate
Copper
Vanadium
Sodium
Nickel
Zinc
Thallium
Beryllium
Cyanide

Organics

Endrin
Lindane
Methoxychlor
Toxaphene
2,4-D
2,4,5-TP
Aldicarb
Chlordane
Dalapon
Diquat
Endothall
Glyphosphate
Carbofuran
Alachlor
Epichlorohydrin
Toluene
Adipates
2,3,7,8-TCDD (Dioxin)
1,1,2-Trichloroethane
Vydate
Simazine
Polynuclear aromatic
 hydrocarbons (PAHs)
Polychlorinated biphenyls
 (PCBs)

Table E.4 Contaminants Regulated Under Safe Drinking Water Act. 1986 Amendments (continued)

        Atrazine
        Phthalates
        Acryamide
        Dibromochloropropane
         (DBCP)
        1,2-Dichloropropane
        Pentachlorophenol
        Pichloram
        Dinoseb
        Ethylene dibromide
        Dibromomethane
        Xylene
        Hexachlorocyclopentadiene

Radionuclides

        Radium 226 and 228
        Beta particle and photon
         radioactivity
        Uranium
        Gross alpha particle activity
        Radon

Table E.5 USEPA Drinking Water Standards And Health Goals

Chemical	MCLG (μg/L)	MCL (μg/L)	SMCL (μg/L)
Volatile Organic Chemicals			
Trichloroethylene	0#	5!	
Carbon tetachloride	0#	5!	
Vinyl chloride	0#	5!	
1,2-Dichloroethane	0#	5!	
Benzene	0#	5!	
1,1-Dichloroethylene	7#	7!	
1,1,1-Trichloroethane	200#	200!	
p-Dichlorobezene	75!	75!	

Table E.5 USEPA Drinking Water Standards And Health Goals (continued)

Chemical	MCLG (µg/L)	MCL (µg/L)	SMCL (µg/L)
Synthetic Organic Chemicals			
Acrylamide	0*		
Alachlor	0*		
Aldicarb (including aldicarb sulfoxide and aldicarb sulfone)	9*		
Carbofuran	36*		
Chlorodane	0*		
cis-1,2-Dichloroethylene	0*		
Dibromochloropropane (DBCP)	0*		
1,2-Dichloropropane	6*		
o-Dichlorobenzene	620*		
2,4-Dichlorophenoxyacetic acid	70*		
Epichlorohydrin	0*		
Ethyl benzene	680*		
Ethylene dibromide (EDB)	0*		
Heptachlor	0*		
Lindane	0.2*		
Methoxychlor	340*		
Monochlorobenzene	60*		
Pentachlorophenol	220*		
Polychlorobiphenyls (PCBs)	0*		
Styrene	140*		
Toluene	2000*		
Toxaphene	0*		
trans-1,2-Dichloroethylene	70*		
2-(2,4,5-Trichlorophenoxy) propionic acid (2,4,5-TP)	52*		
Xylene	440*		
Inorganic Chemicals			
Arsenic	50*		
Asbestos	7.1×10^6 long fibers per liter		
Barium	1500*		

Table E.5 USEPA Drinking Water Standards And Health Goals (continued)

Chemical	MCLG (μg/L)	MCL (μg/L)	SMCL (μg/L)
Cadmium	5*		
Chromium	120*		
Copper	1300*		
Fluoride	4000#	40000*	2000*
Lead	20*		
Mercury	3*		
Nitrate	10000*		
Nitrite	1000*		
Selenium	45*		

Microbiological Parameters

Giardia	0 organisms*		
Total coliforms	0 organisms*		
Turbidity	0.1 turbidity units*		
Viruses	0 organisms		

\# Final value (Federal Register, Nov. 13, 1985)
* Proposed value (Federal Register, Nov. 13, 1985)
! Final value (Federal Register, Jul. 8, 1987)

Table E.6 Industrial Use Criteria

Use	Turbidity, units	Color units	Taste and odor threshold
Air-conditioning	—	—	low
Baking	10	10	none-low
Brewing	0—10	0—10	none-low
Carbonated beverages	1—2	5—10	none-low
Confectionery	—	—	low
Dairy	—	none	none
Drinking	5	15	3
Food canning and freezing	1—10	—	none-low

Table E.6 Industrial Use Criteria

Use	Turbidity, units	Color units	Taste and odor threshold
Food equipment, washing	1	5—20	none
Food processing, general	1—10	5—10	low
Ice manufacture	5	5	low
Paper and pulp, fine	10	5	—
Paper, groundwood	50	30	—
Paper, kraft, bleached	40	25	—
Paper, kraft, unbleached	100	100	—
Paper, soda and sulfate pulps	25	5	—
Rayon and acetate fiber pulp production	5	5	—
Rayon manufacture	0.3	—	—
Tanning	20	10—100	—
Textile	0.3—25	0—70	—

Use	Dissolved solids	Hardness, $CaCO_3$	Alkalinity, as $CaCO_3$
Brewing	500—1500	—	75—80
Carbonated beverages	850	200—250	50—130
Confectionery	50—100	soft	—
Dairy	500	180	—
Drinking	500	—	—
Food canning and freezing	850	—	30—250
Food equipment, washing	850	10	—
Food processing, general	850	10—250	30—250
Ice manufacture	170—1300	—	—
Laundering	—	0—50	60
Paper and pulp, fine	200	100	75
Paper, groundwood	500	200	150
Paper, kraft, bleached	300	100	75
Paper, kraft, unbleached	500	200	150
Paper, soda and sulfate pulps	250	100	75
Rayon and acetate fiber pulp production	100	8	50—75
Rayon manufacture	—	55	—
Sugar	low	low	—
Tanning	—	50—500	130

Table E.6 Industrial Use Criteria

Use	Dissolved solids	Hardness, $CaCO_3$	Alkalinity, as $CaCO_3$
Textile	—	0—50	—

Use	pH units	Chlorides, as Cl	Sulfates, as SO_4
Brewing	6.5—7.0	60—100	—
Carbonated beverages	—	250	250
Confectionery	>70	—	—
Dairy	—	30	60
Drinking	—	250	250
Food canning and freezing	>7.5	—	—
Food equipment, washing	—	250	—
Laundering	6.0—6.8	—	—
Paper, groundwood	—	75	—
Paper, kraft, bleached	—	200	—
Paper, kraft, unbleached	—	200	—
Paper, soda, and sulfate pulps	—	75	—
Rayon manufacture	7.8—8.3	—	—
Sugar	—	20	20
Tanning	6.0—8.0	—	—
Textile	—	100	100

Use	Iron, as Fe	Manganese as Mn	Iron plus manganese
Air conditioning	0.5	0.5	0.5
Baking	0.2	0.2	0.2
Brewing	0.1	0.1	0.1
Carbonated beverages	0.1—0.2	0.2	0.1—0.4
Confectionery	0.2	0.2	0.2
Dairy	0.1—0.3	0.03—0.1	—
Drinking	0.3	0.05	—
Food canning and freezing	0.2	0.2	0.2—0.3
Food equipment, washing	—	—	0.1
Food processing, general	0.2	0.2	0.2—0.3
Ice manufacture	0.2	0.2	0.2
Laundering	0.2—1.0	0.2	0.2—1.0
Paper and pulp, fine	0.1	0.05	—

Table E.6 Industrial Use Criteria (continued)

Use	Iron, as Fe	Manganese as Mn	Iron plus manganese
Paper, groundwood	0.3	0.1	—
Paper, kraft, bleached	0.2	0.1	—
Paper, kraft, unbleached	1.0	0.5	—
Paper, soda, and sulfate pulps	0.1	0.05	—
Rayon and acetate fiber pulp production	0.05	0.03	0.05
Rayon manufacture	0.0	0.0	0.0
Sugar	0.1	—	—
Tanning	0.1—0.2	0.1—0.2	0.2
Textile	0.1—1.0	0.05—1.0	0.2—1.0

Use	Hydrogen sulfide	Fluorides as F
Baking	0.2	—
Brewing	0.2	1.0
Carbonated beverages	0—0.2	0.2—1.0
Confectionery	0.2	—
Drinking	—	1.4—2.4
Food canning and freezing	1.0	1.0
Food equipment, washing	—	1.0
Food processing, general	—	1.0

Use	Other Requirements
Air conditioning	not corrosive or slime promoting
Baking	some calcium is needed for yeast action, too much hardness retards fermentation, too little softens gluten to produce soggy bread, water of zero hardness required for some cakes and crackers, potable
Brewing	not more than 300 mg/L of any one substance, $CaSO_4$

Table E.6 Industrial Use Criteria (continued)

Use	Other Requirements
	less than 100 to 500 mg/L, $MgSO_4$ less than 50 to 200 mg/L, for dark beer alkalinity as $CaCO_3$ may be 80 to 150 mg/L, potable
Carbonated beverages	potable, COD 1.5, organic matter infinitesimal, algae and protozoa none
Confectionery	potable
Dairy	potable, NO_3-N 5.5, NO_2-N 0.0, NH_3N trace only, COD as $KMnO_4$ 12
Drinking	potable
Food canning and freezing	potable, free from saprophytic organisms, NaCl 1000-1500, NO_3-N 2.8, NH_3-N 0.4,
Food equipment, washing	potable, organic matter infinitesimal
Food processing, general	potable
Ice manufacture	potable, SiO_2 10
Paper and pulp, fine	soluable SiO_2 20, free CO_2 10, residual Cl_2 2
Paper groundwood	soluable SiO_2 50, free CO_2 10
Paper, kraft, bleached	soluable SiO_2 50, free CO_2 10
Paper, kraft, unbleached	soluable SiO_2 100, free CO_2 10
Paper, soda, and sulfate pulps	soluable SiO_2 20, free CO_2 10
Rayon and acetate fiber pulp manufacture	Al_2O_3 8, Si 25, Cu 5
Sugar	Ca 20, Mg 10, Bicarbonate as $CaCO_3$ 100, sterile, no saprophytic organisms
Textile	COD 8, heavy metals none, Ca 10, Mg 5, bicarbonate as $CaCO_3$ 200

Table E.7 Criteria For Farm Animal Drinking Water

Constituent	Maximum Concentration (mg/L)
Total dissolved solids	3000
Aluminum	5
Arsenic	0.2
Boron	5
Cadmium	0.05
Chromium	1
Cobalt	1
Copper	0.5
Fluorine	2
Lead	0.1
Mercury	0.01
NO_3-N plus NO_2-N	100
NO-N	10
Selenium	0.05
Vanadium	0.1
Zinc	25

(from National Academy of Sciences and National Academy of Engineering, 1972)

Table E. 8 Crop Tolerances To Boron

Water Class	Sensitive Crops (mg/L)	Semitolerant Crops (mg/L)	Tolerant Crops (mg/L)
Excellent	<0.33	<0.67	<1.00
Good	0.33—0.67	0.67—1.33	1.00—2.00
Permissible	0.67—1.00	1.33—2.00	2.00—3.00
Doubtful	1.00—1.25	2.00—2.50	3.00—3.75
Unsuitable	>1.25	>2.50	>3.75

(after Wilcox, 1955)

Table E.9 Relative Tolerances Of Plants To Boron

(from Salinity Laboratory, 1954)

Sensitive	Semitolerent	Tolerant
Lemon	Lima bean	Carrot
Grapefruit	Sweet potato	Lettuce
Avocado	Bell pepper	Cabbage
Orange	Pumpkin	Turnip
Thornless blackberry	Zinnia	Onion
Apricot	Oat	Broadbean
Peach	Milo	Gladiolus
Cherry	Corn	Alfalfa
Persimmon	Wheat	Garden beet
Kadota fig	Barley	Mangel
Grape	Olive	Sugar beet
Apple	Ragged robin rose	Date palm
Pear	Field pea	Palm
Plum	Radish	Asparagus
American elm	Sweetpea	Athel
Navy bean	Tomato	
Jerusalem artichoke	Cotton	
Engish walnut	Potato	
Black walnut	Sunflower	
Pecan		

Table E.10 Tolerance Of Crops To Water Salinity

Crop	Low Salt Tolerance	Medium Salt Tolerance	High Salt Tolerance
Fruit	Avocado	Cantaloupe	Date palm
		Lemon	Date
		Strawberry	Olive
		Peach	Fig
		Apricot	Pomegranate
		Almond	
		Plum	
		Prune	
		Grapefuit	
		Orange	

Table E.10 Tolerance Of Crops To Water Salinity (continued)

Crop	Low Salt Tolerance	Medium Salt Tolerance	High Salt Tolerance
Vegetable	Apple Pear 3000 S/cm Green bean Celery Radish 4000 S/cm	4000 S/cm Cucumber Squash Peas Onion Carrot Potato Sweet corn Lettuce Cauliflower Bell pepper Cabbage Broccoli Tomato 10000 S/cm	10000 S/cm Spinach Asparagus Kale Garden beet 12000 S/cm

Forage	2000 S/cm
	Burnet
	
	Ladino
	  clover
	Red
	  clover
	Alsike
	  clover
	
	Meadow
	  foxtail
	White
	  dutch
	  clover
	
	4000 S/cm
		4000 S/cm
		Sickle
		  milkvetch
		Sour
		  clover
		Cicer
		  milkvetch
		Tall
		  meadow
		  oat
		Grass
		
		Smooth
		  brome
		Big trefoil
		
		Reed canary
		
		Meadow fescue
		Blue grame
		Orchard grass
			12000 S/cm
			Bird's-foot
			  trefoil
			Barley(hay)
			
			Western
			  wheat grass
			Canada
			  wild rye
			
			Rescue grass
			
			Rhodes grass
			
			Bermuda
			  grass
			Nattall alkali
			  grass
			Salt grass
			Alkali sacaton
			18000 S/cm

Table E.10 Tolerance Of Crops To Water Salinity (continued)

Crop	Low Salt Tolerance	Medium Salt Tolerance	High Salt Tolerance
		Oat(hay)	
		Wheat(hay)	
		Rye(hay)	
		Tall fescue	
		Alfalfa	
		Hubam clover	
		Sudan grass	
		Dallis grass	
		Strawberry clover	
		Mountain brome	
		Perennial rye grass	
		Yellow sweet clover	
		White sweet clover	
		12000 S/cm	
Field	4000 S/cm	6000 S/cm	10000 S/cm
	Field bean	Caster bean	Cotton
		Sunflower	Rape
		Flax	Sugar beet
		Corn(field)	Barley(grain)
		Sorghum	16000 S/cm
		Rice	
		Oat(grain)	
		Wheat(grain)	
		Rye(grain)	
		10000 S/cm	

(after Salinity Laboratory, 1954)

Appendix F
Industrial Contaminants

Table F.1 Chemicals And Their Uses

Chemicals	Industrial Use
Metals and cations	
Aluminum	Alloys, foundry, paints, protective coatings, electrical, packaging, building and construction, machinery, and equipment
Antimony	Hardening alloys, solders, sheet and pipe, pyrotechnics
Arsenic	Alloys, dyestuffs, medicine, solders, electronic devices, insectices, rodenticides, herbicides, preservatives
Barium	Alloys, lubricant
Beryllium	Structural material in space technology, inertial guidance systems, additive to rocket fuels, moderator and reflector of neutrons in nuclear reactors
Cadmium	Alloys, coatings, batteries, electrical equipment, fire protection systems, paints, fungicides, photography

Table F.1 Chemicals And Their Uses (continued)

Chemicals	Industrial Use
Calcium	Alloys, fertilizers, reducing agent
Chromium	Alloys, protective coatings, paints, nuclear and high-temperature research
Cobalt	Alloys, ceramics, drugs, paints, glass, printing, catalyst, electroplating, lamp filaments
Copper	Alloys, paints, electrical wiring, machinery, constr-uction materials, electro-plating, piping, insecticides
Iron	Alloys, machinery, magnets
Lead	Alloys, batteries, gasoline additive, sheet and pipe, paints, radiation shielding
Lithium	Alloys, pharmaceuticals, coolant, batteries, solders, propellants
Magnesium	Alloys, batteries, pyrotech-nics, precision instruments, optical mirrors
Manganese	Alloys, purifying agent
Mercury	Alloys, electrical apparatus, fungicides, bactericides, mildew-proofing paper, pharm-aceuticals
Molybdenum	Alloys, pigments, lubricant
Palladium	Alloys, catalyst, jewelry, pro-tective coatings, electrical equipment
Potassium	Alloys, catalyst
Selenium	Alloys, electronics, ceramics, catalyst
Silver	Alloys, photography, chemical, manufacturing, mirrors, elec-tronic equipment, jewelry, equipment, catalyst, pharm-aceuticals

Table F.1 Chemicals And Their Uses (continued)

Chemicals	Industrial Use
Sodium	Chemical manufacturing, catalyst, coolant, non-glare lighting for highways, laboratory, reagent
Thallium	Alloys, glass, pesticides, photoelectric applications
Titanium	Alloys, structural materials, abrasives, coatings
Vanadium	Alloys, catalysts, target material for x-rays
Zinc	Alloys, electronics, automobile parts, fungicides, roofing, cable wrappings, nutrition

Nonmetals and anions

Chemicals	Industrial Use
Ammonia	Fertilizers, chemical manufacturing, refrigerants, synthetic fibers, fuels, dyestuffs
Boron	Alloys, fibers and filaments, semi-conductors, propellants
Chlorides	Chemical manufacturing, water purification, shrink-proofing, flame-retardants, food processing
Cyanides	Polymer production (heavy duty tires), coatings, metallurgy, pesticides
Fluorides	Toothpastes and other dentrifices, additive to drinking water
Nitrates	Fertilizers, food perservatives
Nitrites	Fertilizers, food perservatives
Phosphates	Detergents, fertilizers, food preservatives
Sulfates	Fertilizers, pesticides
Sulfites	Pulp production and processing food preservatives

Table F.1 Chemicals And Their Uses (continued)

Chemicals	Industrial Use
Radionuclides	
Cesium 137	Gamma radiation source for certain foods
Chromium 51	Diagnosis of blood volume, blood cell life, cardiac output
Cobalt 60	Radiation therapy, irradiation, radiographic testing, research
Iodine 131	Medical diagnosis, therapy, leak detection, tracers, measuring film thickness
Iron 59	Medicine, tracer
Phosophorous 32	Tracer, medical treatment, industrial measurements
Plutonium 238, 243	Energy source, weaponry
Radium 226	Medical treatment, radiography
Radon 222	Medicine, leak detection, radiography, flow measurement
Ruthenium 106	Catalyst
Scandium 46	Tracer, leak detection, semi-conductors
Strontium 90	Medicine, industrial applications
Tritium	Tracer, luminous instrument dials
Uranium 238	Nuclear reactors
Zinc 65	Industrial tracers
Aromatic hydrocarbons	
Acetanilide	Intermediate manufacturing, pharmaceuticals, dyestuffs
Alkyl benzene sulfonates	Detergents
Aniline	Dyestuffs, intermediate manufacturing, photographic chemicals, pharmaceuticals, herbicides, fungicides, petroleum, refining, explosives
Anthracene	Dyestuffs, intermediate manu-

Table F.1 Chemicals And Their Uses (continued)

Chemicals	Industrial Use
	facturing, semiconductor research
Benzene	Detergents, intermediate manufacturing, solvents, antiknock gasoline
Benzidine	Dyestuffs, reagent, stiffening agent in rubber compounding
Benzyl alcohol	Solvent, perfumes and flavors, photographic developer inks, dyestuffs, intermediate manufacturing
Chrysene	Organic synthesis
Creosote mixture	Wood preservatives, disinfectants
Dihydrotrimethylquiunoline	Rubber antioxidant
Ethylbenzene	Intermediate manufacturing, solvent
Fluorene	Resinous products, dyestuffs, insecticides
Fluorescein	Dyestuffs
Isopropyyl benzene	Solvent, chemical manufacturing
4,4'-Methylene-bis-2-chloroaniline	Curing agent for polyurethanes and epoxy resins
Napthalene	Solvent, lubricant, explosives, preservatives, intermediate manufacturing, fungicide, moth repellant
o-Nitroaniline	Dyestuffs, intermediate, interior paint pigments, chemical manufacturing
Nitrobenzene	Solvent, polishes, chemical manufacturing
4-Nitrophenol	Chemical manufacturing
n-Nitrosodiphenylamine	Pesticides, retarder of vulcanization of rubber
Phenanthrene	Dyestuffs, explosives, syn-

Table F.1 Chemicals And Their Uses (continued)

Chemicals	Industrial Use
	thesis of drugs, biochemical research
n-Propylbenzene	Dyestuffs, solvent
Pyrene	Biochemical research
Styrene (vinyl benzene)	Plastics, resins, protective coatings, intermediate manufacturing
Toluene	Adhesive solvent in plastics, solvent, aviation and high octane blending stock, dilutent and thinner, chemicals, explosives, detergents
1,2,4-Trimethylbenzene	Manufacture of dyestuffs, pharmaceuticals, chemical manufacturing
Xylenes	Aviation gasoline, protective coatings, solvent, synthesis of organic chemicals
Oxygenated hydrocarbons	
Acetic acid	Food additives, plastics, dyestuffs, pharmaceuticals, photographic chemicals, insecticides
Acetone	Dyestuffs, solvent, chemical manufacturing, cleaning and drying of precision equipment
Benzophenone	Organic synthesis, odor fixative, flavoring, pharmaceuticals
Butyl	Solvent
N-Butyl-benzylphthalate	Plastic, intermediate manufacturing
Di-n-butyl phthalate	Pasticizer, solvent, adhesives, insecticides, safety glass, inks, paper coatings
Diethyl ether	Chemical manufacturing, solvent, analytical chemistry, anesthetic, perfumes

Table F.1 Chemicals And Their Uses (continued)

Chemicals	Industrial Use
Diethyl phthalate	Plastics, explosives, solvents, insecticides, perfumes
Dilsopropyl ether	Solvent, rubber cements, paint and varnish removers
2,4-Dimethyl-3-hexanol	Intermediate maufacturing, solvent, lubricant
2,4-Dimethyl phenol	Pharmaceuticals, plastics, disinfectants, solvent, dye-stuffs, insecticides, fungicides, additives to lubricants and gasolines
Di-n-octyl phthalate	Plasticizer for polyvinyl chloride and other vinyls
1,4 Dioxane	Solvent, lacquers, paints, varnishes, cleaning and detergent preparations, fumigants, paint and varnish removers, wetting agent, cosmetics
Ethyl acrylate	Polymers, acrylic paints, intermediate manufacturing
Formic acid	Dyeing and finishing, chemicals, manufacture of fumigants, insecticides, solvents, plastics, refrigerants
Methanol	Chemical manufacturing, solvents, automobile antifreeze, fuels
Methylcyclohexanone	Solvent, lacquers
Methyl ethyl ketone	Solvent, paint removers, cements and adhesives, cleaning fluids, printing, acrylic coatings
Phenols	Resins, solvent, pharmaceuticals, reagent, dyestuffs, indicators, germicidal paints
Phthalic acid	Dyestuffs, medicine, perfumes, reagent

Table F.1 Chemicals And Their Uses (continued)

Chemicals	Industrial Use
2-Propanol	Chemical manufacturing, solvent, deicing agent, pharmaceuticals, perfumes, lacquers, dehydrating agent, preservatives
2-Propyl-1-heptanol	Solvent
Tetrahydrofuran	Solvent
Varsol	Paint and varnish thinner

Hydrocarbons with specific elements

Acetyl chloride	Dyestuffs, pharmaceuticals, organic preparations
Alachlor (Lasso)	Herbicides
Aldicarb	Insecticide, nematocide
Aldrin	Insecticides
Atrazine	Herbicides, plant growth regulator, wed control agent
Benzoyl chloride	Medicine, intermediate manufacturing
Bromacil	Herbicides
Bromobenzene	Solvent, motor oils, organic synthesis
Bromomochloromehtane	Fire extinguishers, organic synthesis
Bromodichloromethane	Solvent, fire extinguisher fluid, mineral and salt separations
Bromoform	Solvent, intermediate manufacturing
Carbofuran	Insecticide, nematocide
Carbon tetrachloride	Degreasers, refrigerants and propellants, fumigants, chemical manufacturing
Chlordane	Insecticides, oil emulsions
Chlorobenzene	Solvent, pesticides, chemical manufacturing

Table F.1 Chemicals And Their Uses (continued)

Chemicals	Industrial Use
Chloroform	Plastics, fumigants, insecticides, refrigerants and propellants
Chloromethane	Refrigerants, medicine, propellants, herbicide, organic synthesis
2-Chloronaphthalene	Oil: plasticizer, solvent for dyestuffs, varnish gums and resins, waxes Wax: moisture-, flame-, acid-, and insect-proofing of fibrous materials; moisture- and flame-proofing of electrical cable; solvent
Chlorthal-methyl	Herbicide
o-Chlorotoluene	Solvent, intermediate manufacturing
p-Chlorotoluene	Solvent, intermediate manufacturing
Dibromochloromethane	Organic synthesis
Dibromochloropropane	Fumigant, nematocide
Dibromoethane	Fumigant, nematocide, solvent, waterproofing preparations, organic synthesis
Dibromomethane	Organic synthesis, solvent
Dichlofenthion	Pesticides
o-Dichlorobenzene	Solvent, fumigants, dyestuffs, insecticides, degreasers, polishes, industrial odor control
p-Dichlorobenzene	Insecticides, moth repellant, germicide, space odorant, fumigants, intermediate manufacturing
Dichlorobenzidine	Intermediate manufacturing, curing agent for resins

Table F.1 Chemicals And Their Uses (continued)

Chemicals	Industrial Use
Dichlorocyclooctadiene	Pesticides
Dichlorodiphenyldichlor-ethane	Insecticides
Dichlorodiphenyldichl-oroethylene	Degration product of DDT, found as an impurity in DDT residues
Dichlorodiphenyltri-chloroethane (DDT)	Pesticides
1,1-Dichloroethane	Solvent, fumigants, medicine
1,2-Dichloroethane	Solvent, degreasers, soaps and scouring compounds, organic synthesis, additive in antiknock gasoline, paint and finish removers
1,1-Dichloroethylene	Saran (used in screens, upholstery, fabrics, carpets), adhesives, synthetic fibers
1,2-Dichloroethylene	Solvent, perfumes, lacquers, themoplastics, dye extraction, organic synthesis, medicine
Dichloroethyl ether	Solvent, organic sysnthesis, paints, varnishes, lacquers, finish removers, drycleaning, fumigants
Dichloroisopropylether	Solvent, paint and varnish removers, cleaning solutions
Dichloromethane	Solvent, plastics, paint removers, propellants, blowing agent in foams
2,4 Dichlorophenol	Organic synthesis
2,4-Dichlorophenoxy-acetic acid	Herbicides
1,2-Dichloropropane	Solvent, intermediate

Table F.1 Chemicals And Their Uses (continued)

Chemicals	Industrial Use
	manufacturing, scouring compounds, fumigant, nematocide, additive for antiknock fluids
Dieldrin	Insectides
Diiodomethane	Organic synthesis
Dimethylformamide	Solvent, organic synthesis
2,4-Dinotrohenol	Herbicides
Dioxins	Impurity in the herbicide 2,4,5-T
Dodecyl mercaptan	Manufacture of synthetic rubber and plastics, pharmaceuticals, insecticides, fungicides
Endosulfan	Insecticides
Endrin	Insecticides
Ethyl chloride	Chemical manufacturing, anesthetic, solvent, refrigerants, insecticides
Bis-2-ethylhexylph- thalate	Plastics
Di-2-ethylhexylph- thalate	Plasticizers
Fluorobenzene	Insecticide and larvicide intermediate manufacturing
Fluoroform	Refrigerants, intermediate manufacturing, blowing agent for forms
Heptachlor	Insecticides
Heptachlorepoxide	Degradation product of heptachlor, also acts as an insecticide
Hexachlorobutadiene	Solvent, transformer and hydraulic fluid, heat-transfer liquid
alpha-Hexachlorocycl-	

Table F.1 Chemicals And Their Uses (continued)

Chemicals	Industrial Use
ohexane	Insecticides
beta-Hexachlorocycl- ohexane	Insecticides
gamma-Hexachlorocy- clohexane	Insecticides
Hexachlorocyclop- entadiene	Intermediate manufac- turing for resins, dyestuffs, pesticides, fungicides, phar- maceuticals
Hexachloroethane	Solvent, pyrotechnics and smoke devices, explosives, organic synthesis
Keopone	Pesticides
Malathion	Insecticides
Methoxychlor	Insecticides
Methyl bromide	Fumigants, pesticides, organic synthesis
Methyl parathion	Insecticides
Parathion	Insecticides
Pentachlorophenol	Insecticides, fung- icides, bactericides, algicides, herb- icides, wood preser- vative
Phorate	Insecticides
Polybrominated biphenyls (PBBs)	Flame retardant for plastics, paper, and textiles
Polychlorinated biphenyls (PCBs)	Heat-exchange and insulating fluids in closed systems
Prometon	Herbicides
RDX (Cyclonite)	Explosives
Simazine	Herbicides

Table F.1 Chemicals And Their Uses (continued)

Chemicals	Industrial Use
Tetrachloroethanes (1,1,1,2&1,1,2,2)	Degreasers, paint removers, varnishes, lacquers, photographic film, organic synthesis, solvent, insecticides, fumigants, weed killer
Tetrachloroethylene	Degreasers, drycleaning, solvent, drying agent, chemical manufacturing, heat-transfer medium, vermifuge
Toxaphene	Insecticides
Triazine	Herbicides
1,2,4-Trichlorobenzene	Solvent, dyestuffs, insecticides, lubricants, heat-transfer medium
Trichlorethanes (1,1,1 and 1,1,2)	Pesticides, degreasers, solvent
1,1,2-Trichloroethylene (TCE)	Degreasers, paints, drycleaning, dyestuffs, textiles, solvent, refrigerant and heat exchange liquid, fumigant, intermediate manufacturing, aerospace operations
Trichlorofluromethane	Solvent, refrigerants, fire extinguishers, intermediate manufacting
2,4,6-Trichlorophenol	Fungicides, herbicides, defoliant
2,4,5-Trichlorophenoxyacetic acid	Herbicides, defoliant

Table F.1 Chemicals And Their Uses (continued)

Chemicals	Industrial Use
2,4,5-Trichlorophen-oxypropionic acid	Herbicies and plant growth regulator
Trichlorotrifluoroethane	Drycleaning, fire extinguishers, refrigerants, intermediate drying agent manufacturing
Trinitrotoluene (TNT)	Explosives, intermediate in dyestuffs and photographic chemicals
Tris-(2,3-dibromopropyl) phospate	Flame retardant
Vinyl chloride	Organic synthesis, polyvinyl chloride and copolymers, adhesives

Other hydrocarbons

Chemicals	Industrial Use
Alkyl sulfonates	Detergents
Cyclohexane	Organic synthesis, solvent, oil extraction
1,3,5,7-Cyclooctatraene	Organic research
Dicyclopentadiene	Intermediate for insecticides, paints and varnishes, flame retardants
Fuel oil	Fuel heating
Gasoline	Fuel
Jet fuels	Fuel
Kerosene	Fuel, heating, solvent, insecticides
Lignin	Newsprint, ceramic binder, dyestuffs, drilling fuel additive, plastics
Methylene blue activated substances	Dyestuffs, analytical chemistry
Propane	Fuel, solvent, refrig-

Table F.1 Chemicals And Their Uses (continued)

Chemicals	Industrial Use
	erants, propellants, organic synthesis,
Tannin	Chemical manufacturing, tanning, textiles, electroplating, inks, pharmaceuticals, photography, paper
Undecane	Petroleum research, organic synthesis

(after OTA, 1984)

Appendix G
Well And Other Function Values

Table G.1 W(u)

u	W(u)	u	W(u)	u	W(u)
1.0×10^{-8}	17.8435	1.0×10^{-7}	15.5409	1.0×10^{-6}	13.2383
1.5×10^{-8}	17.4380	1.5×10^{-7}	15.1354	1.5×10^{-6}	12.8328
2.0×10^{-8}	17.1503	2.0×10^{-7}	14.8477	2.0×10^{-6}	12.5451
2.5×10^{-8}	16.9272	2.5×10^{-7}	14.6246	2.5×10^{-6}	12.3220
3.0×10^{-8}	16.7449	3.0×10^{-7}	14.4423	3.0×10^{-6}	12.1397
3.5×10^{-8}	16.5907	3.5×10^{-7}	14.2881	3.5×10^{-6}	11.9855
4.0×10^{-8}	16.4572	4.0×10^{-7}	14.1546	4.0×10^{-6}	11.8520
5.0×10^{-8}	16.2340	5.0×10^{-7}	13.9314	5.0×10^{-6}	11.6280
6.0×10^{-8}	16.0517	6.0×10^{-7}	13.7491	6.0×10^{-6}	11.4465
7.0×10^{-8}	15.8976	7.0×10^{-7}	13.5950	7.0×10^{-6}	11.2924
8.0×10^{-8}	15.7640	8.0×10^{-7}	13.4614	8.0×10^{-6}	11.1589
9.0×10^{-8}	15.6462	9.0×10^{-7}	13.3437	9.0×10^{-6}	11.0411
1.0×10^{-5}	10.9357	1.0×10^{-4}	8.6332	1.0×10^{-3}	6.3315
1.5×10^{-5}	10.5303	1.5×10^{-4}	8.2278	1.5×10^{-3}	5.9266
2.0×10^{-5}	10.2426	2.0×10^{-4}	7.9402	2.0×10^{-3}	5.6394
2.5×10^{-5}	10.0194	2.5×10^{-4}	7.7172	2.5×10^{-3}	5.4167
3.0×10^{-5}	9.8371	3.0×10^{-4}	7.5348	3.0×10^{-3}	5.2349
3.5×10^{-5}	9.6830	3.5×10^{-4}	7.3807	3.5×10^{-3}	5.0813
4.0×10^{-5}	9.5495	4.0×10^{-4}	7.2472	4.0×10^{-3}	4.9482
5.0×10^{-5}	9.3263	5.0×10^{-4}	7.0242	5.0×10^{-3}	4.7261
6.0×10^{-5}	9.1440	6.0×10^{-4}	6.8420	6.0×10^{-3}	4.5448
7.0×10^{-5}	8.9899	7.0×10^{-4}	6.6879	7.0×10^{-3}	4.3916
8.0×10^{-5}	8.8563	8.0×10^{-4}	6.5545	8.0×10^{-3}	4.2591
9.0×10^{-5}	8.7386	9.0×10^{-4}	6.4368	9.0×10^{-3}	4.1423
1.0×10^{-2}	4.0379	1.0×10^{-1}	1.8229	1.0×10^{0}	0.2194
1.5×10^{-2}	3.6374	1.5×10^{-1}	1.4645	1.5×10^{0}	0.1000
2.0×10^{-2}	3.3547	2.0×10^{-1}	1.2227	2.0×10^{0}	0.0489

Table G.1 W(u) (continued)

u	W(u)	u	W(u)	u	W(u)
2.5×10^{-2}	3.1365	2.5×10^{-1}	1.0443	2.5×10^{0}	0.0249
3.0×10^{-2}	2.9591	3.0×10^{-1}	0.9057	3.0×10^{0}	0.0131
3.5×10^{-2}	2.8099	3.5×10^{-1}	0.7942	3.5×10^{0}	0.0070
4.0×10^{-2}	2.6813	4.0×10^{-1}	0.7024	4.0×10^{0}	0.0038
5.0×10^{-2}	2.4679	5.0×10^{-1}	0.5598	5.0×10^{0}	0.0011
6.0×10^{-2}	2.2953	6.0×10^{-1}	0.4544	6.0×10^{0}	0.0004
7.0×10^{-2}	2.1508	7.0×10^{-1}	0.3738	7.0×10^{0}	0.0001
8.0×10^{-2}	2.0269	8.0×10^{-1}	0.3106	8.0×10^{0}	
9.0×10^{-2}	1.9187	9.0×10^{-1}	0.2602	9.0×10^{0}	

(after Ferris, et al, 1962; pp. 96–97)

Table G.2 W(u, ρ) For Production Well
(ρ = 1)

1/u	W(u, ρ)	1/u	W(u, ρ)	1/u	W(u, ρ)
0.00001	0.00356	0.01	0.1106	10	2.1881
0.00002	0.00504	0.02	0.1552	20	2.7249
0.00004	0.00713	0.04	0.2159	50	3.4966
0.00006	0.00873	0.06	0.2623	100	4.1241
0.00008	0.01008	0.08	0.3013	200	4.7767
0.0001	0.01127	0.10	0.3353	500	5.6636
0.0002	0.01591	0.20	0.4603	1000	6.3451
0.0004	0.02246	0.40	0.6308	2000	7.0317
0.0006	0.02748	0.60	0.7500	5000	7.9436
0.0008	0.03176	0.80	0.8496	10000	8.6351
0.001	0.03542	1.00	0.9346	20000	9.3273
0.002	0.04998	2.00	1.2328	50000	10.2426
0.004	0.07043	4.00	1.6000	100000	10.9359
0.006	0.08602	6.00	1.8451	1000000	13.2383
0.008	0.09909	8.00	2.0366	10000000	15.5409

(after Streltsova, 1988, p. 48)

Table G.3 W(u, ρ, α)

ρ = 1

u	1.0×10^{-1}	1.0×10^{-2}	α 1.0×10^{-3}	1.0×10^{-4}	1.0×10^{-5}
2.0×10^{0}	4.88×10^{-2}	4.99×10^{-3}	5.00×10^{-4}	5.00×10^{-5}	5.00×10^{-6}
1.0×10^{0}	9.19×10^{-2}	9.91×10^{-3}	9.99×10^{-4}	1.00×10^{-4}	1.00×10^{-5}
5.0×10^{-1}	1.77×10^{-1}	1.97×10^{-2}	2.00×10^{-3}	2.00×10^{-4}	2.00×10^{-5}
2.0×10^{-1}	4.06×10^{-1}	4.89×10^{-2}	4.99×10^{-3}	5.00×10^{-4}	5.00×10^{-5}
1.0×10^{-1}	7.34×10^{-1}	9.67×10^{-2}	9.97×10^{-3}	1.00×10^{-3}	1.00×10^{-4}
5.0×10^{-2}	1.26×10^{0}	1.90×10^{-1}	1.99×10^{-2}	2.00×10^{-3}	2.00×10^{-4}
2.0×10^{-2}	2.30×10^{0}	4.53×10^{-1}	4.95×10^{-2}	5.00×10^{-3}	5.00×10^{-4}
1.0×10^{-2}	3.28×10^{0}	8.52×10^{-1}	9.83×10^{-2}	9.98×10^{-3}	1.00×10^{-3}
5.0×10^{-3}	4.26×10^{0}	1.54×10^{0}	1.95×10^{-1}	1.99×10^{-2}	2.00×10^{-3}
2.0×10^{-3}	5.42×10^{0}	3.04×10^{0}	4.73×10^{-1}	4.97×10^{-2}	5.00×10^{-3}
1.0×10^{-3}	6.21×10^{0}	4.55×10^{0}	9.07×10^{-1}	9.90×10^{-2}	9.99×10^{-3}
5.0×10^{-4}	6.96×10^{0}	6.03×10^{0}	1.69×10^{0}	1.97×10^{-1}	2.00×10^{-2}
2.0×10^{-4}	7.87×10^{0}	7.56×10^{0}	3.52×10^{0}	4.81×10^{-1}	4.98×10^{-2}
1.0×10^{-4}	8.57×10^{0}	8.44×10^{0}	5.53×10^{0}	9.34×10^{-1}	9.93×10^{-2}
5.0×10^{-5}	9.32×10^{0}	9.23×10^{0}	7.63×10^{0}	1.77×10^{0}	1.98×10^{-1}
2.0×10^{-5}	1.02×10^{1}	1.02×10^{1}	9.68×10^{0}	3.83×10^{0}	4.86×10^{-1}
1.0×10^{-5}		1.09×10^{1}	1.07×10^{1}	6.25×10^{0}	9.49×10^{-1}
5.0×10^{-6}		1.16×10^{1}	1.15×10^{1}	8.99×10^{0}	1.82×10^{0}
2.0×10^{-6}		1.25×10^{1}	1.25×10^{1}	1.17×10^{1}	4.03×10^{0}
1.0×10^{-6}		1.32×10^{1}	1.32×10^{1}	1.29×10^{1}	6.78×10^{0}
5.0×10^{-7}			1.39×10^{1}	1.38×10^{1}	1.01×10^{1}
2.0×10^{-7}			1.48×10^{1}	1.48×10^{1}	1.37×10^{1}
1.0×10^{-7}			1.55×10^{1}	1.55×10^{1}	1.51×10^{1}
5.0×10^{-8}				1.62×10^{1}	1.61×10^{1}
2.0×10^{-8}				1.71×10^{1}	1.71×10^{1}
1.0×10^{-8}				1.78×10^{1}	1.78×10^{1}
5.0×10^{-9}					1.85×10^{1}
2.0×10^{-9}					1.94×10^{1}
1.0×10^{-9}					2.02×10^{1}

ρ = 10

u	1.0×10^{-1}	1.0×10^{-2}	1.0×10^{-3}	1.0×10^{-4}	1.0×10^{-5}
2.0×10^{0}	2.41×10^{-2}	3.52×10^{-3}	3.70×10^{-4}	3.73×10^{-5}	4.19×10^{-6}
1.0×10^{0}	1.41×10^{-1}	2.69×10^{-2}	2.95×10^{-3}	2.98×10^{-4}	3.07×10^{-5}
5.0×10^{-1}	4.44×10^{-1}	1.21×10^{-1}	1.42×10^{-2}	1.45×10^{-3}	1.47×10^{-4}
2.0×10^{-1}	1.13×10^{0}	5.12×10^{-1}	7.24×10^{-2}	7.54×10^{-3}	7.61×10^{-4}
1.0×10^{-1}	1.76×10^{0}	1.12×10^{0}	2.01×10^{-1}	2.16×10^{-2}	2.18×10^{-3}

Table G.3 $W(u, \rho, \alpha)$ (continued)

u	1.0×10^{-1}	1.0×10^{-2}	α 1.0×10^{-3}	1.0×10^{-4}	1.0×10^{-5}
5.0×10^{-2}	2.43×10^{0}	1.95×10^{0}	4.87×10^{-1}	5.55×10^{-2}	5.65×10^{-3}
2.0×10^{-2}	3.34×10^{0}	3.11×10^{0}	1.31×10^{0}	1.74×10^{-1}	1.80×10^{-2}
1.0×10^{-2}	4.03×10^{0}	3.90×10^{0}	2.38×10^{0}	3.86×10^{-1}	4.09×10^{-2}
5.0×10^{-3}	4.72×10^{0}	4.65×10^{0}	3.68×10^{0}	8.13×10^{-1}	9.03×10^{-2}
2.0×10^{-3}	5.64×10^{0}	5.61×10^{0}	5.23×10^{0}	1.97×10^{0}	2.47×10^{-1}
1.0×10^{-3}		6.31×10^{0}	6.13×10^{0}	3.44×10^{0}	5.15×10^{-1}
5.0×10^{-4}		7.01×10^{0}	6.92×10^{0}	5.26×10^{0}	1.04×10^{0}
2.0×10^{-4}		7.94×10^{0}	7.90×10^{0}	7.33×10^{0}	2.45×10^{0}
1.0×10^{-4}			8.61×10^{0}	8.37×10^{0}	4.28×10^{0}
5.0×10^{-5}			9.31×10^{0}	9.20×10^{0}	6.63×10^{0}
2.0×10^{-5}			1.02×10^{1}	1.02×10^{1}	9.36×10^{0}
1.0×10^{-5}				1.09×10^{1}	1.06×10^{1}
5.0×10^{-6}				1.16×10^{1}	1.15×10^{1}
2.0×10^{-6}				1.25×10^{1}	1.25×10^{1}
1.0×10^{-6}				1.32×10^{1}	1.32×10^{1}
5.0×10^{-7}					1.39×10^{1}
2.0×10^{-7}					1.49×10^{1}

$\rho = 50$

u	1.0×10^{-1}	1.0×10^{-2}	1.0×10^{-3}	1.0×10^{-4}	1.0×10^{-5}
2.0×10^{0}	4.24×10^{-2}	2.03×10^{-2}	3.05×10^{-3}	3.16×10^{-4}	3.21×10^{-5}
1.0×10^{0}	2.09×10^{-1}	1.42×10^{-1}	2.81×10^{-2}	3.23×10^{-3}	3.27×10^{-4}
5.0×10^{-1}	5.49×10^{-1}	4.65×10^{-1}	1.54×10^{-1}	1.80×10^{-2}	1.84×10^{-3}
2.0×10^{-1}	1.22×10^{0}	1.16×10^{0}	6.59×10^{-1}	1.03×10^{-1}	1.08×10^{-2}
1.0×10^{-1}		1.78×10^{0}	1.38×10^{0}	2.97×10^{-1}	3.30×10^{-2}
5.0×10^{-2}		2.44×10^{0}	2.27×10^{0}	7.30×10^{-1}	8.90×10^{-2}
2.0×10^{-2}		3.34×10^{0}	3.22×10^{0}	1.87×10^{0}	2.89×10^{-1}
1.0×10^{-2}		4.03×10^{0}	3.96×10^{0}	3.08×10^{0}	6.49×10^{-1}
5.0×10^{-3}		4.72×10^{0}	4.69×10^{0}	4.25×10^{0}	1.35×10^{0}
2.0×10^{-3}		5.64×10^{0}	5.63×10^{0}	5.47×10^{0}	3.03×10^{0}
1.0×10^{-3}			6.32×10^{0}	6.24×10^{0}	4.75×10^{0}
5.0×10^{-4}			7.02×10^{0}	6.98×10^{0}	6.31×10^{0}
2.0×10^{-4}				7.92×10^{0}	7.71×10^{0}
1.0×10^{-4}				8.62×10^{0}	8.52×10^{0}
5.0×10^{-5}				9.32×10^{0}	9.21×10^{0}
2.0×10^{-5}				1.02×10^{1}	1.02×10^{1}
1.0×10^{-5}					1.09×10^{1}
5.0×10^{-6}					1.16×10^{1}
2.0×10^{-6}					1.25×10^{1}
1.0×10^{-6}					1.32×10^{1}

Table G.3 $W(u, \rho, \alpha)$ (continued)

| | α | | | | |
u	1.0×10^{-1}	1.0×10^{-2}	1.0×10^{-3}	1.0×10^{-4}	1.0×10^{-5}
$\rho = 100$					
2.0×10^{0}	4.48×10^{-2}	3.44×10^{-2}	8.38×10^{-3}	9.56×10^{-4}	9.77×10^{-5}
1.0×10^{0}	2.14×10^{-1}	1.91×10^{-1}	7.56×10^{-2}	1.01×10^{-2}	1.04×10^{-3}
5.0×10^{-1}	5.55×10^{-1}	5.31×10^{-1}	3.23×10^{-1}	5.62×10^{-2}	6.02×10^{-3}
2.0×10^{-1}		1.20×10^{0}	1.02×10^{0}	3.04×10^{-1}	3.61×10^{-2}
1.0×10^{-1}		1.81×10^{0}	1.70×10^{0}	7.92×10^{-1}	1.10×10^{-1}
5.0×10^{-2}		2.46×10^{0}	2.40×10^{0}	1.62×10^{0}	2.92×10^{-1}
2.0×10^{-2}		3.35×10^{0}	3.32×10^{0}	2.95×10^{0}	8.91×10^{-1}
1.0×10^{-2}			4.02×10^{0}	3.84×10^{0}	1.80×10^{0}
5.0×10^{-3}			4.72×10^{0}	4.63×10^{0}	3.14×10^{0}
2.0×10^{-3}			5.64×10^{0}	5.60×10^{0}	5.01×10^{0}
1.0×10^{-3}				6.31×10^{0}	6.06×10^{0}
5.0×10^{-4}				7.01×10^{0}	6.90×10^{0}
2.0×10^{-4}				7.94×10^{0}	7.89×10^{0}
1.0×10^{-4}					8.61×10^{0}
5.0×10^{-5}					9.31×10^{0}
2.0×10^{-5}					1.02×10^{1}
$\rho = 200$					
2.0×10^{0}	4.50×10^{-2}	4.35×10^{-2}	1.50×10^{-2}	3.83×10^{-3}	3.15×10^{-4}
1.0×10^{0}	2.15×10^{-1}	2.11×10^{-1}	1.47×10^{-1}	3.42×10^{-2}	3.44×10^{-3}
5.0×10^{-1}	5.59×10^{-1}	5.51×10^{-1}	4.78×10^{-1}	1.75×10^{-1}	2.00×10^{-2}
2.0×10^{-1}		1.22×10^{0}	1.17×10^{0}	7.10×10^{-1}	1.19×10^{-1}
1.0×10^{-1}			1.79×10^{0}	1.43×10^{0}	3.50×10^{-1}
5.0×10^{-2}			2.45×10^{0}	2.24×10^{0}	8.57×10^{-1}
2.0×10^{-2}			3.35×10^{0}	3.28×10^{0}	2.12×10^{0}
1.0×10^{-2}				4.02×10^{0}	3.34×10^{0}
5.0×10^{-3}				4.71×10^{0}	4.40×10^{0}
2.0×10^{-3}				5.63×10^{0}	5.52×10^{0}
1.0×10^{-3}				6.33×10^{0}	6.27×10^{0}
5.0×10^{-4}					6.99×10^{0}
2.0×10^{-4}					7.93×10^{0}
1.0×10^{-4}					8.63×10^{0}

(after Reed, 1980; pp. 41-43)

Table G.4 W(ar/m,L/m,d/m, z/m)

L/m = 1.0, d/m = 0.8   ar/m

z/m	0.1	0.2	0.4	0.6
0.0	-3.42×10^0	-2.10×10^0	-9.44×10^{-1}	-4.51×10^{-1}
0.1	-3.37×10^0	-2.06×10^0	-9.16×10^{-1}	-4.34×10^{-1}
0.2	-3.32×10^0	-1.93×10^0	-8.29×10^{-1}	-3.83×10^{-1}
0.4	-2.57×10^0	-1.35×10^0	-4.67×10^{-1}	-1.84×10^{-1}
0.6	-8.77×10^{-1}	-5.70×10^{-2}	1.68×10^{-1}	1.14×10^{-1}
0.8	4.28×10^0	2.40×10^0	9.39×10^{-1}	4.10×10^{-1}
1.0	8.22×10^0	3.97×10^0	1.32×10^0	2.41×10^{-1}

z/m	0.8	1.0	1.2	1.5
0.0	-2.19×10^{-1}	-1.08×10^{-1}	-5.30×10^{-2}	-1.90×10^{-2}
0.1	-2.10×10^{-1}	-1.03×10^{-1}	-5.10×10^{-2}	-1.80×10^{-2}
0.2	-1.82×10^{-1}	-8.90×10^{-2}	-4.40×10^{-2}	-1.50×10^{-2}
0.4	-7.90×10^{-2}	-3.60×10^{-2}	-1.70×10^{-2}	-6.00×10^{-3}
0.6	6.20×10^{-2}	3.20×10^{-2}	1.60×10^{-2}	6.00×10^{-3}
0.8	1.89×10^{-1}	9.00×10^{-2}	4.40×10^{-2}	1.50×10^{-2}
1.0	2.41×10^{-1}	1.13×10^{-1}	5.50×10^{-2}	1.90×10^{-2}

L/m = 1.0, d/m = 0.5   ar/m

z/m	0.1	0.2	0.4	0.6
0.0	-2.06×10^0	-9.93×10^{-1}	-3.31×10^{-1}	-1.35×10^{-1}
0.1	-1.82×10^0	-8.86×10^{-1}	-3.05×10^{-1}	-1.26×10^{-1}
0.2	-1.07×10^0	-6.00×10^{-1}	-2.35×10^{-1}	-1.02×10^{-1}
0.4	2.19×10^{-1}	1.40×10^{-2}	-4.20×10^{-2}	-2.80×10^{-2}
0.6	6.43×10^{-1}	3.38×10^{-1}	1.17×10^{-1}	4.60×10^{-2}
0.8	8.08×10^{-1}	4.82×10^{-1}	2.07×10^{-1}	9.60×10^{-2}
1.0	8.54×10^{-1}	5.24×10^{-1}	2.36×10^{-1}	1.13×10^{-1}

z/m	0.8	1.0	1.2	1.5
0.0	-1.56×10^{-1}	-7.50×10^{-2}	-3.70×10^{-2}	-1.30×10^{-2}
0.1	-1.49×10^{-1}	-7.20×10^{-2}	-3.50×10^{-2}	-1.20×10^{-2}
0.2	-1.27×10^{-1}	-6.10×10^{-2}	-3.00×10^{-2}	-1.00×10^{-2}
0.4	-4.80×10^{-2}	-2.30×10^{-2}	-1.10×10^{-2}	-4.00×10^{-3}
0.6	4.80×10^{-2}	2.30×10^{-2}	1.10×10^{-2}	4.00×10^{-3}

Table G.4 W(ar/m,L/m,d/m, z/m) (continued)

0.8	1.27×10^{-1}	6.10×10^{-2}	3.00×10^{-2}	1.00×10^{-2}
1.0	1.56×10^{-1}	7.50×10^{-2}	3.70×10^{-2}	1.30×10^{-2}

$L/m = 1.0$, $d/m = 0.2$ ⟶ ar/m

z/m	0.1	0.2	0.4	0.6
0.0	-2.06×10^{0}	-9.93×10^{-1}	-3.31×10^{-1}	-1.35×10^{-1}
0.1	-1.82×10^{0}	-8.86×10^{-1}	-3.05×10^{-1}	-1.26×10^{-1}
0.2	-1.07×10^{0}	-6.00×10^{-1}	-2.35×10^{-1}	-1.02×10^{-1}
0.4	2.19×10^{-1}	1.40×10^{-2}	-4.20×10^{-2}	-2.80×10^{-2}
0.6	6.43×10^{-1}	3.38×10^{-1}	1.17×10^{-1}	4.60×10^{-2}
0.8	8.08×10^{-1}	4.82×10^{-1}	2.07×10^{-1}	9.60×10^{-2}
1.0	8.54×10^{-1}	5.24×10^{-1}	2.36×10^{-1}	1.13×10^{-1}

	0.8	1.0	1.2	1.5
0.0	-6.00×10^{-2}	-2.80×10^{-2}	-1.40×10^{-2}	-5.00×10^{-3}
0.1	-5.70×10^{-2}	-2.70×10^{-2}	-1.30×10^{-2}	-5.00×10^{-3}
0.2	-4.70×10^{-2}	-2.30×10^{-2}	-1.10×10^{-2}	-4.00×10^{-3}
0.4	-1.50×10^{-2}	-8.00×10^{-3}	-4.00×10^{-3}	-1.00×10^{-3}
0.6	2.00×10^{-2}	9.00×10^{-3}	4.00×10^{-3}	1.00×10^{-3}
0.8	4.60×10^{-2}	2.20×10^{-2}	1.10×10^{-2}	4.00×10^{-3}
1.0	5.50×10^{-2}	2.70×10^{-2}	1.30×10^{-2}	5.00×10^{-3}

(after Weeks, 1969, pp. 196–214)

Table G.5 W(λ)

λ	W(λ)	λ	W(λ)	λ	W(λ)
1×10^{-4}	56.9	5×10^{2}	0.274	9×10^{8}	0.0932
2	40.4	6	0.268	1×10^{9}	0.0927
3	33.1	7	0.263	2	0.0899
4	28.7	8	0.258	3	0.0883
5	25.7	9	0.254	4	0.0872
6	23.5	1×10^{3}	0.251	5	0.0864
7	21.8	2	0.232	6	0.0857
8	20.4	3	0.222	7	0.0851

Table G.5 $W(\lambda)$ **(continued)**

λ	$W(\lambda)$	λ	$W(\lambda)$	λ	$W(\lambda)$
9	19.3	4	0.215	8	0.0846
1×10^{-3}	18.34	5	0.210	9	0.0842
2	13.11	6	0.206	1×10^{10}	0.0838
3	10.79	7	0.203	2	0.0814
4	9.41	8	0.200	3	0.0801
5	8.47	9	0.198	4	0.0792
6	7.77	1×10^{4}	0.1964	5	0.0785
7	7.23	2	0.1841	6	0.0779
8	6.79	3	0.1777	7	0.0774
9	6.43	4	0.1733	8	0.0770
1×10^{-2}	6.13	5	0.1701	9	0.0767
2	4.47	6	0.1675	1×10^{11}	0.0764
3	3.74	7	0.1654	2	0.0744
4	3.30	8	0.1636	3	0.0733
5	3.00	9	0.1621	4	0.0726
6	2.78	1×10^{5}	0.1608	5	0.0720
7	2.60	2	0.1524	6	0.0716
8	2.46	3	0.1479	7	0.0712
9	2.35	4	0.1449	8	0.0709
1×10^{-1}	2.249	5	0.1426	9	0.0706
2	1.716	6	0.1408	1×10^{12}	0.0704
3	1.477	7	0.1393	2	0.0686
4	1.333	8	0.1380	3	0.0677
5	1.234	9	0.1369	4	0.0671
6	1.160	1×10^{6}	0.1360	5	0.0666
7	1.103	2	0.1299	6	0.0662
8	1.057	3	0.1266	7	0.0658
9	1.018	4	0.1244	8	0.0655
1×10^{0}	0.985	5	0.1227	9	0.0653
2	0.803	6	0.1213	1×10^{13}	0.0651
3	0.719	7	0.1202	2	0.0636
4	0.667	8	0.1192	3	0.0628
5	0.630	9	0.1184	4	0.0622
6	0.602	1×10^{7}	0.1177	5	0.0618
7	0.580	2	0.1131	6	0.0615
8	0.562	3	0.1106	7	0.0612
9	0.547	4	0.1089	8	0.0609
1×10^{1}	0.534	5	0.1076	9	0.0607
2	0.461	6	0.1066	1×10^{14}	0.0605
3	0.427	7	0.1057	2	0.0593

Table G.5 $W(\lambda)$ **(continued)**

λ	$W(\lambda)$	λ	$W(\lambda)$	λ	$W(\lambda)$
4	0.405	8	0.1049	3	0.0586
5	0.389	9	0.1043	4	0.0581
6	0.377	1×10^8	0.1037	5	0.0577
7	0.367	2	0.1002	6	0.0574
8	0.359	3	0.0982	7	0.0572
9	0.352	4	0.0968	8	0.0569
1×10^2	0.346	5	0.0958	9	0.0567
2	0.311	6	0.0950	1×10^{15}	0.0566
3	0.294	7	0.0943	2	0.0555
4	0.283	8	0.0937	3	0.0549

(after Reed, 1080, p. 19)

Table G.6 $W(\lambda , \rho)$

λ	\rho 5	20	100	200	500	1000
1×10^0	0.002					
2	0.022					
3	0.049					
4	0.076					
5	0.101					
7	0.142					
1×10^1	0.188	0.000				
2	0.277	0.001				
3	0.325	0.004				
4	0.358	0.009				
5	0.381	0.016				
7	0.414	0.031				
1×10^2	0.446	0.053				
1.5	0.479	0.085				
2	0.500	0.110				
3	0.528	0.146				
5	0.559	0.194	0.000			
7	0.578	0.223	0.001			
1×10^3	0.596	0.254	0.004			
1.5	0.615	0.287	0.012			

Table G.6 $W(\lambda, \rho)$ **(continued)**

λ	5	20	ρ 100	200	500	1000
2	0.627	0.309	0.021	0.000		
3	0.644	0.338	0.039	0.001		
5	0.662	0.372	0.068	0.006		
7	0.673	0.392	0.089	0.014		
1×10^4	0.685	0.413	0.114	0.025		
1.5	0.696	0.435	0.142	0.043	0.000	
2	0.704	0.450	0.161	0.058	0.001	
3	0.715	0.469	0.188	0.081	0.005	
5	0.727	0.492	0.221	0.113	0.014	0.000
7	0.734	0.506	0.242	0.134	0.025	0.001
1×10^5	0.742	0.520	0.263	0.156	0.039	0.002
1.5	0.750	0.532	0.285	0.180	0.058	0.007
2	0.755	0.544	0.300	0.197	0.072	0.013
3	0.762	0.558	0.321	0.220	0.094	0.024
5	0.771	0.574	0.345	0.247	0.122	0.044
7	0.776	0.584	0.360	0.264	0.141	0.059
1×10^6	0.782	0.594	0.376	0.282	0.160	0.076
1.5	0.788	0.604	0.392	0.301	0.181	0.096
2	0.792	0.612	0.403	0.314	0.196	0.111
3	0.797	0.622	0.418	0.331	0.216	0.132
5	0.803	0.633	0.436	0.352	0.240	0.157
7	0.807	0.641	0.448	0.365	0.255	0.173
1×10^7	0.811	0.648	0.459	0.378	0.270	0.190
1.5	0.815	0.656	0.472	0.392	0.287	0.208
2	0.818	0.662	0.480	0.402	0.299	0.221
3	0.822	0.669	0.492	0.415	0.314	0.238
5	0.827	0.678	0.506	0.431	0.333	0.258
7	0.830	0.684	0.514	0.441	0.344	0.271
1×10^8	0.833	0.690	0.523	0.452	0.357	0.285
1.5	0.837	0.696	0.533	0.463	0.370	0.300
2	0.839	0.701	0.540	0.470	0.379	0.310
3	0.842	0.706	0.549	0.481	0.391	0.323
5	0.846	0.714	0.560	0.494	0.406	0.340
7	0.849	0.718	0.567	0.502	0.415	0.350
1×10^9	0.851	0.723	0.574	0.510	0.425	0.361
1.5	0.854	0.728	0.582	0.519	0.435	0.372
2	0.856	0.731	0.587	0.525	0.443	0.380
3	0.858	0.736	0.594	0.533	0.452	0.392
5	0.861	0.742	0.603	0.544	0.464	0.405
7	0.863	0.746	0.609	0.550	0.472	0.413

Table G.6   W(λ, ρ) (continued)

λ	5	20	ρ 100	200	500	1000
1×10^{10}	0.865	0.749	0.615	0.557	0.480	0.422
2	0.869	0.756	0.625	0.569	0.494	0.438
3	0.871	0.760	0.631	0.576	0.502	0.447
5	0.874	0.765	0.638	0.584	0.512	0.457
7	0.875	0.768	0.643	0.589	0.518	0.464
1×10^{11}	0.877	0.770	0.647	0.594	0.524	0.471

(after Reed, 1980, p. 20)

Table G.7   W(u,b)

u	b 1.0×10^{-3}	5.0×10^{-3}	1.0×10^{-2}	2.5×10^{-2}	5.0×10^{-2}
1.0×10^{-6}	13.0031	10.8283	9.4425	7.6111	6.2285
2.0×10^{-6}	12.4240	10.8174	9.4425	7.6111	6.2285
3.0×10^{-6}	12.0581	10.7849	9.4425	7.6111	6.2285
4.0×10^{-6}	11.7905	10.7374	9.4422	7.6111	6.2285
5.0×10^{-6}	11.5795	10.6822	9.4413	7.6111	6.2285
6.0×10^{-6}	11.4053	10.6240	9.4394	7.6111	6.2285
7.0×10^{-6}	11.2570	10.5652	9.4361	7.6111	6.2285
8.0×10^{-6}	11.1279	10.5072	9.4313	7.6111	6.2285
9.0×10^{-6}	11.0135	10.4508	9.4251	7.6111	6.2285
1.0×10^{-5}	10.9109	10.3963	9.4176	7.6111	6.2285
2.0×10^{-5}	10.2301	9.9530	9.2961	7.6111	6.2285
3.0×10^{-5}	9.8288	9.6392	9.1499	7.6101	6.2285
4.0×10^{-5}	9.5432	9.3992	9.0102	7.6069	6.2285
5.0×10^{-5}	9.3213	9.2052	8.8827	7.6000	6.2285
6.0×10^{-5}	9.1398	9.0426	8.7673	7.5894	6.2285
7.0×10^{-5}	8.9863	8.9027	8.6625	7.5754	6.2285
8.0×10^{-5}	8.8532	8.7798	8.5669	7.5589	6.2284
9.0×10^{-5}	8.7358	8.6703	8.4792	7.5402	6.2283
1.0×10^{-4}	8.6308	8.5717	8.3983	7.5199	6.2282
2.0×10^{-4}	7.9390	7.9092	7.8192	7.2898	6.2173

Table G.7 W(u,b) (continued)

u	1.0×10^{-3}	5.0×10^{-3}	b 1.0×10^{-2}	2.5×10^{-2}	5.0×10^{-2}
3.0×10^{-4}	7.5340	7.5141	7.4534	7.0759	6.1848
4.0×10^{-4}	7.2466	7.2317	7.1859	6.8929	6.1373
5.0×10^{-4}	7.0237	7.0118	6.9750	6.7357	6.0821
6.0×10^{-4}	6.8416	6.8316	6.8009	6.5988	6.0239
7.0×10^{-4}	6.6876	6.6790	6.6527	6.4777	5.9652
8.0×10^{-4}	6.5542	6.5467	6.5237	6.3695	5.9073
9.0×10^{-4}	6.4365	6.4299	6.4094	6.2716	5.8509
1.0×10^{-3}	6.3313	6.3253	6.3069	6.1823	5.7965
2.0×10^{-3}	5.6393	5.6363	5.6271	5.5638	5.3538
3.0×10^{-3}	5.2348	5.2329	5.2267	5.1845	5.0408
4.0×10^{-3}	4.9482	4.9467	4.9421	4.9105	4.8016
5.0×10^{-3}	4.7260	4.7249	4.7212	4.6960	4.6084
6.0×10^{-3}	4.5448	4.5438	4.5407	4.5197	4.4467
7.0×10^{-3}	4.3916	4.3908	4.3882	4.2404	4.1857
8.0×10^{-3}	4.2590	4.2583	4.2561	4.3702	4.3077
9.0×10^{-3}	4.1423	4.1416	4.1396	4.1258	4.0772
1.0×10^{-2}	4.0379	4.0373	4.0356	4.0231	3.9795
2.0×10^{-2}	3.3547	3.3544	3.3536	3.3476	3.3264
3.0×10^{-2}	2.9591	2.9589	2.9584	2.9545	2.9409
4.0×10^{-2}	2.6812	2.6811	2.6807	2.6779	2.6680

u	1.0×10^{-3}	5.0×10^{-3}	b 1.0×10^{-2}	2.5×10^{-2}	5.0×10^{-2}
5.0×10^{-2}	2.4679	2.4678	2.4675	2.4653	2.4576
6.0×10^{-2}	2.2953	2.2952	2.2950	2.2932	2.2870
7.0×10^{-2}	2.1508	2.1508	2.1506	2.1491	2.1439
8.0×10^{-2}	2.0269	2.0269	2.0267	2.0255	2.0210
9.0×10^{-2}	1.9187	1.9187	1.9185	1.9174	1.9136
1.0×10^{-1}	1.8229	1.8229	1.8227	1.8218	1.8184
2.0×10^{-1}	1.2226	1.2226	1.2226	1.2222	1.2209
3.0×10^{-1}	0.9057	0.9057	0.9056	0.9054	0.9047
4.0×10^{-1}	0.7024	0.7024	0.7074	0.7022	0.7016
5.0×10^{-1}	0.5598	0.5598	0.5598	0.5597	0.5594
6.0×10^{-1}	0.4544	0.4544	0.4544	0.4543	0.4541
7.0×10^{-1}	0.3738	0.3738	0.3738	0.3737	0.3735
8.0×10^{-1}	0.3106	0.3106	0.3106	0.3106	0.3104

Table G.7   W(u,b) (continued)

9.0×10^{-1}	0.2602	0.2602	0.2602	0.2602	0.2601
1.0×10^{0}	0.2194	0.2194	0.2194	0.2194	0.2193
2.0×10^{0}	0.0489	0.0489	0.0489	0.0489	0.0489
3.0×10^{0}	0.0130	0.0130	0.0130	0.0130	0.0130
4.0×10^{0}	0.0038	0.0038	0.0038	0.0038	0.0038
5.0×10^{0}	0.0011	0.0011	0.0011	0.0011	0.0011
6.0×10^{0}	0.0004	0.0004	0.0004	0.0004	0.0004
7.0×10^{0}	0.0001	0.0001	0.0001	0.0001	0.0001

b

u	7.5×10^{-2}	1.5×10^{-1}	3.0×10^{-1}	5.0×10^{-1}	7.0×10^{-1}
1.0×10^{-4}	5.4228	4.0601	2.7449	1.8488	1.3210
2.0×10^{-4}	5.4227	4.0601	2.7449	1.8488	1.3210
3.0×10^{-4}	5.4212	4.0601	2.7449	1.8488	1.3210
4.0×10^{-4}	5.4160	4.0601	2.7449	1.8488	1.3210
5.0×10^{-4}	5.4062	4.0601	2.7449	1.8488	1.3210
6.0×10^{-4}	5.3921	4.0601	2.7449	1.8488	1.3210
7.0×10^{-4}	5.3745	4.0600	2.7449	1.8488	1.3210
8.0×10^{-4}	5.3542	4.0599	2.7449	1.8488	1.3210
9.0×10^{-4}	5.3317	4.0598	2.7449	1.8488	1.3210
1.0×10^{-3}	5.3078	4.0595	2.7449	1.8488	1.3210
2.0×10^{-3}	5.0517	4.0435	2.7449	1.8488	1.3210
3.0×10^{-3}	4.8243	4.0092	2.7448	1.8488	1.3210
4.0×10^{-3}	4.6335	3.9551	2.7444	1.8488	1.3210
5.0×10^{-3}	4.4713	3.8821	2.7428	1.8488	1.3210
6.0×10^{-3}	4.3311	3.8384	2.7398	1.8488	1.3210
7.0×10^{-3}	4.2078	3.7529	2.7350	1.8488	1.3210
8.0×10^{-3}	4.0980	3.6903	2.7284	1.8488	1.3210
9.0×10^{-3}	3.9991	3.6302	2.7202	1.8487	1.3210
1.0×10^{-2}	3.9091	3.5725	2.7104	1.8486	1.3210
2.0×10^{-2}	3.2917	3.1158	2.5688	1.8379	1.3207

b

u	7.5×10^{-2}	1.5×10^{-1}	3.0×10^{-1}	5.0×10^{-1}	7.0×10^{-1}
3.0×10^{-2}	2.9183	2.8017	2.4110	1.8062	1.3177
4.0×10^{-2}	2.6515	2.5655	2.2661	1.7603	1.3094
5.0×10^{-2}	2.4448	2.3776	2.1371	1.7075	1.2955

Table G.7 W(u,b) (continued)

6.0×10^{-2}	2.2766	2.2218	2.0227	1.6524	1.2770
7.0×10^{-2}	2.1352	2.0894	1.9206	1.5973	1.2551
8.0×10^{-2}	2.0136	1.9745	1.8290	1.5436	1.2310
9.0×10^{-2}	1.9072	1.8732	1.7460	1.4918	1.2054
1.0×10^{-1}	1.8128	1.7829	1.6704	1.4422	1.1791
2.0×10^{-1}	1.2186	1.2066	1.1602	1.0592	0.9629
3.0×10^{-1}	0.9035	0.8969	0.8713	0.8142	0.7362
4.0×10^{-1}	0.7010	0.6969	0.6809	0.6446	0.5943
5.0×10^{-1}	0.5588	0.5561	0.5453	0.5206	0.4860
6.0×10^{-1}	0.4537	0.4518	0.4441	0.4266	0.4018
7.0×10^{-1}	0.3733	0.3719	0.3663	0.3534	0.3351
8.0×10^{-1}	0.3102	0.3092	0.3050	0.2953	0.2815
9.0×10^{-1}	0.2599	0.2591	0.2559	0.2485	0.2378
1.0×10^{0}	0.2191	0.2186	0.2161	0.2103	0.2020
2.0×10^{0}	0.0489	0.0488	0.0485	0.0477	0.0467
3.0×10^{0}	0.0130	0.0130	0.0130	0.0128	0.0126
4.0×10^{0}	0.0038	0.0038	0.0038	0.0037	0.0037
5.0×10^{0}	0.0011	0.0011	0.0011	0.0011	0.0011
6.0×10^{0}	0.0004	0.0004	0.0004	0.0004	0.0004
7.0×10^{0}	0.0001	0.0001	0.0001	0.0001	0.0001

b

u	8.5×10^{-1}	1.0×10^{0}	1.5×10^{0}	2.0×10^{0}	2.5×10^{0}
1.0×10^{-2}	1.0485	0.8420	0.4276	0.2278	0.1247
2.0×10^{-2}	1.0484	0.8420	0.4276	0.2278	0.1247
3.0×10^{-2}	1.0481	0.8420	0.4276	0.2278	0.1247
4.0×10^{-2}	1.0465	0.8418	0.4276	0.2278	0.1247
5.0×10^{-2}	1.0426	0.8409	0.4276	0.2278	0.1247
6.0×10^{-2}	1.0362	0.8391	0.4276	0.2278	0.1247
7.0×10^{-2}	1.0272	0.8360	0.4276	0.2278	0.1247
8.0×10^{-2}	1.0161	0.8316	0.4275	0.2278	0.1247
9.0×10^{-2}	1.0032	0.8259	0.4274	0.2278	0.1247
1.0×10^{-1}	0.9890	0.8190	0.4271	0.2278	0.1247
2.0×10^{-1}	0.8216	0.7148	0.4135	0.2268	0.1247
3.0×10^{-1}	0.6706	0.6010	0.3812	0.2211	0.1240
4.0×10^{-1}	0.5501	0.5024	0.3411	0.2096	0.1217
5.0×10^{-1}	0.4550	0.4210	0.3007	0.1944	0.1174
6.0×10^{-1}	0.3793	0.3543	0.2630	0.1774	0.1112
7.0×10^{-1}	0.3183	0.2996	0.2292	0.1602	0.1040

Table G.7 W(u,b) (continued)

8.0×10^{-1}	0.2687	0.2543	0.1994	0.1436	0.0961
9.0×10^{-1}	0.2280	0.2168	0.1734	0.1281	0.0881
1.0×10^{0}	0.1943	0.1855	0.1509	0.1139	0.0803
2.0×10^{0}	0.0456	0.0444	0.0394	0.0335	0.0271
3.0×10^{0}	0.0124	0.0122	0.0112	0.0100	0.0086
4.0×10^{0}	0.0036	0.0036	0.0034	0.0031	0.0027
5.0×10^{0}	0.0011	0.0011	0.0010	0.0010	0.0009
6.0×10^{0}	0.0004	0.0004	0.0003	0.0003	0.0003
7.0×10^{0}	0.0001	0.0001	0.0001	0.0001	0.0001

Table G.8 W(λ ,b)

	b			
λ	6×10^{-3}	1×10^{-2}	2×10^{-2}	6×10^{-2}
1×10^{-1}	2.24	2.24	2.24	2.25
2	1.71	1.71	1.71	1.71
5	1.23	1.23	1.23	1.23
1×10^{0}	0.983	0.983	0.983	0.984
2	0.800	0.800	0.800	0.801
5	0.628	0.628	0.628	0.629
1×10^{1}	0.534	0.534	0.534	0.535
2	0.461	0.461	0.461	0.462
5	0.389	0.389	0.389	0.390
1×10^{2}	0.346	0.346	0.346	0.349
2	0.311	0.311	0.312	0.316
5	0.274	0.275	0.276	0.284
1×10^{3}	0.251	0.252	0.255	0.266
2	0.232	0.234	0.239	0.255
5	0.210	0.215	0.222	0.249
1×10^{4}	0.196	0.204	0.216	0.248
2	0.185	0.197	0.213	
5	0.170	0.192	0.212	
1×10^{5}	0.161	0.191		
2	0.152			
5	0.143			
1×10^{6}	0.136			
2	0.130			

Table G.8 $W(\lambda ,b)$ (continued)

| 5 | 0.123 | 0.191 | 0.212 | 0.248 |

| | | b | | |

λ	1×10^{-1}	2×10^{-1}	6×10^{-1}	1×10^{0}
1×10^{-1}	2.25	2.26	2.31	2.43
2	1.72	1.73	1.81	1.96
5	1.24	1.25	1.38	1.61
1×10^{0}	0.990	1.01	1.18	1.49
2	0.809	0.834	1.07	1.44
5	0.642	0.682	1.01	1.43
1×10^{1}	0.554	0.611		
2	0.491	0.569		
5	0.438	0.548		
1×10^{2}	0.417	0.545		
2	0.408			
5	0.406	0.545	1.01	1.43

| | | b | | |

λ	1×10^{-5}	2×10^{-5}	6×10^{-5}	1×10^{-4}
1×10^{4}	0.196	0.196	0.196	0.196
2	0.185	0.185	0.185	0.185
5	0.170	0.170	0.170	0.170
1×10^{5}	0.161	0.161	0.161	0.161
2	0.152	0.152	0.152	0.152
5	0.143	0.143	0.143	0.143
1×10^{6}	0.136	0.136	0.136	0.136
2	0.130	0.130	0.130	0.130
5	0.123	0.123	0.123	0.123
1×10^{7}	0.118	0.118	0.118	0.118
2	0.114	0.114	0.114	0.114
5	0.108	0.108	0.108	0.108
1×10^{8}	0.104	0.104	0.104	0.105
2	0.100	0.100	0.101	0.103
5	0.0959	0.0958	0.0966	0.102
1×10^{9}	0.0927	0.0930	0.0943	
2	0.0899	0.0906	0.0927	
5	0.0864	0.0880	0.0916	
1×10^{10}	0.0838	0.0867	0.0914	

Table G.8 W(λ ,b) (continued)

2	0.0814	0.0862		
5	0.0785	0.0860		
1×10^{11}	0.0764	0.0860	0.0914	0.102

		b		
λ	2×10^{-4}	6×10^{-4}	1×10^{-3}	2×10^{-3}
1×10^4	0.196	0.196	0.196	0.197
2	0.185	0.185	0.185	0.185
5	0.170	0.170	0.170	0.173
1×10^5	0.161	0.162	0.162	0.167
2	0.152	0.153	0.155	0.163
5	0.143	0.144	0.148	0.161
1×10^6	0.137	0.139	0.144	0.159
2	0.131	0.135	0.143	0.159
5	0.124	0.133	0.142	0.158
1×10^7	0.120			
2	0.116	0.133	0.142	0.158

(after Reed, 1980, p. 36)

Table G.9 W(u, τ)

			τ		
u	1.0×10^{-2}	5.0×10^{-2}	1.0×10^{-1}	2.0×10^{-1}	5.0×10^{-1}
1.0×10^{-6}	9.9259	8.3395	7.6497	6.9590	6.0463
2.0×10^{-6}	9.5677	7.9908	7.3024	6.6126	5.7012
3.0×10^{-6}	9.3561	7.7864	7.0991	6.4100	5.4996
4.0×10^{-6}	9.2047	7.6412	6.9547	6.2663	5.3567
5.0×10^{-6}	9.0866	7.5284	6.8427	6.1548	5.2459
6.0×10^{-6}	8.9894	7.4362	6.7512	6.0637	5.1555
7.0×10^{-6}	8.9069	7.3581	6.6737	5.9867	5.0790
8.0×10^{-6}	8.8350	7.2904	6.6066	5.9200	5.0129
9.0×10^{-6}	8.7714	7.2306	6.5474	5.8611	4.9545
1.0×10^{-5}	8.7142	7.1771	6.4944	5.8085	4.9024
2.0×10^{-5}	8.3315	6.8238	6.1453	5.4623	4.5598

Table G.9 W(u, τ) (continued)

3.0×10^{-5}	8.1013	6.6159	5.9406	5.2597	4.3600
4.0×10^{-5}	7.9346	6.4677	5.7951	5.1160	4.2185
5.0×10^{-5}	7.8031	6.3523	5.6821	5.0045	4.1090
6.0×10^{-5}	7.6941	6.2576	5.5896	4.9134	4.0196
7.0×10^{-5}	7.6007	6.1773	5.5113	4.8364	3.9442
8.0×10^{-5}	7.5190	6.1076	5.4434	4.7697	3.8789
9.0×10^{-5}	7.4461	6.0459	5.3834	4.7108	3.8214
1.0×10^{-4}	7.3803	5.9906	5.3297	4.6581	3.7700
2.0×10^{-4}	6.9321	5.6226	4.9747	4.3115	3.4334
3.0×10^{-4}	6.6563	5.4035	4.7655	4.1086	3.2379
4.0×10^{-4}	6.4541	5.2459	4.6161	3.9645	3.0999
5.0×10^{-4}	6.2934	5.1223	4.4996	3.8527	2.9933
6.0×10^{-4}	6.1596	5.0203	4.4040	3.7612	2.9065
7.0×10^{-4}	6.0447	4.9333	4.3228	3.6838	2.8334
8.0×10^{-4}	5.9439	4.8573	4.2523	3.6167	2.7702
9.0×10^{-4}	5.8539	4.7898	4.1898	3.5575	2.7146
1.0×10^{-3}	5.7727	4.7290	4.1337	3.5045	2.6650
2.0×10^{-3}	5.2203	4.3184	3.7598	3.1549	2.3419
3.0×10^{-3}	4.8837	4.0683	3.5363	2.9494	2.1559
4.0×10^{-3}	4.6396	3.8859	3.3750	2.8030	2.0253
5.0×10^{-3}	4.4474	3.7415	3.2483	2.6891	1.9250
6.0×10^{-3}	4.2888	3.6214	3.1436	2.5957	1.8437
7.0×10^{-3}	4.1536	3.5185	3.0542	2.5165	1.7754
8.0×10^{-3}	4.0357	3.4282	2.9762	2.4478	1.7166
9.0×10^{-3}	3.9313	3.3478	2.9068	2.3870	1.6651
1.0×10^{-2}	3.8374	3.2752	2.8443	2.3325	1.6193
2.0×10^{-2}	3.2133	2.7829	2.4227	1.9714	1.3239
3.0×10^{-2}	2.8452	2.4844	2.1680	1.7579	1.1570
4.0×10^{-2}	2.5842	2.2691	1.9841	1.6056	1.0416
5.0×10^{-2}	2.3826	2.1007	1.8401	1.4872	0.9540
6.0×10^{-2}	2.2188	1.9626	1.7217	1.3905	0.8838
7.0×10^{-2}	2.0812	1.8458	1.6213	1.3088	0.8255
8.0×10^{-2}	1.9630	1.7448	1.5343	1.2381	0.7758
9.0×10^{-2}	1.8595	1.6559	1.4577	1.1760	0.7327
1.0×10^{-1}	1.7677	1.5768	1.3893	1.1207	0.6947
2.0×10^{-1}	1.1895	1.0714	0.9497	0.7665	0.4603
3.0×10^{-1}	0.8825	0.7986	0.7103	0.5739	0.3390
4.0×10^{-1}	0.6850	0.6218	0.5543	0.4482	0.2619
5.0×10^{-1}	0.5463	0.4969	0.4436	0.3591	0.2083
6.0×10^{-1}	0.4437	0.4041	0.3613	0.2927	0.1688
7.0×10^{-1}	0.3651	0.3330	0.2980	0.2415	0.1386
8.0×10^{-1}	0.3035	0.2770	0.2481	0.2012	0.1151

Table G.9 W(u, τ) **(continued)**

u					
9.0×10^{-1}	0.2543	0.2323	0.2082	0.1690	0.0010
1.0×10^{0}	0.2144	0.1961	0.1758	0.1427	0.0008
2.0×10^{0}	0.0005	0.0004	0.0004	0.0003	0.0002
3.0×10^{0}	0.0001	0.0001	0.0001	0.0001	

τ

u	1.0×10^{0}	2.0×10^{0}	5.0×10^{0}	1.0×10^{1}	2.0×10^{1}
1.0×10^{-6}	5.3575	4.6721	3.7756	3.1110	2.4671
2.0×10^{-6}	5.0141	4.3312	3.4412	2.7857	2.1568
3.0×10^{-6}	4.8136	4.1327	3.2474	2.5984	1.9801
4.0×10^{-6}	4.6716	3.9922	3.1109	2.4671	1.8571
5.0×10^{-6}	4.5617	3.8836	3.0055	2.3661	1.7633
6.0×10^{-6}	4.4719	3.7951	2.9199	2.2844	1.6877
7.0×10^{-6}	4.3962	3.7204	2.8478	2.2158	1.6246
8.0×10^{-6}	4.3306	3.6558	2.7856	2.1568	1.5706
9.0×10^{-6}	4.2728	3.5989	2.7309	2.1050	1.5234
1.0×10^{-5}	4.2212	3.5481	2.6822	2.0590	1.4816
2.0×10^{-5}	3.8827	3.2162	2.3660	1.7632	1.2170
3.0×10^{-5}	3.6858	3.0241	2.1850	1.5965	1.0716
4.0×10^{-5}	3.5468	2.8889	2.0588	1.4815	0.9730
5.0×10^{-5}	3.4394	2.7848	1.9622	1.3943	0.8994
6.0×10^{-5}	3.3519	2.7002	1.8841	1.3244	0.8412
7.0×10^{-5}	3.2781	2.6290	1.8189	1.2664	0.7934
8.0×10^{-5}	3.2143	2.5677	1.7629	1.2169	0.7530
9.0×10^{-5}	3.1583	2.5138	1.7139	1.1739	0.7182
1.0×10^{-4}	3.1082	2.4658	1.6704	1.1359	0.6878
2.0×10^{-4}	2.7819	2.1549	1.3937	0.8992	0.5044
3.0×10^{-4}	2.5937	1.9778	1.2401	0.7721	0.4111
4.0×10^{-4}	2.4617	1.8545	1.1352	0.6875	0.3514
5.0×10^{-4}	2.3601	1.7604	1.0564	0.6252	0.3089
6.0×10^{-4}	2.2778	1.6846	0.9937	0.5765	0.2766
7.0×10^{-4}	2.2087	1.6212	0.9420	0.5370	0.2510
8.0×10^{-4}	2.1492	1.5670	0.8982	0.5040	0.2300
9.0×10^{-4}	2.0971	1.5196	0.8603	0.4758	0.2125
1.0×10^{-3}	2.0506	1.4776	0.8271	0.4513	0.1976
2.0×10^{-3}	1.7516	1.2116	0.6238	0.3084	0.1164
3.0×10^{-3}	1.5825	1.0652	0.5182	0.2394	0.0008
4.0×10^{-3}	1.4656	0.9658	0.4496	0.1970	0.0006
5.0×10^{-3}	1.3767	0.8915	0.4001	0.1677	0.0005
6.0×10^{-3}	1.3054	0.8327	0.3620	0.1460	0.0004
7.0×10^{-3}	1.2460	0.7843	0.3315	0.1292	0.0003
8.0×10^{-3}	1.1953	0.7435	0.3064	0.1158	0.0003

Table G.9 W(u, τ) (continued)

9.0×10^{-3}	1.1512	0.7083	0.2852	0.1047	0.0003
1.0×10^{-2}	1.1122	0.6775	0.2670	0.0010	0.0002
2.0×10^{-2}	0.8677	0.4910	0.1653	0.0005	0.0001
3.0×10^{-2}	0.7353	0.3965	0.1197	0.0004	
4.0×10^{-2}	0.6467	0.3357	0.0009	0.0003	
5.0×10^{-2}	0.5812	0.2923	0.0008	0.0002	
6.0×10^{-2}	0.5298	0.2593	0.0006	0.0001	
7.0×10^{-2}	0.4880	0.2332	0.0005	0.0001	
8.0×10^{-2}	0.4530	0.2119	0.0005	0.0001	
9.0×10^{-2}	0.4230	0.1941	0.0004	0.0001	
1.0×10^{-1}	0.3970	0.1789	0.0004	0.0001	
2.0×10^{-1}	0.2452	0.0010	0.0001		
3.0×10^{-1}	0.1729	0.0006	0.0001		
4.0×10^{-1}	0.1296	0.0004	0.0001		
5.0×10^{-1}	0.1006	0.0003			
6.0×10^{-1}	0.0008	0.0002			
7.0×10^{-1}	0.0006	0.0002			
8.0×10^{-1}	0.0005	0.0002			
9.0×10^{-1}	0.0004	0.0001			
1.0×10^{0}	0.0004	0.0001			
2.0×10^{0}	0.0001	0.0001			
3.0×10^{0}	0.0001				

Table G.10 $W_c(u,z/m)$

			z/m				
$1/u$	**0**	**0.10**	**0.20**	**0.30**	**0.40**	**0.50**	**0.60**
0.01							0
0.02						0	0.04
0.03					0	0.04	0.10
0.04	0	0	0	0	0.03	0.07	0.15
0.06	0.01	0.01	0.02	0.03	0.07	0.13	0.23
0.08	0.03	0.03	0.04	0.08	0.13	0.20	0.29
0.10	0.06	0.06	0.07	0.12	0.17	0.26	0.36
0.15	0.15	0.15	0.17	0.21	0.27	0.35	0.44
0.20	0.22	0.24	0.26	0.30	0.36	0.44	0.52
0.30	0.38	0.40	0.42	0.45	0.49	0.55	0.62

Table G.10 W$_c$(u,z/m) (continued)

1/u	0	0.10	0.20	0.30	0.40	0.50	0.60
0.40	0.52	0.54	0.55	0.58	0.61	0.66	0.71
0.60	0.71	0.71	0.72	0.73	0.76	0.79	0.82
0.80	0.82	0.82	0.82	0.83	0.84	0.86	0.88
1.00	0.88	0.88	0.88	0.88	0.89	0.90	0.92
1.50	0.96	0.96	0.96	0.96	0.96	0.96	0.97

z/m

1/u	0.70	0.80	0.85	0.90	0.95	1.00
0.005	0	0.05	0.17	0.34	0.60	1.00
0.01	0.03	0.17	0.33	0.50	0.73	1.00
0.02	0.13	0.30	0.44	0.60	0.78	1.00
0.03	0.21	0.40	0.53	0.65	0.81	1.00
0.04	0.28	0.45	0.58	0.70	0.83	1.00
0.06	0.36	0.54	0.63	0.74	0.85	1.00
0.08	0.42	0.58	0.67	0.77	0.87	1.00
0.10	0.48	0.63	0.72	0.80	0.88	1.00
0.15	0.55	0.68	0.75	0.82	0.90	1.00
0.20	0.61	0.72	0.78	0.85	0.91	1.00
0.30	0.70	0.78	0.83	0.87	0.93	1.00
0.40	0.78	0.84	0.88	0.92	0.96	1.00
0.60	0.86	0.90	0.92	0.94	0.97	1.00
0.80	0.90	0.93	0.94	0.96	0.97	1.00
1.00	0.94	0.96	0.96	0.97	0.98	1.00
1.50	0.97	0.98	0.98	0.99	0.99	1.00

(based on graph presented by Domenico and Mifflin, 1965, p. 568)

Table G.11 W$_c$(u) (continued)

1/u	W$_c$(u)	1/u	W$_c$(u)
0.05	0.17	0.55	0.80
0.10	0.38	0.60	0.82
0.15	0.42	0.65	0.84

Table G.11 $W_c(u)$ (continued)

1/u	$W_c(u)$	1/u	$W_c(u)$
0.20	0.52	0.70	0.86
0.25	0.58	0.75	0.88
0.30	0.62	0.80	0.90
0.35	0.67	0.85	0.91
0.40	0.72	0.90	0.92
0.45	0.76	0.95	0.93
0.50	0.78	1.00	0.93

(based on graph presented by Domenico and Mifflin, 1965, p. 574)

Table G.12 $W(u_A, u_B, \beta, \sigma)$

	β			
$1/u_A$	1.0×10^{-3}	1.0×10^{-2}	6.0×10^{-2}	2.0×10^{-1}
4.0×10^{-1}	2.48×10^{-2}	2.41×10^{-2}	2.30×10^{-2}	2.14×10^{-2}
8.0×10^{-1}	1.45×10^{-1}	1.40×10^{-1}	1.31×10^{-1}	1.19×10^{-1}
1.4×10^{0}	3.58×10^{-1}	3.45×10^{-1}	3.18×10^{-1}	2.79×10^{-1}
2.4×10^{0}	6.62×10^{-1}	6.33×10^{-1}	5.70×10^{-1}	4.83×10^{-1}
4.0×10^{0}	1.02×10^{0}	9.63×10^{-1}	8.49×10^{-1}	6.88×10^{-1}
8.0×10^{0}	1.57×10^{0}	1.46×10^{0}	1.23×10^{0}	9.18×10^{-1}
1.4×10^{1}	2.05×10^{0}	1.88×10^{0}	1.51×10^{0}	1.03×10^{0}
2.4×10^{1}	2.52×10^{0}	2.27×10^{0}	1.73×10^{0}	1.07×10^{0}
4.0×10^{1}	2.97×10^{0}	2.61×10^{0}	1.85×10^{0}	1.08×10^{0}
8.0×10^{1}	3.56×10^{0}	3.00×10^{0}	1.92×10^{0}	1.08×10^{0}
1.4×10^{2}	4.01×10^{0}	3.23×10^{0}	1.93×10^{0}	1.08×10^{0}
2.4×10^{2}	4.42×10^{0}	3.37×10^{0}	1.94×10^{0}	1.08×10^{0}
4.0×10^{2}	4.77×10^{0}	3.43×10^{0}	1.94×10^{0}	1.08×10^{0}
8.0×10^{2}	5.16×10^{0}	3.45×10^{0}	1.94×10^{0}	1.08×10^{0}
1.4×10^{3}	5.40×10^{0}	3.46×10^{0}	1.94×10^{0}	1.08×10^{0}
2.4×10^{3}	5.54×10^{0}	3.46×10^{0}	1.94×10^{0}	1.08×10^{0}
4.0×10^{3}	5.59×10^{0}	3.46×10^{0}	1.94×10^{0}	1.08×10^{0}
8.0×10^{3}	5.62×10^{0}	3.46×10^{0}	1.94×10^{0}	1.08×10^{0}
1.4×10^{4}	5.62×10^{0}	3.46×10^{0}	1.94×10^{0}	1.08×10^{0}

Table G.12 W(u_A,u_B, β, σ) (continued)

	β			
$1/u_A$	6.0×10^{-1}	1.0×10^0	2.0×10^0	4.0×10^0
4.0×10^{-1}	1.88×10^{-2}	1.70×10^{-2}	1.38×10^{-2}	9.33×10^{-3}
8.0×10^{-1}	9.88×10^{-2}	8.49×10^{-2}	6.03×10^{-2}	3.17×10^{-2}
1.4×10^0	2.17×10^{-1}	1.75×10^{-1}	1.07×10^{-1}	4.45×10^{-2}
2.4×10^0	3.43×10^{-1}	2.56×10^{-1}	1.33×10^{-2}	4.76×10^{-2}
4.0×10^0	4.38×10^{-1}	3.00×10^{-1}	1.40×10^{-1}	4.78×10^{-2}
8.0×10^0	4.97×10^{-1}	3.17×10^{-1}	1.41×10^{-1}	4.78×10^{-2}
1.4×10^1	5.07×10^{-1}	3.17×10^{-1}	1.41×10^{-1}	4.78×10^{-2}
2.4×10^1	5.07×10^{-1}	3.17×10^{-1}	1.41×10^{-1}	4.78×10^{-2}

Values obtained by setting σ = 10^{-9}

	β			
$1/u_B$	1.0×10^{-3}	1.0×10^{-2}	6.0×10^{-2}	2.0×10^{-2}
4.0×10^{-2}	5.62×10^0	3.46×10^0	1.94×10^0	1.09×10^0
8.0×10^{-2}	5.62×10^0	3.46×10^0	1.94×10^0	1.09×10^0
1.4×10^{-1}	5.62×10^0	3.46×10^0	1.94×10^0	1.10×10^0
2.4×10^{-1}	5.62×10^0	3.46×10^0	1.95×10^0	1.11×10^0
4.0×10^{-1}	5.62×10^0	3.46×10^0	1.96×10^0	1.13×10^0
8.0×10^{-1}	5.62×10^0	3.46×10^0	1.98×10^0	1.18×10^0
1.4×10^0	5.63×10^0	3.47×10^0	2.01×10^0	1.24×10^0
2.4×10^0	5.63×10^0	3.49×10^0	2.06×10^0	1.35×10^0
4.0×10^0	5.63×10^0	3.51×10^0	2.13×10^0	1.50×10^0
8.0×10^0	5.64×10^0	3.56×10^0	2.31×10^0	1.85×10^0
1.4×10^1	5.65×10^0	3.63×10^0	2.55×10^0	2.23×10^0
2.4×10^1	5.67×10^0	3.74×10^0	2.86×10^0	2.68×10^0
4.0×10^1	5.70×10^0	3.90×10^0	3.24×10^0	3.15×10^0
8.0×10^1	5.76×10^0	4.22×10^0	3.85×10^0	3.82×10^0
1.4×10^2	5.85×10^0	4.58×10^0	4.38×10^0	4.37×10^0
2.4×10^2	5.99×10^0	5.00×10^0	4.91×10^0	4.91×10^0
4.0×10^2	6.16×10^0	5.46×10^0	5.42×10^0	5.42×10^0
8.0×10^2	6.47×10^0	6.11×10^0	6.11×10^0	6.11×10^0
1.4×10^3	6.67×10^0	6.67×10^0	6.67×10^0	6.67×10^0
2.4×10^3	7.21×10^0	7.21×10^0	7.21×10^0	7.21×10^0
4.0×10^3	7.72×10^0	7.72×10^0	7.72×10^0	7.72×10^0
8.0×10^3	8.41×10^0	8.41×10^0	8.41×10^0	8.41×10^0

Table G.12 $W(u_A, u_B, \beta, \sigma)$ (continued)

β

1.4×10^4	8.97×10^0	8.97×10^0	8.97×10^0	8.97×10^0
2.4×10^4	9.51×10^0	9.51×10^0	9.51×10^0	9.51×10^0
4.0×10^4	1.94×10^1	1.94×10^1	1.94×10^1	1.94×10^1
$1/u_B$	6.0×10^{-1}	1.0×10^0	2.0×10^0	4.0×10^0
4.0×10^{-4}	5.08×10^{-1}	3.18×10^{-1}	1.42×10^{-1}	4.79×10^{-2}
8.0×10^{-4}	5.08×10^{-1}	3.18×10^{-1}	1.42×10^{-1}	4.80×10^{-2}
1.4×10^{-3}	5.08×10^{-1}	3.18×10^{-1}	1.42×10^{-1}	4.81×10^{-2}
2.4×10^{-3}	5.08×10^{-1}	3.18×10^{-1}	1.42×10^{-1}	4.84×10^{-2}
4.0×10^{-3}	5.08×10^{-1}	3.18×10^{-1}	1.42×10^{-1}	4.88×10^{-2}
8.0×10^{-3}	5.09×10^{-1}	3.19×10^{-1}	1.43×10^{-1}	4.96×10^{-2}
1.4×10^{-2}	5.10×10^{-1}	3.21×10^{-1}	1.45×10^{-1}	5.09×10^{-2}
2.4×10^{-2}	5.12×10^{-1}	3.23×10^{-1}	1.47×10^{-1}	5.32×10^{-2}
4.0×10^{-2}	5.16×10^{-1}	3.27×10^{-1}	1.52×10^{-1}	5.68×10^{-2}
8.0×10^{-2}	5.24×10^{-1}	3.37×10^{-1}	1.62×10^{-1}	6.61×10^{-2}
1.4×10^{-1}	5.37×10^{-1}	3.50×10^{-1}	1.78×10^{-1}	8.06×10^{-2}
2.4×10^{-1}	5.57×10^{-1}	3.74×10^{-1}	2.05×10^{-1}	1.06×10^{-1}
4.0×10^{-1}	5.89×10^{-1}	4.12×10^{-1}	2.48×10^{-1}	1.49×10^{-1}
8.0×10^{-1}	6.67×10^{-1}	5.06×10^{-1}	3.57×10^{-1}	2.66×10^{-1}
1.4×10^0	7.80×10^{-1}	6.42×10^{-1}	5.17×10^{-1}	4.45×10^{-1}
2.4×10^0	9.54×10^{-1}	8.50×10^{-1}	7.63×10^{-1}	7.18×10^{-1}
4.0×10^0	1.20×10^0	1.13×10^0	1.08×10^0	1.06×10^0
8.0×10^0	1.68×10^0	1.65×10^0	1.63×10^0	1.63×10^0
1.4×10^1	2.15×10^0	2.14×10^0	2.14×10^0	2.14×10^0
2.4×10^1	2.65×10^0	2.65×10^0	2.64×10^0	2.64×10^0
4.0×10^1	3.14×10^0	3.14×10^0	3.14×10^0	3.14×10^0
8.0×10^1	3.82×10^0	3.82×10^0	3.82×10^0	3.82×10^0
1.4×10^2	4.37×10^0	4.37×10^0	4.37×10^0	4.37×10^0
2.4×10^2	4.91×10^0	4.91×10^0	4.91×10^0	4.91×10^0
4.0×10^2	5.42×10^0	5.42×10^0	5.42×10^0	5.42×10^0
8.0×10^2	6.11×10^0	6.11×10^0	6.11×10^0	6.11×10^0
1.4×10^3	6.67×10^0	6.67×10^0	6.67×10^0	6.67×10^0
2.4×10^3	7.21×10^0	7.21×10^0	7.21×10^0	7.21×10^0
4.0×10^3	7.72×10^0	7.72×10^0	7.72×10^0	7.72×10^0
8.0×10^3	8.41×10^0	8.41×10^0	8.41×10^0	8.41×10^0

1.4×10^4	8.97×10^0	8.97×10^0	8.97×10^0	8.97×10^0
2.4×10^4	9.51×10^0	9.51×10^0	9.51×10^0	9.51×10^0
4.0×10^4	1.94×10^1	1.94×10^1	1.94×10^1	1.94×10^1

Table G.13 $W(u_w, c)$

	c			
$1/u_w$	**1**	**.1**	**.01**	**.001**
7.692×10^3	8.1515	6.5479	4.3329	2.0393
5.917	7.8891	6.2856	4.0706	1.7770
3.501	7.3645	5.7610	3.5460	1.2524
1.226	6.3156	4.7121	2.4971	0.2034
9.430×10^2	6.0535	4.4499	2.2349	
5.580	5.5295	3.9259	1.7109	
3.302	5.0060	3.4025	1.1874	
1.156	3.9621	2.3586	0.1436	
8.893×10^1	3.7024	2.0988		
5.262	3.1853	1.5818		
3.114	2.6736	1.0700		
1.090	1.6819	0.0784		
8.386×10^0	1.4457			
4.962	0.9970			
2.259	0.4163			
1.028	0.0103			

(after Moench and Prickett, 1972, p. 497)

Table G.14 $W(u_i, \sigma_i)$

	ρ_i				
u_i	**10**	**5**	**2**	**1**	**0.5**
1.00×10^{-6}	0.9923	0.9948	0.9968	0.9977	0.9984
2.00×10^{-6}	0.9823	0.9927	0.9955	0.9968	0.9977
4.00×10^{-6}	0.9853	0.9898	0.9936	0.9955	0.9968

Table G.14 $W(u_i, \sigma_i)$ (continued)

u_i	ρ_i				
	10	5	2	1	0.5
6.00×10^{-6}	0.9822	0.9876	0.9922	0.9945	0.9961
8.00×10^{-6}	0.9796	0.9857	0.9910	0.9936	0.9955
1.00×10^{-5}	0.9773	0.9841	0.9900	0.9929	0.9949
2.00×10^{-5}	0.9683	0.9776	0.9858	0.9900	0.9929
4.00×10^{-5}	0.9558	0.9687	0.9801	0.9858	0.9899
6.00×10^{-5}	0.9464	0.9619	0.9757	0.9827	0.9877
8.00×10^{-5}	0.9387	0.9562	0.9720	0.9800	0.9858
1.00×10^{-4}	0.9318	0.9512	0.9688	0.9777	0.9841
2.00×10^{-4}	0.9059	0.9321	0.9562	0.9687	0.9776
4.00×10^{-4}	0.8711	0.9061	0.9389	0.9560	0.9685
6.00×10^{-4}	0.8458	0.8869	0.9258	0.9465	0.9615
8.00×10^{-4}	0.8253	0.8711	0.9151	0.9385	0.9557
1.00×10^{-3}	0.8079	0.8576	0.9057	0.9315	0.9505
2.00×10^{-3}	0.7450	0.8075	0.8702	0.9048	0.9307
4.00×10^{-3}	0.6684	0.7439	0.8232	0.8686	0.9031
6.00×10^{-3}	0.6178	0.7001	0.7896	0.8419	0.8825
8.00×10^{-3}	0.5797	0.6662	0.7626	0.8202	0.8654
1.00×10^{-2}	0.5492	0.6384	0.7400	0.8017	0.8505
2.00×10^{-2}	0.4517	0.5450	0.6595	0.7336	0.7947
4.00×10^{-2}	0.3556	0.4454	0.5654	0.6489	0.7214
6.00×10^{-2}	0.3030	0.3872	0.5055	0.5919	0.6697
8.00×10^{-2}	0.2682	0.3469	0.4618	0.5486	0.6289
1.00×10^{-1}	0.2428	0.3168	0.4276	0.5137	0.5951
2.00×10^{-1}	0.1740	0.2313	0.3234	0.4010	0.4799
4.00×10^{-1}	0.1207	0.1612	0.2292	0.2902	0.3566
6.00×10^{-1}	0.09616	0.1280	0.1817	0.2311	0.2864
8.00×10^{-1}	0.08134	0.1077	0.1521	0.1931	0.2397
1.00×10^{0}	0.17120	0.09375	0.1315	0.1663	0.2061
2.00×10^{0}	0.04620	0.05940	0.08044	0.09912	0.1202
4.00×10^{0}	0.02908	0.03621	0.04668	0.05521	0.06420
6.00×10^{0}	0.02185	0.02663	0.03326	0.03830	0.04331
8.00×10^{0}	0.01771	0.02125	0.02594	0.02933	0.03254
1.00×10^{1}	0.01499	0.01776	0.02130	0.02376	0.02600
2.00×10^{1}	0.00872	0.00994	0.01133	0.01219	0.01288
4.00×10^{1}	0.00490	0.00540	0.00617	0.00637	0.00657
6.00×10^{1}	0.00345	0.00373	0.00400	0.00413	0.00423
8.00×10^{1}	0.00267	0.99285	0.00302	0.00311	0.00316

Table G.14 $W(u_i, \sigma_i)$ (continued)

1.00×10^2	0.00218	0.00231	0.00249	0.00253	0.00256
2.00×10^2	0.00115	0.00120	0.00125	0.00126	0.00127
4.00×10^2	0.00060	0.00061	0.00062	0.00063	0.00063
6.00×10^2	0.00040	0.00041	0.00042	0.00042	0.00042
8.00×10^2	0.00030	0.00031	0.00031	0.00031	0.00031
1.00×10^2	0.00024	0.00025	0.00025	0.00025	0.00025

			ρ_i		
u_i	10^{-1}	10^{-2}	10^{-3}	10^{-4}	10^{-5}
1.00×10^{-3}	0.9771	0.9920	0.9969	0.9985	0.9992
2.15×10^{-3}	0.9658	0.9876	0.9949	0.9974	0.9985
4.64×10^{-3}	0.9490	0.9807	0.9914	0.9954	0.9970
1.00×10^{-2}	0.9238	0.9693	0.9853	0.9915	0.9942
2.15×10^{-2}	0.8860	0.9505	0.9744	0.9841	0.9883
4.64×10^{-2}	0.8293	0.9187	0.9545	0.9701	0.9781
1.00×10^{-1}	0.7460	0.8655	0.9183	0.9434	0.9572
2.15×10^{-1}	0.6289	0.7782	0.8538	0.8935	0.9167
4.64×10^{-1}	0.4782	0.6436	0.7436	0.8031	0.8410
1.00×10^0	0.3117	0.4598	0.5729	0.6520	0.7080
2.15×10^0	0.1665	0.2597	0.3543	0.4364	0.5038
4.64×10^0	0.07415	0.1086	0.1554	0.2082	0.2620
7.00×10^0	0.04625	0.06204	0.08519	0.1161	0.1521
1.00×10^1	0.03065	0.03780	0.04821	0.06355	0.08378
1.40×10^1	0.02092	0.02414	0.02844	0.03492	0.04426
2.15×10^1	0.01297	0.01414	0.01545	0.01723	0.01999
3.00×10^1	0.009070	0.009615	0.01016	0.01083	0.01169
4.64×10^1	0.005711	0.004919	0.006111	0.006319	0.006554
7.00×10^1	0.003722	0.003809	0.003884	0.003962	0.004046
1.00×10^2	0.002577	0.002618	0.002653	0.002688	0.002725
2.15×10^2	0.001179	0.001187	0.001194	0.001201	0.001208

			ρ_i		
u_i	10^{-6}	10^{-7}	10^{-8}	10^{-9}	10^{-10}
1.00×10^{-3}	0.9994	0.9996	0.9996	0.9997	0.9997
2.00×10^{-3}	0.9989	0.9992	0.9993	0.9994	0.9995
4.00×10^{-3}	0.9980	0.9985	0.9987	0.9989	0.9991
6.00×10^{-3}	0.9972	0.9978	0.9982	0.9984	0.9986
8.00×10^{-3}	0.9964	0.9971	0.9976	0.9980	0.9982
1.00×10^{-2}	0.9956	0.9965	0.9971	0.9975	0.9978
2.00×10^{-2}	0.9919	0.9934	0.9944	0.9952	0.9958
4.00×10^{-2}	0.9848	0.9875	0.9894	0.9908	0.9919

Table G.14 $W(u_i, \rho_i)$ (continued)

u_i	ρ_i				
	10^{-6}	10^{-7}	10^{-8}	10^{-9}	10^{-10}
6.00×10^{-2}	0.9782	0.9819	0.9846	0.9866	0.9881
8.00×10^{-2}	0.9718	0.9765	0.9799	0.9824	0.9844
1.00×10^{-1}	0.9655	0.9712	0.9753	0.9784	0.9807
2.00×10^{-1}	0.9361	0.9459	0.9532	0.9587	0.9631
4.00×10^{-1}	0.8828	0.8995	0.9122	0.9220	0.9298
6.00×10^{-1}	0.8345	0.8569	0.8741	0.8875	0.8964
8.00×10^{-1}	0.7901	0.8173	0.8383	0.8550	0.8686
1.00×10^{0}	0.7489	0.7801	0.8045	0.8240	0.8401
2.00×10^{0}	0.5800	0.6235	0.6591	0.6889	0.7139
3.00×10^{0}	0.4554	0.5033	0.5442	0.5792	0.6096
4.00×10^{0}	0.3613	0.4093	0.4517	0.4891	0.5222
5.00×10^{0}	0.2893	0.3351	0.3768	0.4146	0.4487
6.00×10^{0}	0.2337	0.2759	0.3157	0.3525	0.3865
7.00×10^{0}	0.1903	0.2285	0.2655	0.3007	0.3337
8.00×10^{0}	0.1562	0.1903	0.2243	0.2573	0.2888
9.00×10^{0}	0.1292	0.1594	0.1902	0.2208	0.2505
1.00×10^{1}	0.1078	0.1343	0.1620	0.1900	0.2178
2.00×10^{1}	0.02720	0.03343	0.04129	0.05071	0.06149
3.00×10^{1}	0.01286	0.01448	0.01667	0.01956	0.02320
4.00×10^{1}	0.008337	0.008898	0.009637	0.01062	0.01190
5.00×10^{1}	0.006209	0.006470	0.006789	0.007192	0.007709
6.00×10^{1}	0.004961	0.005111	0.005283	0.005487	0.005735
8.00×10^{1}	0.003547	0.003617	0.003691	0.003773	0.003863
1.00×10^{2}	0.002763	0.002803	0.002845	0.002890	0.002938
2.00×10^{2}	0.001313	0.001322	0.001330	0.001339	0.001348

(after Reed, 1980, p. 47; Bredehoeft and Papadopulos, 1980, p. 235)

Table G.15 s_c/s

u_c	u				
	2.5×10^{-1}	2.5×10^{-2}	2.5×10^{-3}	2.5×10^{-4}	2.5×10^{-5}
2.0×10^{-4}	0.9710	0.9780	0.9800	0.9810	0.9820
4.0×10^{-4}	0.9580	0.9680	0.9720	0.9730	0.9740

Table G.15 s_c/s

			u		
u_c	2.5×10^{-1}	2.5×10^{-2}	2.5×10^{-3}	2.5×10^{-4}	2.5×10^{-5}
6.0×10^{-4}	0.9490	0.9610	0.9660	0.9680	0.9690
8.0×10^{-4}	0.9420	0.9550	0.9600	0.9630	0.9640
2.0×10^{-3}	0.9090	0.9300	0.9370	0.9410	0.9430
4.0×10^{-3}	0.8730	0.9010	0.9120	0.9170	0.9200
6.0×10^{-3}	0.8460	0.8800	0.8930	0.8990	0.9020
8.0×10^{-3}	0.8240	0.8620	0.8770	0.8830	0.8870
2.0×10^{-2}	0.7320	0.7870	0.8080	0.8180	0.8230
4.0×10^{-2}	0.6370	0.7070	0.7340	0.7470	0.7540
6.0×10^{-2}	0.5710	0.6490	0.6800	0.6940	0.7020
8.0×10^{-2}	0.5200	0.6030	0.6360	0.6510	0.6600
2.0×10^{-1}	0.3360	0.4260	0.4640	0.4830	0.4930
4.0×10^{-1}	0.1950	0.2750	0.3110	0.3280	0.3380
6.0×10^{-1}	0.1230	0.1890	0.2200	0.2360	0.2440
8.0×10^{-1}	0.0818	0.1350	0.1610	0.1740	0.1810
2.0×10^{0}	0.0103	0.0233	0.0311	0.0352	0.0376
4.0×10^{0}	0.0001	0.0018	0.0028	0.0033	0.0036
6.0×10^{0}		0.0002	0.0003	0.0004	0.0004

(after Witherspoon and Neuman, 1972; p. 267)

Table G.16 $W_h(a,b)$

			b			
a	0.1	0.2	0.4	0.6	0.8	1.0
0	1.00	1.00	1.00	1.00	1.00	1.00
5.0×10^{-2}	0.83	0.93	0.95	0.97	0.97	0.98
1.0×10^{-1}		0.80	0.90	0.93	0.94	0.95
1.5×10^{-1}		0.60	0.84	0.88	0.91	0.92
2.0×10^{-1}			0.75	0.82	0.87	0.89
2.5×10^{-1}			0.61	0.75	0.83	0.85
3.0×10^{-1}			0.42	0.68	0.78	0.80

Table G.16 $W_h(a,b)$ (continued)

a				b			
4.0×10^{-1}					0.50	0.67	0.71
5.0×10^{-1}					0.22	0.52	0.60
6.0×10^{-1}						0.32	0.49
7.0×10^{-1}						0.10	0.36
8.0×10^{-1}							0.20
9.0×10^{-1}							0.05

a	1.5	2.0	3.0	4.0	6.0	10.0
0	1.00	1.00	1.00	1.00	1.00	1.00
5.0×10^{-2}	0.98	0.98	0.99	0.99	0.99	0.99
1.0×10^{-1}	0.96	0.96	0.97	0.97	0.98	0.98
1.5×10^{-1}	0.93	0.94	0.95	0.96	0.97	0.97
2.0×10^{-1}	0.90	0.92	0.93	0.94	0.95	0.96
2.5×10^{-1}	0.88	0.90	0.91	0.93	0.94	0.95
3.0×10^{-1}	0.84	0.87	0.89	0.91	0.92	0.94
4.0×10^{-1}	0.78	0.82	0.86	0.88	0.90	0.93
5.0×10^{-1}	0.70	0.78	0.82	0.86	0.88	0.91
6.0×10^{-1}	0.63	0.72	0.78	0.82	0.85	0.89
7.0×10^{-1}	0.56	0.67	0.75	0.79	0.82	0.87
8.0×10^{-1}	0.48	0.60	0.71	0.76	0.80	0.85
9.0×10^{-1}	0.40	0.54	0.66	0.72	0.77	0.83
1.0×10^{0}	0.30	0.48	0.62	0.68	0.74	0.80
1.1×10^{0}	0.20	0.40	0.57	0.64	0.72	0.78
1.2×10^{0}	0.11	0.34	0.52	0.61	0.69	0.76
1.3×10^{0}	0.04	0.27	0.48	0.57	0.66	0.75
1.4×10^{0}		0.20	0.43	0.53	0.63	0.72
1.5×10^{0}		0.13	0.38	0.49	0.61	0.70
1.6×10^{0}		0.06	0.33	0.45	0.59	0.68
1.7×10^{0}		0.02	0.28	0.42	0.56	0.66
1.8×10^{0}			0.23	0.38	0.53	0.64
1.9×10^{0}			0.18	0.34	0.50	0.62
2.0×10^{0}			0.15	0.30	0.47	0.60

(based on graph presented by Spillette, 1972, p. 22)

Table G.17 $W_h(\alpha, \lambda)$

α	0.3	1	λ 3	6	15	30
1.5			0.04	0.07	0.15	0.20
2×10^0		0.03	0.10	0.15	0.22	0.28
4×10^0	0.02	0.12	0.22	0.30	0.36	0.41
6×10^0	0.06	0.18	0.28	0.36	0.42	0.47
1×10^1	0.11	0.24	0.35	0.42	0.48	0.53
2×10^1	0.19	0.32	0.42	0.50	0.55	0.60
4×10^1	0.26	0.40	0.50	0.56	0.61	0.65
6×10^1	0.32	0.45	0.54	0.60	0.64	0.68
1×10^2	0.35	0.49	0.58	0.63	0.68	0.71
2×10^2	0.44	0.55	0.62	0.67	0.72	0.75
4×10^2	0.50	0.60	0.66	0.71	0.75	0.78
6×10^2	0.54	0.63	0.69	0.73	0.76	0.79
1×10^3	0.58	0.66	0.72	0.75	0.78	0.81
2×10^3	0.63	0.70	0.75	0.78	0.81	0.85
4×10^3	0.68	0.74	0.78	0.81	0.83	0.85
6×10^3	0.70	0.75	0.79	0.82	0.84	0.86
1×10^4	0.72	0.77	0.81	0.83	0.85	0.87

α	60	150	300	1×10^3	1×10^4	∞
1.5	0.23	0.24	0.26	0.28	0.30	0.30
2×10^0	0.31	0.33	0.35	0.37	0.40	0.40
4×10^0	0.44	0.47	0.49	0.51	0.53	0.53
6×10^0	0.50	0.53	0.56	0.58	0.59	0.59
1×10^1	0.56	0.60	0.62	0.64	0.66	0.66
2×10^1	0.63	0.66	0.68	0.70	0.71	0.72
4×10^1	0.62	0.71	0.74	0.75	0.77	0.78
6×10^1	0.70	0.74	0.76	0.78	0.79	0.81
1×10^2	0.73	0.76	0.79	0.81	0.82	0.84
2×10^2	0.77	0.80	0.82	0.84	0.85	0.87
4×10^2	0.80	0.83	0.85	0.86	0.88	0.90
6×10^2	0.82	0.84	0.86	0.88	0.90	0.92
1×10^3	0.83	0.86	0.87	0.89	0.91	0.93
2×10^3	0.85	0.87	0.89	0.91	0.93	0.95
4×10^3	0.87	0.88	0.90	0.92	0.94	0.96
6×10^3	0.88	0.89	0.91	0.93	0.95	0.97
1×10^4	0.89	0.90	0.92	0.94	0.96	0.97

Table G.18 W(f,0)

f	W(f,0)	f	W(f,0)
0.0	0.0	5.6	1.5690
0.2	0.5471	5.8	1.5694
0.4	0.8256	6.0	1.5696
0.6	1.0119	6.2	1.5699
0.8	1.1447	6.4	1.5700
1.0	1.2425	6.6	1.5702
1.2	1.3457	6.8	1.5703
1.4	1.3718	7.0	1.5704
1.6	1.4146	7.2	1.5705
1.8	1.4478	7.4	1.5705
2.0	1.4737	7.6	1.5706
2.2	1.4939	7.8	1.5706
2.4	1.5098	8.0	1.5707
2.6	1.5223	8.2	1.57068
2.8	1.5321	8.4	1.57071
3.0	1.5400	8.6	1.57072
3.2	1.5462	8.8	1.57074
3.4	1.5511	9.0	1.57075
3.6	1.5550	9.2	1.57076
3.8	1.5582	9.4	1.57076
4.0	1.5606	9.6	1.57077
4.2	1.5626	9.8	1.57078
4.4	1.5642	10.0	1.57078
4.6	1.5655	10.2	1.57078
4.8	1.5666	10.4	1.57078
5.0	1.5674	10.6	1.57078
5.2	1.5681	10.8	1.57079
5.4	1.5686	11.0	1.57079

(after Hantush, 1964, p. 319)

Table G.19   D(u) (continued)

u	u^2	D(u)
0.03162	0.0010	0.9643
0.04000	0.0016	0.9549
0.05000	0.0025	0.9436
0.06325	0.0040	0.9287
0.07746	0.0060	0.9128
0.08944	0.0080	0.8994
0.1000	0.010	0.8875
0.1265	0.016	0.8580
0.1581	0.025	0.8231
0.2000	0.040	0.7730
0.2449	0.060	0.7291
0.2828	0.080	0.6892
0.3162	0.10	0.6548
0.4000	0.16	0.5716
0.5000	0.25	0.4795
0.6325	0.40	0.3711
0.7746	0.60	0.2733
0.8944	0.80	0.2059
1.000	1.00	0.1573
1.140	1.30	0.1069
1.265	1.60	0.0736
1.378	1.90	0.0513
1.483	2.20	0.0359
1.581	2.50	0.0254
1.643	2.70	0.0202
1.732	3.00	0.0143
1.789	3.20	0.0114

(after Ferris, et al., 1962, p. 127)

Table G.20 $K_0(x)$ and $I_0(x)$

x	$K_0(x)$	$I_0(x)$	x	$K_0(x)$	$I_0(x)$
0.00		1.0000	0.70	0.6605	1.1263

Table G.20 $K_0(x)$ and $I_0(x)$ (continued)

x	$K_0(x)$	$I_0(x)$	x	$K_0(x)$	$I_0(x)$
0.01	4.7212	1.0000	0.71	0.6501	1.1301
0.02	4.0285	1.0001	0.72	0.6399	1.1339
0.03	3.6235	1.0002	0.73	0.6300	1.1377
0.04	3.3365	1.0004	0.74	0.6202	1.1417
0.05	3.1142	1.0006	0.75	0.6106	1.1457
0.06	2.9329	1.0009	0.76	0.6012	1.1497
0.07	2.7798	1.0012	0.77	0.5920	1.1538
0.08	2.6475	1.0016	0.78	0.5829	1.1580
0.09	2.5310	1.0020	0.79	0.5741	1.1622
0.10	2.4271	1.0025	0.80	0.5654	1.1665
0.11	2.3333	1.0030	0.81	0.5569	1.1709
0.12	2.2479	1.0036	0.82	0.5484	1.1753
0.13	2.1695	1.0042	0.83	0.5402	1.1798
0.14	2.0972	1.0049	0.84	0.5322	1.1843
0.15	2.0300	1.0056	0.85	0.5242	1.1890
0.16	1.9674	1.0064	0.86	0.5165	1.1936
0.17	1.9088	1.0072	0.87	0.5088	1.1984
0.18	1.8537	1.0081	0.88	0.5013	1.2032
0.19	1.8018	1.0091	0.89	0.4940	1.2081
0.20	1.7527	1.0100	0.90	0.4867	1.2130
0.21	1.7062	1.0111	0.91	0.4796	1.2180
0.22	1.6620	1.0121	0.92	0.4727	1.2231
0.23	1.6194	1.0133	0.93	0.4658	1.2282
0.24	1.5798	1.0145	0.94	0.4591	1.2334
0.25	1.5415	1.0157	0.95	0.4525	1.2387
0.26	1.5048	1.0170	0.96	0.4459	1.2440
0.27	1.4697	1.0183	0.97	0.4396	1.2496
0.28	1.4360	1.0197	0.98	0.4333	1.2549
0.29	1.4036	1.0211	0.99	0.4271	1.2605
0.30	1.3725	1.0226	1.00	0.4210	1.2661
0.31	1.3425	1.0242	1.10	0.3656	1.3262
0.32	1.3136	1.0258	1.20	0.3185	1.3937
0.33	1.2857	1.0274	1.30	0.2783	1.4693
0.34	1.2587	1.0291	1.40	0.2437	1.5534
0.35	1.2327	1.0309	1.50	0.2138	1.6467
0.36	1.2075	1.0327	1.60	0.1880	1.7500
0.37	1.1832	1.0345	1.70	0.1655	1.8640
0.38	1.1596	1.0364	1.80	0.1459	1.9896
0.39	1.1367	1.0384	1.90	0.1289	2.1277

Table G.20 $K_0(x)$ and $I_0(x)$ (continued)

x	$K_0(x)$	$I_0(x)$	x	$K_0(x)$	$I_0(x)$
0.40	1.1145	1.0404	2.00	0.1139	2.2796
0.41	1.0930	1.0425	2.10	0.1008	2.4463
0.42	1.0721	1.0446	2.20	0.0893	2.6291
0.43	1.0518	1.0468	2.30	0.0791	2.8296
0.44	1.0321	1.0490	2.40	0.0702	3.0493
0.45	1.0129	1.0513	2.50	0.0624	3.2898
0.46	0.9943	1.0536	2.60	0.0554	3.5258
0.47	0.9761	1.0560	2.70	0.0493	3.8417
0.48	0.9584	1.0584	2.80	0.0438	4.1573
0.49	0.9412	1.0609	2.90	0.0390	4.5028
0.50	0.9244	1.0635	3.00	0.0347	4.8808
0.51	0.9081	1.0661	3.10	0.0310	5.2945
0.52	0.8921	1.0688	3.20	0.0276	5.7472
0.53	0.8766	1.0715	3.30	0.0246	6.2426
0.54	0.8614	1.0742	3.40	0.0220	6.7848
0.55	0.8466	1.0771	3.50	0.0196	7.3782
0.56	0.8321	1.0800	3.60	0.0177	8.0277
0.57	0.8180	1.0829	3.70	0.0156	8.7386
0.58	0.8042	1.0859	3.80	0.0140	9.5169
0.59	0.7970	1.0889	3.90	0.0125	10.369
0.60	0.7775	1.0921	4.00	0.0112	11.302
0.61	0.7646	1.0952	5.00	0.0037	27.240
0.62	0.7520	1.0984	6.00	0.0012	67.234
0.63	0.7397	1.1017	7.00	0.0004	168.59
0.64	0.7277	1.1051	8.00	0.0001	427.56
0.65	0.7159	1.1085	9.00	0.0001	1093.6
0.66	0.7043	1.1119	10.0	0.0000	2815.7
0.67	0.6930	1.1154	11.0		7288.5
0.68	0.6820	1.1190	12.0		18949
0.69	0.6711	1.1226	13.0		49444

(after Huisman and Olsthoorn, 1983, pp. 306 and 307)

Table G.21 $K_1(x)$ and $I_1(x)$

x	$K_1(x)$	$I_1(x)$	x	$K_1(x)$	$I_1(x)$
0.01	99.974	0.0050	0.69	1.0722	0.3659
0.02	49.955	0.0100	0.70	1.0503	0.3719
0.03	33.271	0.0150	0.71	1.0290	0.3778
0.04	24.923	0.0200	0.72	1.0083	0.3838
0.05	19.910	0.0250	0.73	0.9882	0.3899
0.06	16.564	0.0300	0.74	0.9686	0.3959
0.07	14.171	0.0350	0.75	0.9496	0.4020
0.08	12.374	0.0400	0.76	0.9311	0.4081
0.09	10.977	0.0451	0.77	0.9131	0.4143
0.10	9.8538	0.0501	0.78	0.8955	0.4204
0.11	8.9353	0.0551	0.79	0.8784	0.4266
0.12	8.1688	0.0601	0.80	0.8618	0.4329
0.13	7.5192	0.0651	0.81	0.8456	0.4391
0.14	6.9615	0.0702	0.82	0.8298	0.4454
0.15	6.4775	0.0752	0.83	0.8144	0.4518
0.16	6.0533	0.0803	0.84	0.7993	0.4582
0.17	5.6784	0.0853	0.85	0.7847	0.4646
0.18	5.3447	0.0904	0.86	0.7704	0.4710
0.19	5.0456	0.0954	0.87	0.7564	0.4775
0.20	4.7760	0.1005	0.88	0.7428	0.4840
0.21	4.5317	0.1056	0.89	0.7295	0.4905
0.22	4.3092	0.1107	0.90	0.7165	0.4971
0.23	4.1058	0.1158	0.91	0.7039	0.5038
0.24	3.9191	0.1209	0.92	0.6915	0.5104
0.25	3.7470	0.1260	0.93	0.6794	0.5171
0.26	3.5880	0.1311	0.94	0.6675	0.5239
0.27	3.4405	0.1362	0.95	0.6560	0.5306
0.28	3.3034	0.1414	0.96	0.6447	0.5375
0.29	3.1755	0.1465	0.97	0.6336	0.5443
0.30	3.0560	0.1517	0.98	0.6228	0.5512
0.31	2.9441	0.1569	0.99	0.6123	0.5582
0.32	2.8390	0.1621	1.0	0.6019	0.5652
0.33	2.7402	0.1673	1.1	0.5098	0.6375
0.34	2.6470	0.1725	1.2	0.4346	0.7147
0.35	2.5591	0.1777	1.3	0.3726	0.7973
0.36	2.4760	0.1829	1.4	0.3208	0.8861
0.37	2.3973	0.1882	1.5	0.2774	0.9817
0.38	2.3227	0.1935	1.6	0.2446	1.0848
0.39	2.2518	0.1987	1.7	0.2094	1.1964
0.40	2.1844	0.2040	1.8	0.1826	1.3172

Table G.21 $K_1(x)$ and $I_1(x)$ (continued)

x	$K_1(x)$	$I_1(x)$	x	$K_1(x)$	$I_1(x)$
0.41	2.1202	0.2093	1.9	0.1597	1.4482
0.42	2.0590	0.2147	2.0	0.1399	1.5906
0.43	2.0006	0.2200	2.1	0.1228	1.7455
0.44	1.9449	0.2254	2.2	0.1079	1.8280
0.45	1.8915	0.2307	2.3	0.0950	2.0978
0.46	1.8405	0.2361	2.4	0.0837	2.2981
0.47	1.7916	0.2416	2.5	0.0739	2.5167
0.48	1.7447	0.2470	2.6	0.0653	2.7554
0.49	1.6997	0.2524	2.7	0.0577	3.0161
0.50	1.6564	0.2579	2.8	0.0511	3.3011
0.51	1.6149	0.2634	2.9	0.0453	3.6126
0.52	1.5749	0.2689	3.0	0.0402	3.9534
0.53	1.5365	0.2744	3.1	0.0356	4.3262
0.54	1.4994	0.2800	3.2	0.0316	4.7343
0.55	1.4637	0.2855	3.3	0.0281	5.1810
0.56	1.4292	0.2911	3.4	0.0250	5.6701
0.57	1.3960	0.2967	3.5	0.0222	6.2058
0.58	1.3639	0.3024	3.6	0.0198	6.7927
0.59	1.3328	0.3080	3.7	0.0176	7.4358
0.60	1.3028	0.3137	3.8	0.0157	8.1404
0.61	1.2738	0.3194	3.9	0.0140	8.9128
0.62	1.2458	0.3251	4.0	0.0125	9.7595
0.63	1.2186	0.3309	5.0	0.0040	24.336
0.64	1.1923	0.3367	6.0	0.0013	61.342
0.65	1.1668	0.3425	7.0	0.0005	156.04
0.66	1.1420	0.3483	8.0	0.0002	399.87
0.67	1.1181	0.3542	9.0	0.0001	1030.9
0.68	1.0948	0.3600	10	0.0000	2671.0

(after Huisman and Olsthoorn, 1983, pp. 306 and 307)

Table G.22 erf(x) and erfc(x)

x	erf(x)	erfc(x)
0	0	1.0
0.05	0.056372	0.943628
0.1	0.112463	0.887537
0.15	0.167996	0.832004
0.2	0.222703	0.777297
0.25	0.276326	0.723674
0.3	0.328627	0.671373
0.35	0.379382	0.620618
0.4	0.428392	0.571608
0.45	0.475482	0.524518
0.5	0.520500	0.479500
0.55	0.563323	0.436677
0.6	0.603856	0.396144
0.65	0.642029	0.357971
0.7	0.677801	0.322199
0.75	0.711156	0.288844
0.8	0.742101	0.257899
0.85	0.770668	0.229332
0.9	0.796908	0.203092
0.95	0.820891	0.179109
1.0	0.842701	0.157299
1.1	0.880205	0.119795
1.2	0.910314	0.089686
1.3	0.934008	0.065992
1.4	0.952285	0.047715
1.5	0.966105	0.033895
1.6	0.976348	0.023652
1.7	0.983790	0.016210
1.8	0.989091	0.010909
1.9	0.992790	0.007210
2.0	0.995322	0.004678
2.1	0.997021	0.002979
2.2	0.998137	0.001863
2.3	0.998857	0.001143
2.4	0.999311	0.000689
2.5	0.999593	0.000407
2.6	0.999764	0.000236
2.7	0.999866	0.000134

Table G.22 erf(x) and erfc(x)* (continued)

$$*\mathrm{erf}(x) = 1 - (2 / \pi^{0.5}) \sum_{n=0}^{\infty} \{[(-1)^n x^{(2n+1)})] / [n!(2n+1)]\}$$

$$*\mathrm{erfc}(x) = 1 - (2x / \pi^{0.5}) \sum_{n=0}^{\infty} \{(-x^2) / [n!(2n+1)]\}$$

(after Carslaw and Jaeger, 1959)

Table G.23 e^x And e^{-x} *

x	e^x	e^{-x}
0.010	1.010	0.990
0.020	1.020	0.980
0.030	1.030	0.970
0.040	1.041	0.961
0.050	1.051	0.951
0.060	1.062	0.942
0.070	1.072	0.932
0.080	1.083	0.923
0.090	1.094	0.914
0.10	1.105	0.905
0.20	1.221	0.819
0.30	1.350	0.741
0.40	1.492	0.670
0.50	1.649	0.606
0.60	1.822	0.549
0.70	2.014	0.497
0.80	2.225	0.449
0.90	2.460	0.407
1.0	2.718	0.368
1.5	4.482	0.223
2.0	7.389	0.135
2.5	1.218×10^1	8.21×10^{-2}
3.0	2.009×10^1	4.98×10^{-2}
3.5	3.312×10^1	3.02×10^{-2}

Table G.23 e^x And e^{-x} *

x	e^x	e^{-x}
4.0	5.460×10^1	1.83×10^{-2}
4.5	9.002×10^1	
5.0	1.484×10^2	

$$*\exp(x) = \sum_{n=0}^{\infty} x^n / n!$$

$x = \exp(x) = e^x$ $x = e^{\ln(n)}$ $\log_{10}(e) = 0.434294482$

$e = 2.71828183$ $\log(x) = \ln(x)\log(e)$

$\log_{10}(x) = \ln(x)\log_{10}(e)$ $\ln(x) = \log(x)\ln(10)$

$x = 10^{\log(x)}$ $\ln(10) = 2.30258509$

Appendix H
Contaminant Characteristic Values

Table H.1 Radionuclide Distribution Coefficients

Deposit	Radionuclide	Distribution Coefficient (ml/g)
Alluvium	Sr	48—2454
	Cs	121—1165
Tuff	Sr	260—4000
	Cs	1020—17800
Carbonate	Sr	0.19
	Cs	13.5
Dolomite	Sr	5—14
Granolierite	Sr	4—21
	Cs	8—1810
Granite	Sr	1.7
	Cs	34.3
Basalt	Sr	16—1220
	Cs	38—9520
Shaly siltstone	Sr	8.3
	Cs	309
Sandstone	Sr	1.4
	Cs	102
Salt	Sr	0.2
	Cs	0.03
Columbia River Basalt	Ru-106	38.0
	Pu	19.8
	Am-241	231
	Sr-85	120.4
	Pu-237	6.5
	Ra-226	731
	Co-60	315

Table H.1 Radionuclide Distribution Coefficients (continued)

Deposit	Radionuclide	Distribution Coefficient (ml/g)
	I-125	6.2
	Se-75	11.1
	Sr-90	5—38
	Cs-137	12—200
	U	<1
	Tc	<<1
	Ru-89	26—750
	Sr-90	5—25
	Pu-239	>1980

Deposit	Radionuclide	pH	Distribution Coefficient (ml/g)
Alluvium	U-238	2.0	100
		4.5	200
		5.75	1000
		7.0	2000
	Th-230	2.0	16
		4.5	5000
		5.75	10000
		7.0	15000
	Ra-226	2.0	0
		4.5	12
		5.75	60
		7.0	100
	Pb-210	2.0	20
		4.5	100
		5.75	1500
		7.0	4000
	Po-210	2.0	0
		4.5	0
		5.75	12
		7.0	25

(after Borg, et al., 1976; Haji-Djafari, 1981, p. 234)

Table H.2 Soil-Water Partition Coefficient And Organic Solubility

Compound	Solubility (mg/l)	K_{oc}
1,4-dioxane	miscible	1
4-hydroxy-4-methyl-2-pentanone	miscible	1
acetone	miscible	
tetrahydrofuran	miscible	1
N,N′-dimethylformamide		1
N,N′-dimethylacetamide		2
2-methyl-2-butanol	140000	6
2-butanol	125000	6
ethyl ether	84300	8
cyclohexanol	56700	10
3-methylbutanoic acid	42000	12
benzyl alcolhol	40000	12
aniline	34000	13
2-hexanone (butylmethylketone)	35000	14
2-hydroxy-triethylamine		15
2-methylphenol (o-cresol)	31000	15
2-methyl-2-propanol		16
4-methylphenol (p-cresol)	24000	17
pentanoic acid	24000	17
cyclohexanone	23000	18
4-methyl-2-pentanone	19000	20
2,4-dimethyl phenol	17000	21
4-methyl-2-pentanol	17000	21
methylene chloride	13200	25
isophorone	12000	26
phenol	82000	27
2-chlorophenol	11087	27
hexanoic acid	11000	28
chloroform	7840	34
1,2-dichloroethane	8450	36
1,2-trans-dichloroethane	6300	39
chloroethane	5700	42
5-methyl-2-hexanone	5400	43
chloromethane	5380	43
1,1-dichloroethane	5100	45
1,1,2-trichloroethane	4420	49
1,2-dichloropropane	3570	51
benzoic acid	2900	64
octanoic acid	2500	70

Table H.2 Soil-Water Partition Coefficient And Organic Solubility (continued)

Compound	Solubility (mg/l)	K_{oc}
heptanoic acid	2410	71
1,1,2,2-tetrachloroethane	3230	88
benzene	1780	97
diethyl phthalate	1000	123
2-nonanol	1000	123
bromodichloromethane	900	131
3-methylbenzoic acid	850	136
trichloroethene	1100	152
1,1,1-trichloroethane	700	155
di-n-butyl phthalate	400	217
1,1-dichloroethene	400	217
carbon tetrachloride	800	232
2-butanone (methylethylketone)	353	235
4-methylbenzoic acid	340	240
toluene	500	242
tetrachloroethylene	200	303
chlorobenzene	448	318
1,2-dichlorobenzene	148	343
o-xylene	170	363
1,2,2-trifluoro-1,1,2-trichloroethane		372
styrene	162	380
1,3-dichlorobenzene	118	463
fluorotrichloromethane	110	476
4,6-dinitro-2-methylphenol		477
p-xylene	156	552
m-xylene	146	588
1,4-dichlorobenzene	79	594
ethyl benzene	150	622
pentachlorophenol	14	900
N-nitrosodiphenylamine	35.1	982
3,5-dimethylphenol		1038
BHC-delta	31.5	1052
2,6-dimethylphenol		1060
1,2,4-trichlorobenzene	30	1080
naphthalene	31.7	1300
4-ethylphenol		1986
dibenzofuran	10	2140
hexachloroethane	8	2450

Table H.2 Soil-Water Partition Coefficient And Organic Solubility (continued)

Compound	Solubility (mg/l)	K_{oc}
acenaphthene	7.4	2580
tri-N-propylamine		2610
BHC-alpha	8.5	2627
BHC-beta	2.7	3619
hexachlorobenzene	0.035	3910
hexachlorobutadiene	3.2	4330
di-n-octyl phthalate	3	4510
butyl benzyl phthalate	2.9	4606
fluorene	1.98	5835
2-methylnaphthalene	25.4	8500
bis(2-ethylhexyl)phthalate	0.6	12200
toxaphene	0.4	15700
heptachlor epoxide	0.35	17087
endosulfan II	0.28	19623
fluoranthene	0.275	19800
1,2-diphenylhydrazene(as azobenzene)	0.252	20947
endosulfan sulfate	0.22	22788
phenanthrene	1.29	23000
dieldrin	0.188	25120
anthracene	0.073	26000
BHC-gamma	0.15	28900
decanoic acid		39610
chlorodane	0.056	53200
pyrene	0.135	63400
PCB-1254	0.042	63914
heptachlor	0.03	78400
endrin	0.024	90000
benzo(a)anthracene	0.014	125719
aldrin	0.013	132000
4,4'-DDE	0.01	155000
4,4'-DDT	0.0017	238000
4,4'-DDD	0.0017	238000
benzo(a)pyrene	0.0038	282185
PCB-1260	0.0027	349462
chrysene	0.022	420108
benzo(b)fluoranthene		1148497
benzo(k)fluoranthene		2020971

(after Roy and Griffin, 1985, pp. 241-247; Fetter, 1988, pp. 403-405).

Table H.3 Radionuclide Half-Life

Element	Radionuclide	Half-life (years)
Tritium	H-3	12.3
Carbon	C-14	5730
Iron	Fe-55	2.7
Cobalt	Co-60	5.3
Nickel	Ni-63	100
Fission products		
Krypton	Kr-85	10.8
Strontium	Sr-90	28
Zirconium	Zr-93	1000000
Technetium	Tc-99	210000
Ruthenium	Ru-106	1
Palladium	Pd-107	7000000
Antimony	Sb-125	2.7
Iodine	I-129	16000000
Cesium	Cs-134	2.1
	Cs-135	2000000
	Cs-137	30.2
Promethium	Pm-147	2.6
Samarium	Sm-151	90
Europium	Eu-154	16
Europium	Eu-155	1.8
Uranics		
Lead	Pb-210	22.3
Radon	Rn-222	0.01
Radium	Ra-226	1622
Thorium	Th-229	7340
Thorium	Th-230	75400
Uranium	U-234	247000
	U-235	710000000
	U-236	24000000
	U-238	45100000000

Table H.3 Radionuclide Half-Life (continued)

Element	Radionuclide	Half-life (years)
Transuranics		
Neptunium	Np-237	2100000
Plutonium	Pu-238	87
	Pu-239	24110
	Pu-240	6580
	Pu-241	14.4
	Pu-242	380000
Americium	Am-241	445
	Am-243	7650
Curium	Cm-244	18
	Cm-245	9300

(after Milnes, 1985, pp. 16-17)

Table H.4 Inorganic Solubility

Compound	Solubility in water (mg/l)
Antimony Sb	Antomonic acid and antimony oxides are very slightly soluble
Arsenic As	As_2O_5-1.5 × 10^6 at 16°C As_2O_3-3.7 × 10^6 at 20°C
Beryllium Be	BeO-0.2 at 30°C
Cadmium Cd	$CdCl_2$-1.40 × 10^6 at 20°C CdS-1.3 at 18°C $Cd(OH)_2$-2.6 at 25°C
Chromium Cr	CrO(as H_2CrO_4)-6.17× 10^5 at 0°C
Copper Cu	$CuCl_2$-7.06 × 10^5 at 0°C
Lead	PbO-17 at 20°C

Table H.4 Inorganic Solubility (continued)

Compound	Solubility in water (mg/l)
Pb	$PbCl_2$-9.9 $\times$ 10^3 at 20°C
Mercury	HgO-53 at 25°C
Hg	Hgs(alpha)-0.01 at 18°C
	HgS(beta) insoluble
	$HgCl_2$-6.9 $\times$ 10^4 at 20°C
Nickel	NiS-3.6 at 18°C
Ni	$NiCl_2$-6.42 $\times$ 10^5 at 20°C
Selenium	SeO_2-3.84 $\times$ 10_5 at 14°C
Se	SeO_3-very soluble
Silver	AgO-13 at 20°C
Ag	AgCl-0.89 at 10°C
Thallium	Tl_2S-2.0 $\times$ 10^2 at 20°C
Tl	TlCl-2.9 $\times$ 10^3 at 16°C
Zinc	ZmO-1.6 at 29°C
Zm	$ZmCl_2$-4.32 $\times$ 10^6 at 25°C

(after ORD, 1981)

Table H.5 Range Of Organic-Carbon Fraction In Selected Deposits

Deposit	Range of OC
Outwash sand and gravel	0.0001—0.01
Alluvial sand and gravel	0.0004—0.0073
Sandstone	0.0001—0.01
Limestone	0.0001—0.1
Granite	0.00001—0.0001
Lacustrine	0.01—0.4

(Oudijk and Mujica, 1989, p. 153)

Appendix I
Well And Mine Design Factor Values

Table I.1 Continuous Slot Screen Open Area

Screen Diameter (in)	Slot Size	Open Area (in^2/ft)
4	20	44
	30	58
	40	72
	50	78
	60	90
	80	102
	90	105
	95	106
	100	112
	120	99
	125	100
6	20	45
	30	61
	40	77
	50	88
	60	100
	90	124
	95	127
	100	131
	120	141
	125	127
8	20	58
	30	80
	40	98
	50	114
	60	135
	95	165

Table I.1 Continuous Slot Screen Open Area (continued)

Screen Diameter (in)	Slot Size	Open Area (in^2/ft)
	100	169
	125	166
10	20	72
	30	100
	40	122
	50	143
	60	135
	90	174
	95	179
	100	186
	120	203
	125	207
12	20	69
	30	77
	40	99
	50	117
	60	135
	90	176
	95	182
	100	189
	120	209
	125	214
16	20	68
	30	97
	40	124
	50	146
	60	169
	80	206
	90	221
	95	228
	100	238
	125	269
18	20	76
	30	109
	40	137
	60	187

(after Driscoll, 1986, pp. 948-949)

Table I.2 Optimum Screen Entrance Velocity

Hydraulic Conductivity (gpd/sq ft)	Optimum Screen Entrance Velocity (feet/min)
>2000	6
2000	6
1500	5
1000	4
500	3
250	2
<250	1

(after Walton, 1962, p. 29)

Table I.3 Optimum Well Diameter

Discharge Rate (gpm)	Optimum Diameter (inch)
<100	6
75—175	8
150—350	10
300—700	12
500—1000	14
800—1800	16
1200—3000	20
2000—3800	24
3000—6000	30

(after Driscoll, 1986, p. 415)

Table I.4 Pump Bowl Diameter

Pump Operating Speed (rpm)	Discharge Rate (gpm)	Nominal Pump Bowl Diameter (in.)
3500	200—1200	8
1800	100—600	
1200	160—400	
1800	200—1500	10
1200	370—670	
1800	400—2300	12
1200	250—1500	
1800	1000—4500	14
1200	700—3000	
1800	2000—5200	16
1200	1300—3400	
1800	3200—4100	18
1200	2200—4000	
900	2800—3000	
1200	3100—4400	20
900	2300—3600	
1200	7500	22
900	5600	

(after Campbell and Lehr, 1973, p. 274)

Table I.5 Values of T_m

Percent Mine Penetration	P_h/P_v Ratio				
	1	2	10	50	100
20	.516	.530	.583	.715	.836
40	.506	.514	.548	.612	.667
60	.504	.506	.519	.549	.572
80	.501	.501	.505	.516	.523
100	.500	.500	.500	.500	.500

Appendix J
Microcomputer Groundwater Program Vendors

Hydro Geo Chem, Inc.
1430 N. 6th Avenue
Tucson, Arizona
85705

Microcode, Inc.
2473 Camino Capitan
Santa Fe, New Mexico
87501

Design Professional
Management Systems
Linbrook Office Park
P.O. Box 2364
Kirkland, Washington
98033

Groundwater Graphics
8670 Lake Ashmere Dr.
San Diego, California
92119

GeoTechnical Graphics
930 Dwight Way
Suite 6
Berkeley, California
94710

Data Services
2960 Sonoma Mountain Rd.
Petaluma, California
94952

In-Situ Inc.
P.O. Box I
Laramie, Wyoming
82070

IRRISCO
P.O. Box 5011
University Park,
New Mexico
88003

HydraLogic
Analtical Software
P.O. Box 4722
Missoula, Montana
59806

Geo Trans, Inc.
209 Eldon Street
#301
Herndon, Virginia
22070

Hall Groundwater
Consultants, Inc.
P.O. Box 189
St. Albert, Alberta
Canada T8N 1N3

Watershed Research, Inc.
962 Monterey Court
Shoreview, Minnesota
55126

Rockware, Inc.
7195 W. 30th Avenue
Denver, Colorado
80215

TechMac
c/o Billings &
Associates, Inc.
5801 Osuna Road,
NE, #102
Albuquerque, NM.
87109

Slotta Engineering
Associates
P.O. Box 1376
Corvallis, Oregon
97339

Hunter/Hydrosoft
63 Sarasota Ctr. Blvd.
Suite 107
Sarasota, Florida
34240

Koch & Associates
1660 S. Fillmore St.
Denver, Colorado
80210

T. A. Prickett & Assoc.
University Inn
302 E. John Street
Champaign, Illinois
61820

J.S. Ulrick
1400 Grandview Dr.
Berkeley, California
94705

K. E. Davis Associates
3121 San Jacinto
Suite 102
Houston, Texas

Geo-Slope Programming
Ltd.
7927 Silver Spring Road NW
Calgary, Alberta
Canada T3B 4K4

Interpex Ltd.
715 14th Street
Golden, Colorada
80401

VATNASKIL
Consulting Engineers
Armuli 11
IS-108 Reykjavik
Iceland

Elsevier Science
Publishing Co., Inc.
P.O. Box 1663
Grand Central Station
New York, NY 10163

National Water Well
Association
6375 Riverside Dr.
Dublin, Ohio 43017

International Ground
Water Modeling Center
Holcomb Research
Institute
Butler University
4600 Sunset Ave.
Indianapolis, Indiana
46208

Scientific Software
Group
P.O. Box 23041
Washington, D.C.
20026

Earthware
Rolling Hills Estate,
California

GEOSOFT
204 Richmond Street W.
Suite 500
Toronto, Ontario
Canada M5V1V6

CSW Data Systems
One Overocker Road
Poughkeepsie,
New York 12603

Sci Plot
Micro Glyph Systems
P.O. Box 474
Lexington, MA 02173

Geraghty and Miller
Modeling Group
1895 Preston White Dr.
Suite 301
Reston, Virginia
22091

Scientific Endeavors
Corporation
508 N. Kentucky Street
Kingston, Tennessee
37763

Environmental Systems
& Technologies, Inc.
Software Development
& Sales
P.O. Box 10457

INDEX